D1655175

Katsunori Iwasaki · Hironobu Kimura
Shun Shimomura · Masaaki Yoshida

From Gauss to Painlevé

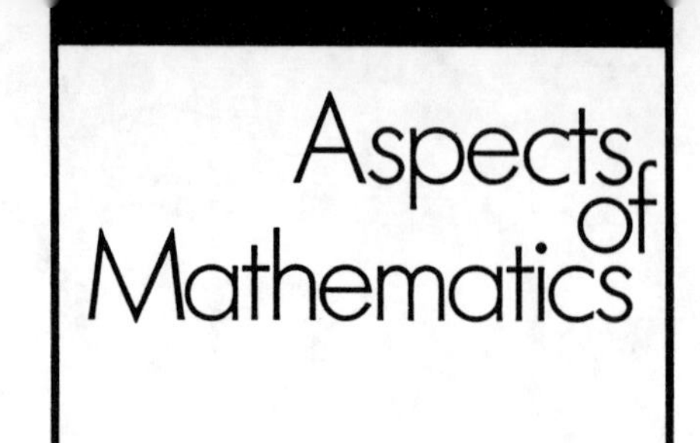

Edited by Klas Diederich

Vol. E 1: G. Hector/U. Hirsch: Introduction to the Geometry of Foliations, Part A

Vol. E 2: M. Knebusch/M. Kolster: Wittrings

Vol. E 3: G. Hector/U. Hirsch: Introduction to the Geometry of Foliations, Part B

Vol. E 4: M. Laska: Elliptic Curves of Number Fields with Prescribed Reduction Type (out of print)

Vol. E 5: P. Stiller: Automorphic Forms and the Picard Number of an Elliptic Surface

Vol. E 6: G. Faltings/G. Wüstholz et al.: Rational Points*

Vol. E 7: W. Stoll: Value Distribution Theory for Meromorphic Maps

Vol. E 8: W. von Wahl: The Equations of Navier-Stokes and Abstract Parabolic Equations (out of print)

Vol. E 9: A. Howard/P.-M. Wong (Eds.): Contributions to Several Complex Variables

Vol. E 10: A. J. Tromba: Seminar of New Results in Nonlinear Partial Differential Equations*

Vol. E 11: M. Yoshida: Fuchsian Differential Equations*

Vol. E 12: R. Kulkarni, U. Pinkall (Eds.): Conformal Geometry*

Vol. E 13: Y. André: G-Function and Geometry*

Vol. E 14: U. Cegrell: Capacities in Complex Analysis

Vol. E 15: J.-P. Serre: Lectures on the Mordell-Weil Theorem

Vol. E 16: K. Iwasaki/H. Kimura/S. Shimomura/M. Yoshida: From Gauss to Painlevé

*A Publication of the Max-Planck-Institut für Mathematik, Bonn

Volumes of the German-language subseries "Aspekte der Mathematik" are listed on page 348.

Aspects of Mathematics

Katsunori Iwasaki · Hironobu Kimura
Shun Shimomura · Masaaki Yoshida

From Gauss to Painlevé

A Modern Theory of Special Functions

Dedicated to Tosihusa Kimura

Die Deutsche Bibliothek – CIP-Einheitsaufnahme

From Gauss to Painlevé: a modern theory of special functions; dedicated to Professor Tosihusa Kimura / Katsunori Iwasaki ... – Braunschweig: Vieweg, 1991
(Aspects of mathematics: E; Vol. 16)
ISBN 3-528-06355-6

NE: Iwasaki, Katunori; Kimura, Tosihusa: Festschrift; Aspects of mathematics / E

Prof. Dr. Katsunori Iwasaki, Departement of Mathematics, University of Tokyo, Japan
Prof. Dr. Hironobu Kimura, Departement of Mathematics, University of Tokyo, Japan
Prof. Dr. Shun Shimomura, Departement of Mathematics, Keio University, Yokohama, Japan
Prof. Dr. Masaaki Yoshida, Departement of Mathematics, Kyushu University, Japan

AMS Subject Classification: 33-XX, 33 A 15, 33 A 30, 34-XX, 34 B 30, 34 A 34

Vieweg is a subsidiary company of the Bertelsmann Publishing Group International.

Cover design by Wolfgang Nieger, Wiesbaden
Printed and bound by W. Langelüddecke, Braunschweig
Printed in Germany

ISSN 0179-2156

ISBN 3-528-06355-6

Preface

The Gamma function, the zeta function, the theta function, the hypergeometric function, the Bessel function, the Hermite function and the Airy function,... are instances of what one calls special functions. These have been studied in great detail. Each of them is brought to light at the right epoch according to both mathematicians and physicists. Note that except for the first three, each of these functions is a solution of a linear ordinary differential equation with rational coefficients which has the same name as the functions. For example, the Bessel equation is the simplest non-trivial linear ordinary differential equation with an irregular singularity which leads to the theory of asymptotic expansion, and the Bessel function is used to describe the motion of planets (Kepler's equation).

Many specialists believe that during the 21st century the Painlevé functions will become new members of the community of special functions. For any case, mathematics and physics nowadays already need these functions. The corresponding differential equations are non-linear ordinary differential equations found by P. Painlevé in 1900 for purely mathematical reasons. It was only 70 years later that they were used in physics in order to describe the correlation function of the two dimensional Ising model. During the last 15 years, more and more people have become interested in these equations, and nice algebraic, geometric and analytic properties were found.

Although many students and researchers in mathematics and in physics need to learn about Painlevé functions, it is not easy for them, even for professional mathematicians not specialized in the theory, to attain the heart of these functions. Indeed, up to now no systematic presentation of the subject was ever written in a western language. In order to fill the gap, the present book was devised as an introductory exposition to the Painlevé functions using only plain language from a modern mathematical view point.

Chapter 1 states elementary theorems in the theory of differential equations needed in the sequel. Most basic theorems are stated without proof. In Chapter 2, the Gauss hypergeometric differential equation is thoroughly studied from various angles. This is not so much because our presentation is intended to be elementary but especially because this differential equation provides the whole theory of linear differential equations with a leading example. In Chapter 3, the monodromy-preserving deformations of second-order Fuchsian equations are presented and the

Hamiltonian structure of the so-called Garnier system is exhibited. The simplest Garnier system turns out to be the celebrated sixth Painlevé equation, of which algebraic properties are explained. Furthermore, it is shown that for some particular values of the parameters, the Garnier system admits solutions which are expressible by the Appell-Lauricella hypergeometric function. Finally in Chapter 4, the behavior of solutions of non-linear differential equations at singular points is investigated. A modern treatment of classical theories is given and a new method to attack the singularities of the Painlevé equations is introduced.

In what follows, only elementary notions of differential equations (the existence theorem for ordinary differential equations, etc.), of function theory (power series, the Cauchy theorem, residues, etc.) and of group theory (normal subgroups, quotient groups, etc.) are supposed to be known, so that any graduate student should be able to understand the presentation.

This book was written in honor of Professor Tosihusa Kimura by his students on the occasion of his sixtieth birthday. The first drafts of Chapters 1,2,3 and 4 were written by M. Yoshida, K. Iwasaki, H. Kimura and S. Shimomura, respectively; then each chapter was improved and polished by the authors and professors Y. Haraoka, Y. Murata, K. Okamoto, T. Sasaki, K. Takano, N. Takayama, and D. Zagier to whom the authors are deeply grateful. Their gratitude goes also to professors H. Majima B. Morin and N. Yamada. Last but not least, we want to thank the members of the department of mathematics of Kobe University, where discussion on the manuscript was held in a comfortable atmosphere, for their tolerance and kindness.

April 1990

Authors

Contents

1 Elements of Differential Equations

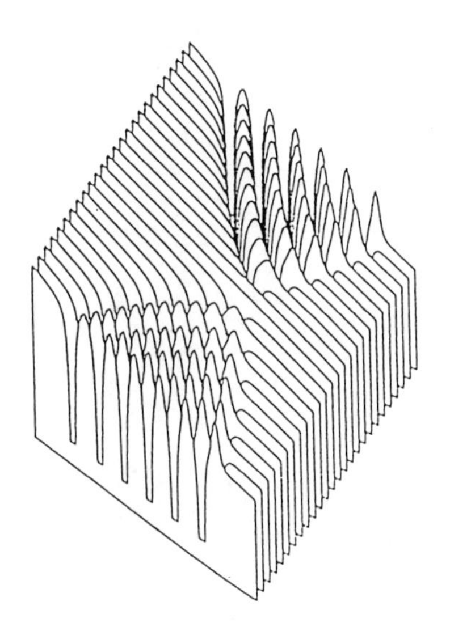

This chapter presents fundamental definitions and theorems on the theory of differential equations which are needed in the course of this book. We do not give proofs for the theorems in Section 1 and 2; the reader may consult any introductory book on differential equations.

1. Cauchy's Existence Theorem

Consider a system of first-order ordinary differential equations

$$\frac{dz_j}{dx} = f_j(x, z) \quad (j = 1, \ldots, r) \tag{1.1}$$

with the independent variable x and the unknown vector $z = (z_1, \ldots, z_r)$, where the vector function $f = (f_1, \ldots, f_r)$ is holomorphic in a domain $\mathcal{D} \subset \mathbb{C} \times \mathbb{C}^r$.

THEOREM 1.1. *For any $(a, b) \in \mathcal{D}$, there is a unique solution z of* (1.1) *holomorphic in a neighborhood of a, such that*

$$z(a) = b. \tag{1.2}$$

THEOREM 1.2. *If the system* (1.1) *and the initial data* (1.2) *depend holomorphically on a system of parameters $s = (s_1, \ldots, s_p)$ then the solution is holomorphic both in x and in s.*

REMARK 1.3. Any system of ordinary differential equations of higher order can be reduced to a system of the form (1.1) by introducing new unknowns.

2. Linear Equations

Consider a linear ordinary differential equation

$$\frac{d^r z}{dx^r} + a_1(x)\frac{d^{r-1} z}{dx^{r-1}} + \cdots + a_r(x) z = 0 \tag{2.1}$$

where the a_j's are holomorphic in a domain $D \subset \mathbb{C}$. Introducing new unknowns, say

$$z_0 = z, \quad z_i = \frac{d^i z}{dx^i} \quad (i = 1, \ldots, r-1),$$

the equation (2.1) can be written in the following form:

$$\frac{dz_i}{dx} = \sum_{j=0}^{r-1} a_i^j(x) z_j \quad (i = 0, \ldots, r-1).$$

THEOREM 2.1. *For any point $a \in D$ and any complex numbers $b_0, \ldots, b_{r-1}$, there is a unique holomorphic solution of* (2.1) *such that*

$$\frac{d^i z}{dx^i}(a) = b_i, \quad i = 0, \ldots, r-1.$$

The solution z can be analytically continued along any curve in D.

A system of r linearly independent solutions of (2.1) at a is called a *fundamental system of solutions* of (2.1) at a.

If the coefficients of the equation (2.1) are holomorphic in $\{x \mid 0 < |x-a| < \epsilon\}$ for some $\epsilon > 0$ and are meromorphic but not all holomorphic in $\{x \mid |x-a| < \epsilon\}$, then the point a is called a *singular point* of (2.1).

DEFINITION 2.2. A singular point a of (2.1) is said to be *regular* if

$$(x-a)^k a_k(x), \quad (k = 1, \ldots, r)$$

are holomorphic at a.

If a is a regular singular point of the equation (2.1), then by putting, for instance,

$$z_1 = z, \quad z_2 = (x-a)\frac{dz_1}{dx}, \ldots, \quad z_r = (x-a)\frac{dz_{r-1}}{dx},$$

the equation can be written in the form

$$\frac{d\mathbf{z}}{dx} = \frac{1}{x-a} A(x)\mathbf{z}, \tag{2.2}$$

where $\mathbf{z} = {}^t(z_1, \ldots, z_r)$ and $A(x)$ is an $r \times r$ matrix holomorphic at a.

THEOREM 2.3. *Consider the equation (2.2) with an $r \times r$ matrix $A(x)$ holomorphic at a. If the matrix $A(a)$ has no eigenvalues that differ from each other by integers, then there exists a matrix $P(x)$ holomorphic at a with $P(a) = I_r$, such that*

$$P(x)(x-a)^{A(a)}$$

satisfies (2.2), where

$$(x-a)^{A(a)} = \exp\{A(a)\log(x-a)\}.$$

Proof. Without loss of generality, we assume $a = 0$. Let us transform the unknown z of (2.2) by

$$z = P(x)w,$$

where $P(x)$ is a holomorphic $r \times r$ matrix with $P(0) = I_r$. Then the equation changes into

$$x\frac{dw}{dx} = B(x)w,$$

with

$$B(x) = P^{-1}(x)A(x)P(x) - xP^{-1}(x)\frac{dP(x)}{dx}.$$

We want to determine $P(x)$ in such a way that $B(x)$ becomes a constant matrix $A(0)$. Our aim is to solve the equation

$$x\frac{dP}{dx} = A(x)P - PA(0).$$

The change of variable $P = I_r + Q$ transforms the equation into the equation

$$x\frac{dQ}{dx} = A(0)Q - QA(0) + \{A(x) - A(0)\}(I_r + Q).$$

Notice that the eigenvalues of the linear map of $r \times r$ matrices

$$X \mapsto A(0)X - XA(0)$$

are the differences of the eigenvalues of $A(0)$ (to see this, take $A(0)$ to be diagonal). Thus, under the assumption for the eigenvalues of $A(0)$, we can apply Proposition 1.1.1 in Chapter 4 to show that the equation for Q has a unique holomorphic solution with $Q(0) = 0$.
The equation

$$x\frac{dw}{dx} = A(0)w$$

is easily solved by the matrix $x^{A(0)}$. Therefore the matrix $z = (I_r + Q(x))x^{A(0)}$ is a solution of (2.2).

3. Local Behavior around Regular Singular Points (Frobenius's Method)

Let $x = 0$ be a regular singular point of the equation (2.1). We introduce the Euler operator

$$\delta = x\frac{d}{dx}.$$

Since δ relates with the operator d/dx as follows:

$$x^k\frac{d^k}{dx^k} = \delta(\delta-1)\cdots(\delta-k+1) \tag{3.0}$$

for $k \geq 1$, we have

$$\begin{aligned} & x^r\left\{\frac{d^r}{dx^r} + a_1(x)\frac{d^{r-1}}{dx^{r-1}} + \cdots + a_r(x)\right\} \\ &= \sum_{k=0}^{r} x^{r-k}a_{r-k}(x)\delta(\delta-1)\cdots(\delta-k+1) \\ &= \delta^r + \left\{xa_1(x) - \frac{(r-1)r}{2}\right\}\delta^{r-1} + \cdots, \end{aligned}$$

where $a_0 = 1$.
Thus, the equation (2.1) can be written in the form

$$Lz = 0,$$

where L is a linear differential operator of the following form:

$$L = \sum_{i=0}^{r} b_i(x)\delta^{r-i},$$

where

$$b_0(x) = 1$$

and $b_1(x), \ldots, b_r(x)$ are given by convergent power series. We set

$$b_i(x) = \sum_{j=0}^{\infty} b_{ij}x^j, \qquad 0 \le i \le r. \tag{3.1}$$

In particular, $b_{00} = 1,\ b_{0j} = 0\ (j \ge 1)$.

Put

$$z = x^s \sum_{k=0}^{\infty} c_k x^k, \qquad c_0 = 1 \tag{3.2}$$

and calculate Lz:

$$\begin{aligned} Lz &= \sum_{i=0}^{r}\sum_{j=0}^{\infty} b_{ij}x^j\delta^{r-i}\sum_{k=0}^{\infty} c_k x^{s+k} \\ &= \sum_{k=0}^{\infty}\sum_{j=0}^{\infty}\sum_{i=0}^{r} b_{ij}(s+k)^{r-i}c_k x^{s+k+j} \\ &= x^s\sum_{n=0}^{\infty}\left\{\sum_{k=0}^{n}\sum_{i=0}^{r} b_{i,n-k}(s+k)^{r-i}c_k\right\}x^n, \end{aligned}$$

where we used the identity $\delta x^\alpha = \alpha x^\alpha$. We put

$$\begin{aligned} f(s) :&= \sum_{i=0}^{r} b_{i,0}s^{r-i} \\ &= \sum_{i=0}^{r} b_i(0)s^{r-i}, \end{aligned}$$

which is equal to $L_{|x=0}$ after replacing δ by s, and put

$$f_j(s) := \sum_{i=0}^{r} b_{ij} s^{r-i}, \quad j \geq 1$$
$$= \sum_{i=1}^{r} b_{ij} s^{r-i},$$

and

$$R_0 := 0,$$

(3.3) $$R_n = R_n(c_1, \ldots, c_{n-1}, s) := \sum_{k=0}^{n-1} f_{n-k}(s+k)c_k, \qquad n \geq 1.$$

Then we have

$$Lz = x^s \sum_{n=0}^{\infty} \{f(s+n)c_n + R_n\} x^n.$$

Thus $Lz = 0$ if and only if

(3.4; n) $$f(s+n)c_n + R_n = 0, \quad n \geq 0.$$

DEFINITION 3.1. The algebraic equation (3.4;0):

(3.5) $$f(s) = 0$$

is called the *characteristic* (or *indicial*) *equation* of the equation (2.1) (and of $Lz = 0$) at the regular singular point $x = 0$. Roots of (3.5) are called the (*characteristic*) *exponents*.

Now consider s as a parameter in a certain domain S, determine the coefficients $c_n = c_n(s)$ ($c_0 = 1$) by the equations (3.4;n) ($n \geq 1$) and put

$$z(s,x) = x^s \sum_{n=0}^{\infty} c_n(s) x^n.$$

If s satisfies

$$f(s+n) \neq 0 \quad \text{for} \quad n \geq 1,$$

then the series $z(s,x)$ converges and represents a holomorphic function in s and x. In fact we have

PROPOSITION 3.2. *Let s be in the domain S and let N be a positive integer such that*

$$f(s+n) \neq 0 \quad \textit{for} \quad s \in S, \quad n \geq N.$$

For arbitrary complex numbers $c_1, \ldots, c_{N-1}$, define $c_n(s)$ $(n \geq N)$ by (3.3) *and* (3.4;n). *Then the series*

$$\sum_{n=N}^{\infty} c_n(s) x^n$$

converges and represents a holomorphic function in $s \in S$ and x.

Proof. The assumption implies that if s is in a compact subset of S then

$$|f(s+n)| \geq F n^r \quad \text{for} \quad n \geq N$$

for some $F > 0$, and the convergence of the series (3.1) yields

$$|b_{ij}| \leq B\epsilon^{-j} \quad 1 \leq i \leq r$$

for some $B > 0$ and $\epsilon > 0$. Since, for sufficiently large t, we have

$$\begin{aligned} |f_j(t)| &\leq \sum_{i=1}^{r} |b_{ij}||t|^{r-i} \quad j \geq 1 \\ &\leq B\epsilon^{-j} \sum_{i=0}^{r-1} |t|^i \\ &\leq C\epsilon^{-j}|t|^{r-1} \end{aligned}$$

for some $C > 0$, we get

$$\begin{aligned} |c_n(s)| &= \frac{|R_n(s)|}{|f(s+n)|} \\ &\leq \frac{1}{Fn^r} \left| \sum_{k=0}^{n-1} f_{n-k}(s+k) c_k(s) \right| \\ &\leq \frac{C}{Fn^r} \sum_{k=0}^{n-1} \epsilon^{k-n} |s+k|^{r-1} |c_k(s)| \end{aligned}$$

and so

$$\epsilon^n |c_n(s)| \leq \frac{D}{Fn} \sum_{k=0}^{n-1} \epsilon^k |c_k(s)|$$

for some $D > 0$; which proves the proposition. ∎

Let $s_1, \ldots, s_r$ be the roots of (3.5): $f(s) = 0$; we can assume

$$\Re s_r \leq \ldots \leq \Re s_1.$$

Then $z(s_1, x)$ is a solution of our differential equation, because we have $f(s_1 + n) \neq 0$ for all $n \geq 1$.

Assume now for simplicity that $r = 2$. If $s_1 - s_2$ is not an integer, then for the same reason, $z(s_2, x)$ is another solution which is linearly independent of $z(s_1, x)$. This is also true when $s_1 - s_2$ is a positive integer, say m, if R_m happens to be zero. In this case we can solve (3.4;n) for $s = s_2$ for all $n \geq 1$ (choose c_m arbitrarily); and the singularity is called *non-logarithmic*.
Otherwise, i.e., when $s_1 = s_2$ or $s_1 - s_2 = m$ is a positive integer and $R_m \neq 0$, differentiate

$$Lz(s, x) = x^s f(s)$$

by s and put $s = s_1$, then we get

$$\frac{\partial}{\partial s}\{Lz(s, x)\}|_{s=s_1} = x^{s_1} f'(s_1).$$

In case $s_1 = s_2$, namely when s_2 is a double root of f, since we have $f'(s_2) = 0$, the following expression

$$\frac{\partial z(s, x)}{\partial s}|_{s=s_1} = z(s_1, x)\log x + x^{s_2} \sum_{j=0}^{\infty} c'_j(s_1) x^j$$

is a solution of $Lz = 0$, because L does not depend on s and so it commutes with $\partial/\partial s$.
Finally consider the remaining case : $s_1 - s_2 = m \in \mathbb{Z}$, $m > 0$ and $R_m \neq 0$. Put

$$u^* = x^{s_2} \sum_{j=0}^{\infty} c_j(s_2) x^j$$

where the c_j's ($j < m$) are determined ($c_0 = 1$) by (3.4; j), while c_m is arbitrarily fixed and the c_j's ($j > m$) are determined again by (3.4; j). Then the series u^* converges (Proposition 3.2) and satisfies the following:

$$Lu^* = R_m(c_1, \dots, c_{m-1}, s_2)x^{s_2+m}.$$

Since $s_2 + m = s_1$, we know that a suitable linear combination of u^* and $\partial z(s,x)/\partial s|_{s=s_1}$, say

$$f'(s_1)u^* - R_m(c_1, \dots, c_{m-1}, s_2)\frac{\partial z(s,x)}{\partial s}|_{s=s_1}$$

is a solution. In the latter two cases, the singularity is said to be *logarithmic.*

REMARK 3.3. By the expressions obtained above, we see that there is a positive number N such that for any solution $z(x)$ and any two real numbers θ_1 and θ_2 ($\theta_1 < \theta_2$), we have

$$x^N z(x) \to 0$$

as x tends to 0 in the sector defined by

$$\theta_1 < \arg x < \theta_2.$$

4. Fuchsian Equations

Before going into the subject of this section, let us give a criterion of regular singularity of differential equations written in terms of the operator δ.

LEMMA 4.1. *A differential equation*

$$\{\delta^r + b_1(x)\delta^{r-1} + \cdots + b_r(x)\}z = 0$$

is regular singular at $x = 0$ if and only if b_j $(1 \leq j \leq r)$ are holomorphic at $x = 0$.

This is a consequence of the identity (3.0) and the following identity

$$\delta^k = x^k \frac{d^k}{dx^k} + \sum_{j=1}^{k-1} c_j^k x^j \frac{d^j}{dx^j},$$

where c_j^k $(1 \le j \le k)$ are constants determined by

$$s^k = \sum_{j=1}^{k} c_j^k s(s-1)\cdots(s-j+1),$$

or by the recurrence formula: $c_j^{k+1} = c_{j-1}^k + jc_j^k$, $1 \le j \le k$, $c_0^k = 0$, $c_k^k = 1$.

The linear equation (2.1) with rational coefficients is said to be *Fuchsian* if every singular point in the x-plane is regular and if, after changing variable x into $t = 1/x$, the transformed equation has a regular singular point at $t = 0$. The exponents at $t = 0$ are called the exponents of (2.1) at infinity. Let us express the equation (2.1) by using the variable t; put

$$\theta = t\frac{d}{dt}.$$

Since $\delta = -\theta$, we have

$$\begin{aligned} & x^r \left\{ \frac{d^r}{dx^r} + a_1(x)\frac{d^{r-1}}{dx^{r-1}} + \cdots + a_r(x) \right\} \\ &= \sum_{k=0}^{r} x^{r-k} a_{r-k}(x)\delta(\delta-1)\cdots(\delta-k+1) \\ &= \sum_{k=0}^{r} t^{-(r-k)} a_{r-k}(\frac{1}{t})(-1)^k\theta(\theta+1)\cdots(\theta+k-1) \\ &= (-1)^r\theta^r + \cdots, \end{aligned}$$

where $a_0 = 1$.

PROPOSITION 4.2 (A characterization of Fuchsian equations). *The equation (2.1) is Fuchsian with regular singularities at $x_1, \ldots, x_m, x_{m+1} = \infty$ if and only if the coefficients have the following form:*

$$a_k(x) = \frac{p_k(x)}{\prod_{i=1}^{m}(x-x_i)^k} \quad (k = 1, \ldots, r)$$

where each $p_k(x)$ is a polynomial of degree at most $k(m-1)$.

Proof. By the definition of regular singularity, at each finite singular point x_j $(1 \leq j \leq m)$, we have the expression of $a_j(x)$ given in the theorem, where $p_j(x)$ are entire functions. Applying Lemma 4.1 at the singular point $t = 0$ of the equation with the variable $t = 1/x$ given above, we conclude that

$$t^{-k} a_k(\frac{1}{t}) \quad (k = 1, \dots, r)$$

are holomorphic at $t = 0$. Therefore each $p_k(x)$ is a polynomial of degree at most $k(m-1)$. ∎

The following table of the regular singular points $x_1, \dots, x_{m+1}$ and the exponents $s_i^1, \dots, s_i^r$ at x_i is called the *Riemann scheme* of the equation (2.1).

$$\begin{pmatrix} x_1 & \dots & x_{m+1} \\ s_1^1 & \dots & s_{m+1}^1 \\ \dots & \dots & \dots \\ s_1^r & \dots & s_{m+1}^r \end{pmatrix}$$

PROPOSITION 4.3 (Fuchs Relation). *The sum of all exponents of* (2.1) *depends only on the number* $m+1$ *of singular points:*

$$\sum_{i=1}^{m+1} \sum_{j=1}^{r} s_i^j = \frac{(m-1)r(r-1)}{2}.$$

Proof. Let $x_1, \dots, x_m, x_{m+1} = \infty$ are regular singular points of the equation (2.1). By the proposition above, $a_1(x)$ admits the following expression:

$$a_1(x) = \sum_{i=1}^{m} \frac{\alpha_i}{x - x_i},$$

where α_i are constants. At each finite singular point x_i, putting

$$\delta_i = (x - x_i)\frac{d}{dx},$$

we have

$$(x-x_i)^r\left\{\frac{d^r}{dx^r}+a_1(x)\frac{d^{r-1}}{dx^{r-1}}+\cdots+a_r(x)\right\}$$

$$=\delta_i^r+\left\{(x-x_i)a_1(x)-\frac{(r-1)r}{2}\right\}\delta_i^{r-1}+\cdots.$$

Thus, by the definition of exponents, we conclude that

$$\sum_{j=1}^r s_i^j=-\alpha_i+\frac{(r-1)r}{2}.$$

At $x_{m+1}=\infty$, by the following expression of the equation with respect to the variable $t=1/x$:

$$x^r\left\{\frac{d^r}{dx^r}+a_1(x)\frac{d^{r-1}}{dx^{r-1}}+\cdots+a_r(x)\right\}$$

$$=\sum_{k=0}^r t^{-(r-k)}a_{r-k}(\frac{1}{t})(-1)^k\theta(\theta+1)\cdots(\theta+k-1)$$

$$=(-1)^r\left[\theta^r+\left\{-\frac{1}{t}a_1(\frac{1}{t})+\frac{(r-1)r}{2}\right\}\theta^{r-1}+\cdots\right],$$

we have

$$\sum_{j=1}^r s_{m+1}^j=\sum_{i=1}^m\alpha_i-\frac{(r-1)r}{2}.$$

Therefore we are led to the desired equality. ∎

5. Pfaffian Systems and Integrability Conditions

Consider a system of first-order partial differential equations with r unknowns $u^1,\ldots,u^r$ and n independent variables $x^1,\ldots,x^n$ in the following form

$$\frac{\partial u^i}{\partial x^j}=a_j^i(x,u),\qquad 1\le i\le r,\ \ 1\le j\le n, \tag{5.1}$$

where the a_j^i's are holomorphic in a domain $\mathcal{D}$ in (x,u)-space. Such a system is called a *Pfaffian system.* The system (5.1) is said to be (*completely*) *integrable* if for any $(x_0,u_0)\in\mathcal{D}$ there is a solution $u(x)$ such that $u(x_0)=u_0$; the manifold consisting of points $(x,u(x))$ is called the *integral manifold* of (5.1) passing a point (x_0,u_0).

THEOREM 5.1 (Frobenius). *The system* (5.1) *is integrable if and only if*

$$\frac{\partial a_j^i}{\partial x^k} + \sum_{s=1}^{r} \frac{\partial a_j^i}{\partial u^s} a_k^s = \frac{\partial a_k^i}{\partial x^j} + \sum_{s=1}^{r} \frac{\partial a_k^i}{\partial u^s} a_j^s, \tag{5.2}$$

$j, k = 1, \ldots, n, \quad i = 1, \ldots, r.$

For the sake of clarity, we prove the theorem for the simplest non-trivial case: $r = 1, n = 2$. Consider the system

$$(5.1)' \qquad \begin{cases} \dfrac{\partial u}{\partial x} = a(x, y, u) \\[2mm] \dfrac{\partial u}{\partial y} = b(x, y, u) \end{cases}$$

defined in a domain $\mathcal{D}$ in (x, y, u)-space. If, for every point of $\mathcal{D}$, there is a solution passing through the point, then by the identity $\partial^2 u/\partial x \partial y = \partial^2 u/\partial x \partial y$, we have

$$(5.2)' \qquad \frac{\partial a}{\partial y} + \frac{\partial a}{\partial u} b = \frac{\partial b}{\partial x} + \frac{\partial b}{\partial u} a.$$

Assuming (5.2)$'$ conversely, for an arbitrary point (x_0, y_0, u_0) in $\mathcal{D}$, we shall construct a solution passing through the point. Let $v(x)$ be the solution of the following ordinary differential equation:

$$(5.3) \qquad \begin{cases} \dfrac{dv}{dx} = a(x, y_0, v) \\[2mm] v|_{x=x_0} = u_0. \end{cases}$$

Let $u(x, y)$ be the solution of the following ordinary differential equation with parameter x:

$$(5.4) \qquad \begin{cases} \dfrac{du}{dy} = b(x, y, u) \\[2mm] u|_{y=y_0} = v(x). \end{cases}$$

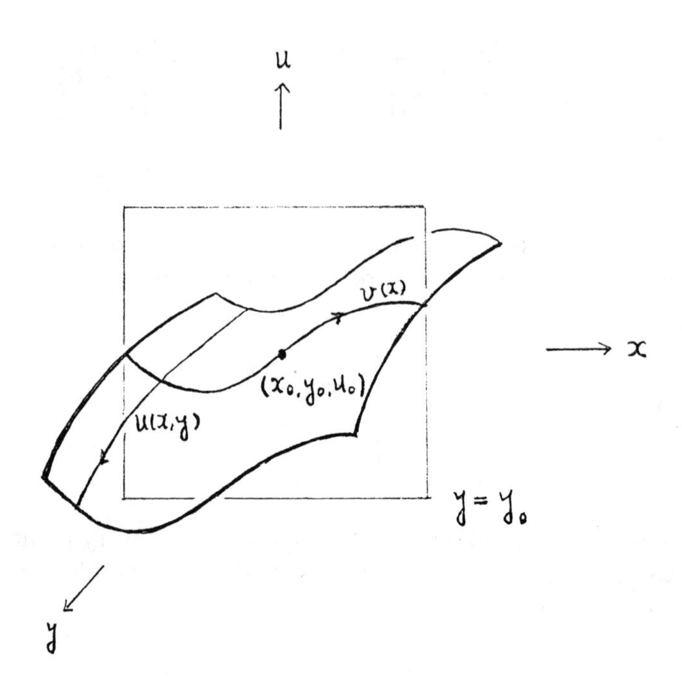

We prove

$$\frac{\partial u(x,y)}{\partial x} = a(x,y,u(x,y)). \tag{5.5}$$

Since the solution $u(x,y)$ of (5.4) depends holomorphically on the parameter x (Theorem 1.2), we have

$$\begin{cases} \dfrac{\partial}{\partial x}\left(\dfrac{\partial u(x,y)}{\partial y}\right) = \dfrac{\partial b}{\partial x} + \dfrac{\partial b}{\partial u}\dfrac{\partial u}{\partial x}, \\ \dfrac{\partial u(x,y)}{\partial x}\Big|_{y=y_0} = \dfrac{du(x,y_0)}{dx} = \dfrac{dv(x)}{dx} = a(x,y_0,v(x)). \end{cases}$$

Thus $\partial u(x,y)/\partial x$ is the solution of the following ordinary differential equation with parameter x:

$$\text{(5.6)} \qquad \begin{cases} \dfrac{dw}{dy} = \dfrac{\partial b}{\partial x} + \dfrac{\partial b}{\partial u} w \\ w|_{y=y_0} = a(x, y_0, v(x)). \end{cases}$$

On the other hand, by (5.2)′, we have

$$\begin{cases} \dfrac{\partial a(x,y,u(x,y))}{\partial y} = \dfrac{\partial a}{\partial y} + \dfrac{\partial a}{\partial u}\dfrac{\partial u}{\partial y} \\ \qquad\qquad = \dfrac{\partial a}{\partial y} + \dfrac{\partial a}{\partial u} b \\ \qquad\qquad = \dfrac{\partial b}{\partial x} + \dfrac{\partial b}{\partial u} a, \\ a(x,y,u(x,y))|_{y=y_0} = a(x, y_0, u(x, y_0)) \\ \qquad\qquad = a(x, y_0, v(x)). \end{cases}$$

Thus $a(x, y, u(x, y))$ also solves (5.6). By the uniqueness of solution, we have the identity (5.5). Since

$$u(x_0, y_0) = v(x_0) = u_0,$$

the function $u(x, y)$ is the solution of (5.1)′ passing through (x_0, y_0, u_0).

The system (5.2) is called the *integrability condition* of the system (5.1). We can paraphrase the theorem by making use of the *exterior (differential) algebra*, consisting of holomorphic p-forms $(0 \leq p \leq n+r)$, over the ring of holomorphic functions in $\mathcal{D}$ (*any left or right ideal of the exterior algebra is a two-sided ideal*). We associate with the Pfaffian system (5.1) the ideal $\mathcal{I}$ of the exterior algebra generated by the 1-forms

$$du^i - \sum_{j=1}^{n} a^i_j(x, u) dx^j, \qquad i = 1, \ldots r.$$

COROLLARY 5.2. *The system (5.1) is integrable if and only if*

$$d\omega \in \mathcal{I} \quad \textit{for all} \quad \omega \in \mathcal{I},$$

in other words,

$$d\omega \equiv 0 \quad \textit{modulo } \mathcal{I} \qquad \textit{for all } \ \omega \in \mathcal{I}.$$

In fact we have

$$
\begin{aligned}
& d\Big(du^i - \sum_{j=1}^{n} a^i_j dx^j\Big) \\
&= - \sum_{j=1}^{n} \Big(\sum_{k=1}^{n} \frac{\partial a^i_j}{\partial x^k} dx^k \wedge dx^j + \sum_{s=1}^{r} \frac{\partial a^i_j}{\partial u^s} du^s \wedge dx^j \Big) \\
&\equiv - \sum_{j=1}^{n} \Big(\sum_{k=1}^{n} \frac{\partial a^i_j}{\partial x^k} dx^k \wedge dx^j + \sum_{s=1}^{r} \frac{\partial a^i_j}{\partial u^s} \sum_{k=1}^{n} a^s_k dx^k \wedge dx^j \Big) \\
&= \sum_{1 \le j < k \le n} \Big\{ \frac{\partial a^i_j}{\partial x^k} - \frac{\partial a^i_k}{\partial x^j} + \sum_{s=1}^{r} \Big(\frac{\partial a^i_j}{\partial u^s} a^s_k - \frac{\partial a^i_k}{\partial u^s} a^s_j \Big) \Big\} dx^j \wedge dx^k
\end{aligned}
$$

modulo $\mathcal{I}$.

When (5.1) is linear, it can be written in the following form:

$$\frac{\partial u^i}{\partial x^j} = \sum_{k=1}^{r} a^i_{jk}(x) u^k, \qquad 1 \le i \le r, \quad 1 \le j \le n, \tag{5.7}$$

where the a^i_{jk}'s are holomorphic in a domain D in x-space. The system can also be written as

$$\frac{\partial \mathbf{u}}{\partial x^j} = A_j(x)\mathbf{u}, \qquad 1 \le j \le n,$$

where

$$\mathbf{u} = {}^t(u^1, \ldots, u^r) \ \text{ and } \ A_j = (a^i_{jk})_{i,k=1,\ldots,r},$$

and, if you like, as

$$d\mathbf{u} = \omega \mathbf{u}, \qquad \omega = \sum_{j=1}^{n} A_j dx^j.$$

COROLLARY 5.3. *The system (5.7) is integrable if and only if*

$$\frac{\partial a^p_{iq}}{\partial x^j} + \sum_{s=1}^{r} a^p_{is} a^s_{jq} = \frac{\partial a^p_{jq}}{\partial x^i} + \sum_{s=1}^{r} a^p_{js} a^s_{iq},$$

$$p, q = 1, \ldots, r, \quad i, j = 1, \ldots, n,$$

or equivalently,

$$\frac{\partial A_i}{\partial x^j} + A_i A_j = \frac{\partial A_j}{\partial x^i} + A_j A_i \qquad i, j = 1, \ldots, n$$

or

$$d\omega - \omega \wedge \omega = 0$$

(where the exterior product $\omega \wedge \omega$ of the r by r matrix-valued 1-form $\omega = (\omega^i_j)$ is the matrix whose (i,j)-th entry is the 2-form $\sum_{k=1}^{r} \omega^k_j \wedge \omega^i_k$).

For any simply connected domain $U \subset D$, the set of solutions on U of the integrable system (5.7) forms an r-dimensional linear space.

6. Hamiltonian Systems

Let $t = (t^1, \ldots, t^m)$ and $x = (x^1, \ldots, x^{2n})$ be variables in $\mathbb{C}^m \times \mathbb{C}^{2n}$. For rational functions f and g in (t, x), we define the *Poisson bracket*:

$$\{f, g\} := \sum_{i=1}^{n} \left(\frac{\partial f}{\partial x^{n+i}} \frac{\partial g}{\partial x^i} - \frac{\partial f}{\partial x^i} \frac{\partial g}{\partial x^{n+i}} \right).$$

The set of rational functions in (t, x) forms a Lie algebra with respect to the Poisson bracket, that is, the Poisson bracket $\{\cdot, \cdot\}$ is a skew symmetric bilinear form over $\mathbb{C}$ which satisfies the *Jacobi identity*:

$$\{f, \{g, h\}\} + \{h, \{f, g\}\} + \{g, \{h, f\}\} = 0.$$

The Pfaffian system:

$$dx^i = \sum_{j=1}^{m} \{H_j, x^i\} dt^j, \quad i = 1, \ldots, 2n \tag{6.1}$$

is called a *Hamiltonian system* with *Hamiltonians* $H_j = H_j(t,x), (j = 1, \ldots, m)$. Note that (6.1) is equivalent to a system of first order partial differential equations:

$$\frac{\partial x^i}{\partial t^j} = \frac{\partial H_j}{\partial x^{n+i}}, \quad \frac{\partial x^{n+i}}{\partial t^j} = -\frac{\partial H_j}{\partial x^i}, \quad i = 1, \ldots, n. \tag{6.2}$$

or to

$$\frac{\partial x}{\partial t} = J\ {}^t(\frac{\partial H}{\partial x}) \tag{6.3}$$

where

$$J = \begin{pmatrix} 0 & I_n \\ -I_n & 0 \end{pmatrix},$$

and $\partial x/\partial t$ and $\partial H/\partial x$ are Jacobian matrices:

$$\frac{\partial x}{\partial t} = \left(\frac{\partial x^i}{\partial t^j}\right), \quad \frac{\partial H}{\partial x} = \left(\frac{\partial H_i}{\partial x^j}\right);$$

that is, for example,

$$\frac{\partial x}{\partial t} = \begin{pmatrix} \frac{\partial x^1}{\partial t^1} & \cdots & \frac{\partial x^1}{\partial t^m} \\ \vdots & & \vdots \\ \frac{\partial x^{2n}}{\partial t^1} & \cdots & \frac{\partial x^{2n}}{\partial t^m} \end{pmatrix}.$$

We associate with a system (6.1) the closed 2-form

$$\Gamma := \sum_{i=1}^{n} dx^i \wedge dx^{n+i} + \sum_{j=1}^{m} dH_j \wedge dt^j,$$

called the *symplectic* 2-*form* or the *fundamental* 2-*form*.

DEFINITION 6.1. A rational transformation $\Phi : (t, x) \mapsto (s, y) := (s(t), y(t, x))$ is said to be *rational symplectic* (with respect to Γ) if $\det(\partial s/\partial t)$ does not vanish identically and if there are rational functions $K_1(s, y), \ldots, K_m(s, y)$ such that

$$\Gamma = \Phi^* \Big(\sum_{i=1}^{n} dy^i \wedge dy^{n+i} + \sum_{j=1}^{m} dK_j \wedge ds^j \Big); \tag{6.4}$$

we denote the transformation by

$$\Phi : (t, x, H) \to (s, y, K).$$

Since we consider only rational transformations of Hamiltonian systems in this book, we call such a transformation simply *symplectic*. Let us study the condition (6.4). We set

$$dx = \begin{pmatrix} dx^1 \\ \vdots \\ dx^{2n} \end{pmatrix}, \quad dt = \begin{pmatrix} dt^1 \\ \vdots \\ dt^m \end{pmatrix}, etc.,$$

$$\frac{\partial y}{\partial x} = \left(\frac{\partial y^i}{\partial x^j} \right), \quad \frac{\partial K}{\partial x} = \left(\frac{\partial K_i}{\partial x^j} \right), etc.,$$

and omit the symbol $\wedge$ for a while. Since we have

$$\sum_{i=1}^{n} dx^i dx^{n+i} = \frac{1}{2}\, {}^t dx J dx,$$

$$\sum_{j=1}^{m} dH_j dt^j = {}^t(\frac{\partial H}{\partial t} dt + \frac{\partial H}{\partial x} dx) dt$$

$$= {}^t dt\, {}^t(\frac{\partial H}{\partial t}) dt + {}^t dx\, {}^t(\frac{\partial H}{\partial x}) dt,$$

$$dy = \frac{\partial y}{\partial t} dt + \frac{\partial y}{\partial x} dx,$$

$$dK = \frac{\partial K}{\partial t} dt + \frac{\partial K}{\partial x} dx,$$

$$ds = \frac{\partial s}{\partial t} dt,$$

the right-hand side of (6.4) equals

$$\begin{aligned}
&\frac{1}{2}\,{}^t(\frac{\partial y}{\partial t}dt + \frac{\partial y}{\partial x}dx)J(\frac{\partial y}{\partial t}dt + \frac{\partial y}{\partial x}dx) \\
&+{}^t(\frac{\partial K}{\partial t}dt + \frac{\partial K}{\partial x}dx)\frac{\partial s}{\partial t}dt \\
= {}^t dt &\left\{\frac{1}{2}\,{}^t(\frac{\partial y}{\partial t})J\frac{\partial y}{\partial t} + {}^t(\frac{\partial K}{\partial t})\frac{\partial s}{\partial t}\right\}dt \\
&+{}^t dx\left\{{}^t(\frac{\partial y}{\partial x})J\frac{\partial y}{\partial t} + {}^t(\frac{\partial K}{\partial x})\frac{\partial s}{\partial t}\right\}dt \\
&+\frac{1}{2}\,{}^t dx\,{}^t(\frac{\partial y}{\partial x})J\frac{\partial y}{\partial x}dx.
\end{aligned}$$

Thus the condition (6.4) is equivalent to the following three conditions:

$$\frac{\partial H}{\partial t} = {}^t(\frac{\partial s}{\partial t})\frac{\partial K}{\partial t} - \frac{1}{2}\,{}^t(\frac{\partial y}{\partial t})J\frac{\partial y}{\partial t},$$

$$J = {}^t(\frac{\partial y}{\partial x})J\frac{\partial y}{\partial x}, \tag{6.5}$$

$$\frac{\partial H}{\partial x} = {}^t(\frac{\partial s}{\partial t})\frac{\partial K}{\partial x} - {}^t(\frac{\partial y}{\partial t})J\frac{\partial y}{\partial x}. \tag{6.6}$$

PROPOSITION 6.2. *By a transformation which satisfies (6.4), the system (6.1) is taken into a Hamiltonian system:*

$$dy^i = \sum_{j=1}^{m}\{K_j, y^i\}ds^j, \quad i = 1, \dots, 2n. \tag{6.7}$$

Proof. We shall derive (6.3) from

$$\frac{\partial y}{\partial s} = J\,{}^t(\frac{\partial K}{\partial y}).$$

Noting

$$\frac{\partial y}{\partial x}\frac{\partial x}{\partial y} = I_{2n}, \quad \frac{\partial s}{\partial t}\frac{\partial t}{\partial s} = I_m, \quad \frac{\partial t}{\partial y}\frac{\partial s}{\partial x} = 0,$$

$$\frac{\partial y}{\partial s} = \frac{\partial y}{\partial t}\frac{\partial t}{\partial s} + \frac{\partial y}{\partial x}\frac{\partial x}{\partial s},$$

$$\frac{\partial K}{\partial y} = \frac{\partial K}{\partial x}\frac{\partial x}{\partial y},$$

we have

$$\frac{\partial y}{\partial s} - J\ {}^t(\frac{\partial K}{\partial y})$$

$$= \frac{\partial y}{\partial x}\left\{\frac{\partial x}{\partial t} + \frac{\partial x}{\partial y}\frac{\partial y}{\partial t} - \frac{\partial x}{\partial y}J\ {}^t(\frac{\partial x}{\partial y})\ {}^t(\frac{\partial K}{\partial x})\frac{\partial s}{\partial t}\right\}\frac{\partial t}{\partial s}.$$

Since (6.6) implies

$$J\ {}^t(\frac{\partial H}{\partial x}) = J\ {}^t(\frac{\partial K}{\partial x})\frac{\partial s}{\partial t} + J\ {}^t(\frac{\partial y}{\partial x})J\frac{\partial y}{\partial t},$$

and (6.5) implies

$$\frac{\partial x}{\partial y} = -J\ {}^t(\frac{\partial y}{\partial x})J,$$

we have

$$\frac{\partial y}{\partial s} - J\ {}^t(\frac{\partial K}{\partial y}) = \frac{\partial y}{\partial x}\left\{\frac{\partial x}{\partial t} - J\ {}^t(\frac{\partial H}{\partial x})\right\}\frac{\partial t}{\partial s}. \quad \blacksquare$$

We shall see how the integrability condition for the system is expressed. Let $\mathcal{I}$ be the ideal of the exterior algebra in the variables t and x generated by

$$\omega_i := dx^i - \sum_{j=1}^{m}\{H_j, x^i\}dt^j, \quad i = 1, \ldots, 2n,$$

and put

$$\Gamma_{ij} := \partial_i H_j - \partial_j H_i + \{H_i, H_j\},$$

where $\partial_i = \partial/\partial t^i$. Notice that for a function $f(t,x)$, we have

$$df(t,x) \equiv \sum_i \left(\frac{\partial f}{\partial t^i} + \{H_i, f\} \right) dt^i \mod \mathcal{I}. \tag{6.8}$$

In fact

$$\begin{aligned} df &= \sum_i \frac{\partial f}{\partial t^i} dt^i + \sum_{j=1}^{2n} \frac{\partial f}{\partial x^j} dx^j \\ &\equiv \sum_i \left\{ \frac{\partial f}{\partial t^i} + \sum_{j=1}^{n} \left(\frac{\partial f}{\partial x^j} \frac{\partial H_i}{\partial x^{j+n}} - \frac{\partial f}{\partial x^{j+n}} \frac{\partial H_i}{\partial x^j} \right) \right\} dt^i \\ &= \sum_i \left(\frac{\partial f}{\partial t^i} + \{H_i, f\} \right) dt^i. \end{aligned}$$

LEMMA 6.3. *For the symplectic 2-form Γ associated with the system (6.1), we have*

$$\Gamma \equiv \sum_{1 \le i < j \le m} \Gamma_{ij} dt^i \wedge dt^j \mod \mathcal{I}.$$

Proof. We have

$$\begin{aligned} \Gamma &= \sum_{i=1}^{n} dx^i \wedge dx^{n+i} + \sum_{k=1}^{m} dH_k \wedge dt^k \\ &\equiv \sum_{i=1}^{n} \sum_{j,k=1}^{m} \{H_j, x^i\}\{H_k, x^{n+i}\} dt^j \wedge dt^k \\ &\quad + \sum_{j,k=1}^{m} (\partial_j H_k + \{H_j, H_k\}) dt^j \wedge dt^k \end{aligned}$$

$$= \sum_{1\le j<k\le m}\left[\sum_{i=1}^{n}(\{H_j,x^i\}\{H_k,x^{n+i}\}-\{H_k,x^i\}\{H_j,x^{n+i}\})\right]dt^j\wedge dt^k$$

$$+ \sum_{1\le j<k\le m}(\partial_j H_k-\partial_k H_j+2\{H_j,H_k\})dt^j\wedge dt^k$$

$$= \sum_{1\le j<k\le m}(\partial_j H_k-\partial_k H_j+\{H_j,H_k\})dt^j\wedge dt^k$$

$$= \sum_{1\le j<k\le m}\Gamma_{jk}dt^j\wedge dt^k$$

modulo $\mathcal{I}$, where we used the following identity to show the last equality:

$$\sum_{i=1}^{n}(\{H_j,x^i\}\{H_k,x^{n+i}\}-\{H_k,x^i\}\{H_j,x^{n+i}\}) = -\{H_j,H_k\}. \blacksquare$$

PROPOSITION 6.4. *A Hamiltonian system (6.1) is completely integrable if and only if Γ_{jk} $(j,k=1,\ldots,m)$ are independent of x, i.e.,*

$$\{\Gamma_{jk},x^i\}=0 \quad i=1,\ldots,2n;\ j,k=1,\ldots,m.$$

Proof. We have the following equalities modulo $\mathcal{I}$:

$$d\omega_i = d(dx^i-\sum_{j=1}^{m}\{H_j,x^i\}dt^j)$$

$$\equiv \sum_{j,k=1}^{m}(\partial_k\{H_j,x^i\}+\{H_k,\{H_j,x^i\}\})dt^j\wedge dt^k$$

$$= \sum_{1\le j<k\le m}(\{\partial_k H_j,x^i\}-\{\partial_j H_k,x^i\}$$

$$+\{H_k,\{H_j,x^i\}\}-\{H_j,\{H_k,x^i\}\})dt^j\wedge dt^k$$

$$= \sum_{1\le j<k\le m}(\{\partial_k H_j,x^i\}-\{\partial_j H_k,x^i\}+\{x^i,\{H_k,H_j\}\})dt^j\wedge dt^k$$

$$= \sum_{1\le j<k\le m}\{\Gamma_{jk},x^i\}dt^j\wedge dt^k$$

modulo $\mathcal{I}$, where we used (6.8) and the Jacobi identity to show the second and the fourth equality, respectively. Thus Corollary 5.2 leads to the conclusion. ∎

2 Hypergeometric Differential Equation

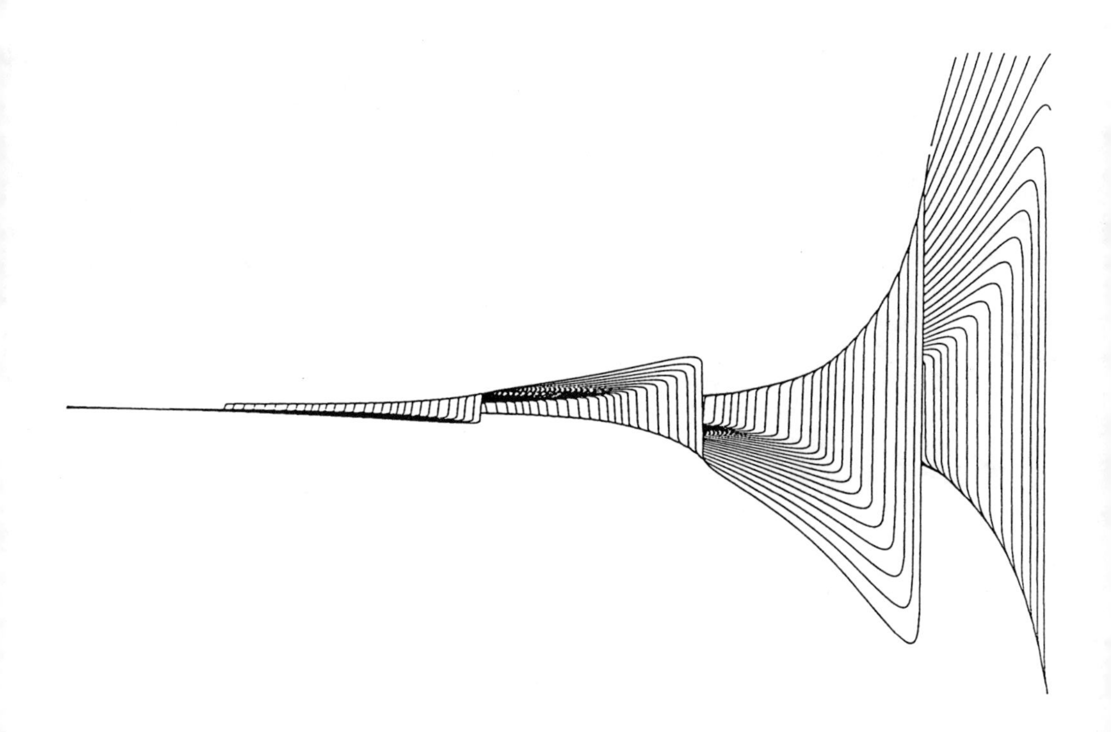

The leading example, in the theory of linear ordinary differential equations with regular singular points of one complex variable, is doubtless the hypergeometric differential equation:

$$x(1-x)\frac{d^2u}{dx^2} + \{\gamma - (\alpha+\beta+1)x\}\frac{du}{dx} - \alpha\beta u = 0,$$

which is a normalized form of linear ordinary differential equations with three regular singular points in the Riemann sphere.

A series of works on this equation by Euler, by Gauss, by Kummer and especially by Riemann inspired L. Fuchs to conceive the concept of regular singularity of linear ordinary differential equations in complex domains. Since then, the theory of Fuchsian differential equations has been developed into various directions, for example, expressing local solutions at singular points, studying the relation between a fundamental system of solutions at a singular point and that at another singular point (the so-called connection problem), studying the monodromy of linear differential equations, and constructing a linear differential equations for a given monodromy (the so-called Riemann-Hilbert problem). The theory has been developed also for various classes of Fuchsian equations, for example, the Jordan-Pochhammer equations, the generalized hypergeometric equations ${}_pF_q$ and hypergeometric equations in several variables, which inherit some of the nice properties of the hypergeometric differential equation presented here. Thus what follows may be considered as a resumé of the whole theory of linear ordinary differential equations with regular singular points of one complex variable. It should be noticed that the equation also appears in many different situations, for instance, in conformal mapping theory, in automorphic function theory, in the theory of representations of Lie algebras, and in the theory of difference equations.

The present chapter is entirely devoted to discussing various aspects of the hypergeometric equation and its solutions by focusing our attention on the connection problem; we exhibit five different methods to find its monodromy groups. We also explain in detail contiguity relations and related topics. On the other hand we do not mention subjects related to conformal mapping theory or automorphic function theory. The reader who is interested in these topics can consult [YosM].

1. Definition and basic facts

1.1 The Gauss hypergeometric equation

Let a *Riemann equation* be a second order Fuchsian equation with three regular singular points on the Riemann sphere $\mathbb{P}^1$, and let a *Riemann scheme* be a table of the following type

$$\begin{pmatrix} a_1 & a_2 & a_3 \\ \sigma_1 & \sigma_2 & \sigma_3 \\ \tau_1 & \tau_2 & \tau_3 \end{pmatrix}, \tag{1.1.1}$$

where a_j are three points on the Riemann sphere $\mathbb{P}^1$ and σ_j, τ_j are complex numbers satisfying the Fuchs relation :

$$\sum_{j=1}^{3}(\sigma_j + \tau_j) = 1. \tag{1.1.2}$$

PROPOSITION 1.1.1. *To each Riemann scheme* (1.1.1) *there is a unique Riemann equation* $E(a;\sigma,\tau)$ *with regular singular points* a_j *and with the characteristic exponents* σ_j, τ_j *at* a_j $(j = 1, 2, 3)$.

Proof. Assume $a_3 = \infty$, then by Proposition 4.2 in Chapter 1 a Fuchsian equation with regular singular points at $\{a_1, a_2, a_3\}$ is of the form

$$\frac{d^2u}{dx^2} + q_1\frac{du}{dx} + q_2 u = 0$$

where

$$q_1 = \frac{p_1}{(x-a_1)(x-a_2)}, \qquad q_2 = \frac{p_2}{(x-a_1)^2(x-a_2)^2}$$

and p_k $(k = 1, 2)$ are polynomials in x with degree at most k. Thus q_1 and q_2 can be written as

$$q_1 = \frac{A_1}{x-a_1} + \frac{A_2}{x-a_2},$$

$$q_2 = \frac{B_1}{(x-a_1)^2} + \frac{B_2}{(x-a_2)^2} + \frac{B_3}{(x-a_1)(x-a_2)},$$

where A_i and B_j are constants. Characteristic equations (Chapter 1 Section 4) at $x = a_1, a_2, \infty$ are given by

$$s^2 + (A_1 - 1)s + B_1 = 0,$$

$$s^2 + (A_2 - 1)s + B_2 = 0,$$

$$s^2 + (-A_1 - A_2 + 1)s + B_1 + B_2 + B_3 = 0,$$

respectively. The exponents $\{\sigma_k, \tau_k\}$ $(k = 1,2,3)$ at a_k are, by definition, solutions of the above three quadratic equations, and hence we have

$$\begin{aligned} A_1 - 1 &= -\sigma_1 - \tau_1, & B_1 &= \sigma_1\tau_1 \\ A_2 - 1 &= -\sigma_2 - \tau_2, & B_2 &= \sigma_2\tau_2 \\ -A_1 - A_2 + 1 &= -\sigma_3 - \tau_3, & B_1 + B_2 + B_3 &= \sigma_3\tau_3. \end{aligned}$$

The Fuchs relation (1.1.2) tells us that the above linear equations with respect to A_i and B_j can be uniquely solved.
When $a_3 \neq \infty$, one has only to apply a linear fractional change of the variable x to send a_3 to infinity. ∎

If we identify a Riemann scheme with those obtained by reordering the columns and by permuting σ_j and $\tau_j (j = 1,2,3)$, then the correspondence is bijective. *The Riemann P-function*

$$P\begin{pmatrix} a_1 & a_2 & a_3 & \\ \sigma_1 & \sigma_2 & \sigma_3 & ;x \\ \tau_1 & \tau_2 & \tau_3 & \end{pmatrix} \tag{1.1.3}$$

is, by definition, the collection of solutions of the Riemann equation $E(a;\sigma,\tau)$ associated with the Riemann scheme (1.1.1).

Now we shall see that the investigation of a Riemann equation can be reduced to that of a more special equation. Since any automorphism T of $\mathbb{P}^1$ transforms $E(a;\sigma,\tau)$ into $E(T(a);\sigma,\tau)$, we may assume, without loss of generality, that $a_1 = 0, a_2 = 1$ and $a_3 = \infty$. Moreover, since

$$P\begin{pmatrix} 0 & 1 & \infty & \\ \sigma_0 & \sigma_1 & \sigma_\infty & ;x \\ \tau_0 & \tau_1 & \tau_\infty & \end{pmatrix} \tag{1.1.4}$$

$$= x^{\sigma_0}(x-1)^{\sigma_1} P\begin{pmatrix} 0 & 1 & \infty & \\ 0 & 0 & \sigma_\infty + \sigma_0 + \sigma_1 & ;x \\ \tau_0 - \sigma_0 & \tau_1 - \sigma_1 & \tau_\infty + \sigma_0 + \sigma_1 & \end{pmatrix},$$

the knowledge of the P-function (1.1.3) reduces to that of

$$P\begin{pmatrix} 0 & 1 & \infty & \\ 0 & 0 & \alpha & ;x \\ 1-\gamma & \gamma-\alpha-\beta & \beta & \end{pmatrix} \tag{1.1.5}$$

by putting $\alpha = \sigma_0 + \sigma_1 + \sigma_\infty, \beta = \tau_\infty + \sigma_0 + \sigma_1$ and $\gamma = 1 + \sigma_0 - \tau_0$. The associated Riemann equation, called the (*Gauss*) *hypergeometric equation*, denoted by $E(\alpha, \beta, \gamma)$, reads as follows

$$x(1-x)\frac{d^2u}{dx^2} + \{\gamma - (\alpha+\beta+1)x\}\frac{du}{dx} - \alpha\beta u = 0. \tag{1.1.6}$$

Notice that $E(\alpha, \beta, \gamma)$ admits another convenient expression

$$\delta_x(\delta_x + \gamma - 1)u - x(\delta_x + \alpha)(\delta_x + \beta)u = 0, \tag{1.1.7}$$

which will be used later, where δ_x is the Euler operator $x(d/dx)$. Thus the investigation of a general Riemann equation boils down to that of the hypergeometric equation with suitable parameters α, β and γ.

REMARK 1.1.2. In other words, in the present context, one may define the hypergeometric equation as a Riemann equation with singular points $0, 1$ and ∞ such that one of the characteristic exponents at each finite singular point is zero.

1.2 Hypergeometric series

The hypergeometric differential equation $E(\alpha,\beta,\gamma)$ admits solutions holomorphic in the unit open disk

$$\Delta = \{x \in \mathbb{C} |\ |x| < 1\}$$

if

$$\gamma \neq 0, -1, -2, \cdots, \tag{1.2.1}$$

since its Riemann scheme is of the form (1.1.5). The Frobenius method gives a solution which admits the following power series expansion:

$$F(\alpha,\beta,\gamma;x) = \sum_{m=0}^{\infty} \frac{(\alpha)_m(\beta)_m}{(\gamma)_m(1)_m} x^m, \tag{1.2.2}$$

called the *hypergeometric series*, where $(\alpha)_m$ denotes the following factorial function

$$(\alpha)_m = \begin{cases} 1 & (m=0), \\ \alpha(\alpha+1)\cdots(\alpha+m-1) & (m=1,2,3,\cdots). \end{cases}$$

This factorial function can be expressed also in terms of the Gamma function $\Gamma(z)$ defined by

$$\Gamma(z) := \int_0^\infty e^{-t} t^{z-1} dt \quad \Re z > 0.$$

It satisfies the difference equation:

$$\Gamma(z+1) = z\Gamma(z),$$

and so we have

$$\Gamma(1) = 1, \quad \Gamma(m+1) = m!$$

and

$$(\alpha)_m = \frac{\Gamma(\alpha+m)}{\Gamma(\alpha)} \qquad \alpha \neq 0, -1, -2, \cdots.$$

By using the above functional equation, Γ is continued analytically to define a meromorphic function in $\mathbb{C}$, which has simple poles at $-m$ ($m = 0,1,2,\dots$) with the residue $(-1)^m/m!$. We quote here the *Stirling formula:*

$$\Gamma(z) \sim (2\pi)^{\frac{1}{2}} z^{z-\frac{1}{2}} e^{-z} \quad (z \to \infty,\ |\arg z| < \pi - \delta)$$

for an arbitrary positive δ, where $\sim$ implies that the ratio of the right and left-hand sides tends to 1.

REMARK 1.2.1. (i) The power series $F(\alpha, \beta, \gamma; x)$ is symmetric in α and β. (ii) $F(\alpha, \beta, \gamma; x)$ is a polynomial in x if and only if either α or β is a non-positive integer.

THEOREM 1.2.2. *If neither α nor β is a non-positive integer, then*
(i) *the radius of convergence of $F(\alpha, \beta, \gamma; x)$ is 1,*
(ii) *if $\Re(\gamma - \alpha - \beta) > 0$, then $F(\alpha, \beta, \gamma; x)$ converges absolutely on the unit circle $\{|x| = 1\}$.*

This theorem will be proved by applying the following two criteria.

LEMMA 1.2.3.
(a) (*the d'Alembert test*). *The series $\sum_{n=0}^{\infty} a_n$ converges (resp. diverges) if*

$$\lim_{n\to\infty} |\frac{a_{n+1}}{a_n}| < 1 \quad (\textit{resp.} > 1).$$

(b) (*the Raabe test*). *The series $\sum_{n=0}^{\infty} a_n$ converges if there exists a constant $C > 1$ and a positive number ε such that*

$$|\frac{a_{n+1}}{a_n}| = 1 - \frac{C}{n} + \mathcal{O}(\frac{1}{n^{1+\varepsilon}}) \quad (n \to \infty),$$

where $\mathcal{O}$ stands for Landau's symbol, and the condition means that there is a positive number M such that

$$\left| |\frac{a_{n+1}}{a_n}| - 1 + \frac{C}{n} \right| \leq \frac{M}{n^{1+\varepsilon}}$$

for all sufficiently large n.

Proof. (a) We have only to recall the fact: The series

$$\sum_{n=0}^{\infty} r^n \quad (r > 0)$$

converges (resp. diverges) if $r < 1$ (resp. $r > 1$).

(b) We recall the fact: The series

$$\sum_{n=1}^{\infty} b_n \quad b_n = \frac{1}{n^s}, \qquad s \in \mathbb{R}$$

converges if $s > 1$. Since

$$\begin{aligned}\frac{b_{n+1}}{b_n} &= (1 + \frac{1}{n})^{-s} \\ &= 1 - \frac{s}{n} + \mathcal{O}(\frac{1}{n^2}) \quad (n \to \infty),\end{aligned}$$

choosing s so that $1 < s < C$, we have

$$|\frac{a_{n+1}}{a_n}| \leq \frac{b_{n+1}}{b_n}$$

for sufficiently large n. ∎

Proof of Theorem 1.2.2. (i) Since

$$\begin{aligned}&\frac{(\alpha)_{n+1}(\beta)_{n+1}}{(\gamma)_{n+1}(1)_{n+1}} x^{n+1} \left(\frac{(\alpha)_n(\beta)_n}{(\gamma)_n(1)_n} x^n \right)^{-1} \\ &= \frac{(\alpha+n)(\beta+n)}{(\gamma+n)(1+n)} x,\end{aligned}$$

the assertion follows from (a) of the lemma.

(ii) We apply (b) of the lemma to

$$a_n = \frac{(\alpha)_n(\beta)_n}{(\gamma)_n(1)_n} x^n.$$

We have, on $|x| = 1$,

$$
\begin{aligned}
\left|\frac{a_{n+1}}{a_n}\right| &= \left|\frac{(\alpha+n)(\beta+n)}{(\gamma+n)(1+n)}\right| \\
&= \left|\frac{(1+\frac{\alpha}{n})(1+\frac{\beta}{n})}{(1+\frac{\gamma}{n})(1+\frac{1}{n})}\right| \\
&= \left|\frac{1+\dfrac{\alpha+\beta}{n}+\mathcal{O}(\dfrac{1}{n^2})}{1+\dfrac{\gamma+1}{n}+\mathcal{O}(\dfrac{1}{n^2})}\right| \\
&= \left|\left(1+\frac{\alpha+\beta}{n}+\mathcal{O}(\frac{1}{n^2})\right)\left(1-\frac{\gamma+1}{n}+\mathcal{O}(\frac{1}{n^2})\right)\right| \\
&= \left|1-\frac{1+\gamma-\alpha-\beta}{n}+\mathcal{O}(\frac{1}{n^2})\right| \\
&= 1-\frac{1+\Re(\gamma-\alpha-\beta)}{n}+\mathcal{O}(\frac{1}{n^2}),
\end{aligned}
$$

which proves (ii). Another proof using the Stirling formula is given as follows: We have

$$\frac{\Gamma(z+n)}{\Gamma(z)} \sim n^z,$$

and so that

$$\frac{(\alpha)_n(\beta)_n}{(\gamma)_n(1)_n} = \frac{\Gamma(\alpha+n)\Gamma(\beta+n)}{\Gamma(\gamma+n)\Gamma(1+n)} \sim \frac{1}{n^{\gamma-\alpha-\beta+1}}. \blacksquare$$

REMARK 1.2.4. From the viewpoint of the theory of differential equations, one can give a satisfactory explanation of the assumption $\Re(\gamma-\alpha-\beta) > 0$ made in Theorem 1.2.2, (ii): Since the hypergeometric differential equation has no singular points on $|x| \leq 1$ except at $x = 0, 1$, the hypergeometric series converges in $|x| \leq 1, x \neq 1$. Moreover, since the hypergeometric differential equation has a regular singular point at $x = 1$ with characteristic exponents 0 and $\gamma - \alpha - \beta$, there is a fundamental system of solutions around $x = 1$ which is of the form $(f_1(x), (x-1)^{\gamma-\alpha-\beta} f_2(x))$, where $f_j(x)$ are holomorphic functions around $x = 1$. Thus, near $x = 1$ in the unit disk, $F(\alpha, \beta, \gamma; x)$ can be written in a linear combination :

$$F(\alpha, \beta, \gamma; x) = c_1 f_1(x) + c_2 (x-1)^{\gamma-\alpha-\beta} f_2(x).$$

If $\Re(\gamma - \alpha - \beta) > 0$ the second term on the right-hand side tends to zero as $x \to 1$. Thus the limit $\lim_{x\to 1} F(\alpha, \beta, \gamma; x)$ exists with value $c_1 f_1(1)$, which will be calculated explicitly in Section 3.7. Hence one can expect that the series $F(\alpha, \beta, \gamma; x)$ is convergent for $x = 1$.

We consider the sum of the hypergeometric series as a function in four variables $(\alpha, \beta, \gamma, x)$. Since the hypergeometric series converges uniformly with respect to α, β, γ and x in any compact subset of $\Omega := \mathbb{C} \times \mathbb{C} \times (\mathbb{C} \backslash \{0, -1, -2, \cdots\}) \times \Delta$, we have

THEOREM 1.2.5. *The sum of the hypergeometric series $F(\alpha, \beta, \gamma; x)$ is holomorphic in $\Omega = \mathbb{C} \times \mathbb{C} \times (\mathbb{C} \backslash \{0, -1, -2, \cdots\}) \times \Delta$.*

The holomorphic function in $|x| < 1$ defined by the hypergeometric series, as well as its analytic continuation, is called the (*Gauss*) *hypergeometric function.*

1.3 Finite group action and Kummer's 24 solutions

Let $E(\alpha,\beta,\gamma)$ be the hypergeometric equation, i.e., the Riemann equation corresponding to the Riemann scheme

$$R(\alpha,\beta,\gamma)=\begin{pmatrix} 0 & 1 & \infty \\ 0 & 0 & \alpha \\ 1-\gamma & \gamma-\alpha-\beta & \beta \end{pmatrix}.$$

A suitable change of the dependent and the independent variables

$$u(x)\mapsto x^{\mu}(1-x)^{\nu}u(h(x)),\quad x\mapsto h(x),$$

$\mu,\nu\in\mathbb{C}$, $h\in \mathrm{Aut}(\mathbb{C}-\{0,1\})$ takes $E(\alpha,\beta,\gamma)$ into another hypergeometric equation $E(a,b,c)$ for some a,b,c. There are 24 of them; it is checked by the following consideration. The group

$$\begin{aligned} H := & \ \mathrm{Aut}(\mathbb{C}-\{0,1\}) \\ = & \ \{x\mapsto x,\frac{1}{x},1-x,\frac{1}{1-x},\frac{x-1}{x},\frac{x}{x-1}\} \end{aligned}$$

acts on the set $\{0,1,\infty\}$ as the full group of permutations. For each $h\in H$, there is a vector (not unique)

$$\boldsymbol{\mu}=(\mu_0,\mu_1,\mu_\infty)\qquad \mu_0+\mu_1+\mu_\infty=0$$

such that

$$R(h,\boldsymbol{\mu}):=\begin{pmatrix} h(0) & h(1) & h(\infty) \\ 0+\mu_{h(0)} & 0+\mu_{h(1)} & \alpha+\mu_{h(\infty)} \\ 1-\gamma+\mu_{h(0)} & \gamma-\alpha-\beta+\mu_{h(1)} & \beta+\mu_{h(\infty)} \end{pmatrix}$$

is in the form $R(a,b,c)$ for some a,b,c. For example, let h be $x\mapsto 1/x$ then we can choose $\boldsymbol{\mu}=(-\alpha,0,\alpha)$ and we have

$$\begin{aligned} R(h,\boldsymbol{\mu}) &= \begin{pmatrix} h(0)=\infty & h(1)=1 & h(\infty)=0 \\ 0+\alpha & 0 & \alpha-\alpha \\ 1-\gamma+\alpha & \gamma-\alpha-\beta & \beta-\alpha \end{pmatrix} \\ &= \begin{pmatrix} 0 & 1 & \infty \\ 0 & 0 & \alpha \\ \beta-\alpha & \gamma-\alpha-\beta & 1-\gamma+\alpha \end{pmatrix} \\ &= R(\alpha,\alpha-\gamma+1,\alpha+1-\beta). \end{aligned}$$

Another example: Let h be

$$x \mapsto \frac{x-1}{x} \text{ and put } \boldsymbol{\mu} = (-(\gamma-\alpha-\beta), -\alpha, \gamma-\beta);$$

then we have

$$\begin{aligned} R(h,\boldsymbol{\mu}) &= \begin{pmatrix} h(0)=\infty & h(1)=0 & h(\infty)=1 \\ 0+(\gamma-\beta) & 0-(\gamma-\alpha-\beta) & \alpha-\alpha \\ 1-\gamma+(\gamma-\beta) & \gamma-\alpha-\beta-(\gamma-\alpha-\beta) & \beta-\alpha \end{pmatrix} \\ &= \begin{pmatrix} 0 & 1 & \infty \\ 0 & 0 & \gamma-\beta \\ -(\gamma-\alpha-\beta) & \beta-\alpha & 1-\beta \end{pmatrix} \\ &= R(\gamma-\beta, 1-\beta, \gamma-\alpha-\beta+1). \end{aligned}$$

One more example: Let h be the identity and put $\boldsymbol{\mu} = (-(1-\gamma), 0, 1-\gamma)$; then we have

$$\begin{aligned} R(h,\boldsymbol{\mu}) &= \begin{pmatrix} 0 & 1 & \infty \\ 0-(1-\gamma) & 0 & \alpha+1-\gamma \\ 1-\gamma-(1-\gamma) & \gamma-\alpha-\beta & \beta+1-\gamma \end{pmatrix} \\ &= R(\alpha-\gamma+1, \beta-\gamma+1, 2-\gamma). \end{aligned}$$

In this way, we can find 24 pairs $(h,\boldsymbol{\mu})$ and the corresponding 24 Riemann scheme $R(h,\boldsymbol{\mu}) = R(a,b,c)$. Passing to the corresponding Riemann equation $E(a,b,c)$, we see that its any solution is expressed as

$$y^{\mu_0}(1-y)^{\mu_1} f(h^{-1}(y)),$$

where f is a solution of the original equation $E(a,b,c)$. Associating the series $F(a,b,c;x)$ with $E(a,b,c)$, conversely, we conclude that

$$h(x)^{-\mu_0}(1-h(x))^{-\mu_1} F(a,b,c;h(x))$$

is a solution of $E(\alpha,\beta,\gamma)$. Therefore we get 24 expressions of solutions of the equation $E(\alpha,\beta,\gamma)$ of the form

$$x^{\mu}(1-x)^{\nu} F(a,b,c;h(x)),$$

which are called *Kummer's 24 solutions.* From the three examples above, we have

$$\left(\frac{1}{x}\right)^{\alpha} F(\alpha, \alpha-\gamma+1, \alpha+1-\beta; \frac{1}{x}),$$

$$\left(\frac{x-1}{x}\right)^{\gamma-\alpha-\beta}\left(1-\frac{x-1}{x}\right)^{\alpha} F(\gamma-\beta, 1-\beta, \gamma-\alpha-\beta+1; \frac{x-1}{x})$$

$$\left(= x^{\beta-\gamma}(x-1)^{\gamma-\alpha-\beta} F(1-\beta, \gamma-\beta, \gamma-\alpha-\beta+1; \frac{x-1}{x})\right),$$

and

$$x^{1-\gamma} F(\alpha-\gamma+1, \beta-\gamma+1, 2-\gamma; x).$$

For later use we name and tabulate the 24 expressions:

$$\begin{aligned} f_0(x;0) := & F(\alpha, \beta, \gamma; x) \\ = & (1-x)^{\gamma-\alpha-\beta} F(\gamma-\alpha, \gamma-\beta, \gamma; x) \\ = & (1-x)^{-\alpha} F(\gamma-\beta, \alpha, \gamma; \frac{x}{x-1}) \\ = & (1-x)^{-\beta} F(\gamma-\alpha, \beta, \gamma; \frac{x}{x-1}). \end{aligned}$$

$$\begin{aligned} f_0(x;1-\gamma) := & x^{1-\gamma} F(\alpha-\gamma+1, \beta-\gamma+1, 2-\gamma; x) \\ = & x^{1-\gamma}(1-x)^{\gamma-\alpha-\beta} F(1-\alpha, 1-\beta, 2-\gamma; x) \\ = & x^{1-\gamma}(1-x)^{\gamma-\alpha-1} F(1-\beta, \alpha+1-\gamma, 2-\gamma; \frac{x}{x-1}) \\ = & x^{1-\gamma}(1-x)^{\gamma-\beta-1} F(1-\alpha, \beta+1-\gamma, 2-\gamma; \frac{x}{x-1}). \end{aligned}$$

$$\begin{aligned} f_1(x;0) := & F(\alpha, \beta, \alpha+\beta-\gamma+1; 1-x) \\ = & x^{1-\gamma} F(\beta+1-\gamma, \alpha+1-\gamma, \alpha+\beta+1-\gamma; 1-x) \\ = & x^{-\alpha} F(\alpha+1-\gamma, \alpha, \alpha+\beta+1-\gamma; \frac{x-1}{x}) \\ = & x^{-\beta} F(\beta+1-\gamma, \beta, \beta+\alpha+1-\gamma; \frac{x-1}{x}). \end{aligned}$$

$$\begin{aligned}
f_1(x;\gamma-\alpha-\beta)&\\
&:=(1-x)^{\gamma-\alpha-\beta}F(\gamma-\alpha,\gamma-\beta,\gamma+1-\alpha-\beta;1-x)\\
&=x^{1-\gamma}(1-x)^{\gamma-\alpha-\beta}F(1-\alpha,1-\beta,\gamma+1-\alpha-\beta;1-x)\\
&=x^{\beta-\gamma}(1-x)^{\gamma-\alpha-\beta}F(1-\beta,\gamma-\beta,\gamma+1-\alpha-\beta;\frac{x-1}{x})\\
&=x^{\alpha-\gamma}(1-x)^{\gamma-\beta-\alpha}F(1-\alpha,\gamma-\alpha,\gamma+1-\alpha-\beta;\frac{x-1}{x}).
\end{aligned}$$

$$\begin{aligned}
f_\infty(x;\alpha)&:=x^{-\alpha}F(\alpha,\alpha-\gamma+1,\alpha+1-\beta;\frac{1}{x})\\
&=(-x)^{\beta-\gamma}(1-x)^{\gamma-\alpha-\beta}F(1-\beta,\gamma-\beta,\alpha+1-\beta;\frac{1}{x})\\
&=(1-x)^{-\alpha}F(\alpha,\gamma-\beta,\alpha+1-\beta;\frac{1}{1-x})\\
&=(-x)^{1-\gamma}(1-x)^{\gamma-\alpha-1}F(\alpha+1-\gamma,1-\beta,\alpha+1-\beta;\frac{1}{1-x}).
\end{aligned}$$

$$\begin{aligned}
f_\infty(x;\beta)&:=x^{-\beta}F(\beta,\beta-\gamma+1,\beta+1-\alpha;\frac{1}{x})\\
&=(-x)^{\alpha-\gamma}(1-x)^{\gamma-\alpha-\beta}F(1-\alpha,\gamma-\alpha,\beta+1-\alpha;\frac{1}{x})\\
&=(1-x)^{-\beta}F(\beta,\gamma-\alpha,\beta+1-\alpha;\frac{1}{1-x})\\
&=(-x)^{\alpha-\gamma}(1-x)^{\gamma-\beta-1}F(\beta+1-\gamma,1-\alpha,\beta+1-\alpha;\frac{1}{1-x}).
\end{aligned}$$

These series make sense if

$$\gamma,\ \gamma-\alpha-\beta,\ \alpha-\beta\notin\mathbb{Z}.$$

The pairs of solutions $(f_0(x;0), f_0(x;1-\gamma))$, $(f_1(x;0), f_1(x;\gamma-\alpha-\beta))$ and $(f_\infty(x;\alpha), f_\infty(x;\beta))$ have characteristic exponents indicated in (1.1.5) at $x=0,1$ and ∞, respectively.

Group-theoretical Remark. The 24 linear transformations

$$(\alpha,\beta,\gamma)\to(a,b,c)$$

together with the transformation

$$n_0 : (\alpha, \beta, \gamma) \to (\beta, \alpha, \gamma)$$

form a group G of order 48: Let

$$\begin{aligned}
n_1 : (\alpha, \beta, \gamma) &\to (\alpha - \gamma + 1, \beta - \gamma + 1, 2 - \gamma) \\
n_2 : (\alpha, \beta, \gamma) &\to (\gamma - \alpha, \gamma - \beta, \gamma) \\
h_1 : (\alpha, \beta, \gamma) &\to (\alpha, \alpha - \gamma + 1, \alpha - \beta + 1) \\
h_2 : (\alpha, \beta, \gamma) &\to (\alpha, \beta, \alpha + \beta - \gamma + 1),
\end{aligned}$$

where n_1 and n_2 are the transformations of the parameters corresponding to $R(h, \boldsymbol{\mu})$:

$$h = \text{identity}, \quad \boldsymbol{\mu} = (-(1-\gamma), 0, 1-\gamma)$$

and

$$h = \text{identity}, \quad \boldsymbol{\mu} = (0, -(\gamma - \alpha - \beta), \gamma - \alpha - \beta),$$

respectively; h_1 and h_2 are the transformations of the parameters corresponding to $R(h, \boldsymbol{\mu})$:

$$h : x \mapsto \frac{1}{x}, \qquad \boldsymbol{\mu} = (-\alpha, 0, \alpha)$$

and

$$h : x \mapsto 1 - x, \quad \boldsymbol{\mu} = (0, 0, 0),$$

respectively.

Put

$$N = \langle n_0, n_1, n_2 \rangle, \quad H' = \langle h_1, h_2 \rangle,$$

where $\langle k_1, k_2, \cdots \rangle$ stands for the group generated by $k_1, k_2, \cdots$. Then N is a normal subgroup of G, and G has the following structure:

$$\begin{aligned}
&N \simeq (\mathbb{Z}/2\mathbb{Z})^3, \\
&H' \simeq H = \mathrm{Aut}(\mathbb{C} - \{0, 1\}) \simeq \text{symmetric group of degree 3}
\end{aligned}$$

and

$$G = NH, \quad N \cap H = \{1\}.$$

2 Contiguity relations

Let us introduce the ring of linear differential operators with coefficients in the field $\mathbb{C}(x)$ of rational functions. Consider the $\mathbb{C}(x)$-vector space with basis:

$$\frac{d^k}{dx^k} \qquad k = 0, 1, 2, \cdots.$$

Define multiplication inductively as follows:

$$\frac{d}{dx}(p\frac{d^k}{dx^k}) = p'\frac{d^k}{dx^k} + p\frac{d^{k+1}}{dx^{k+1}} \qquad p \in \mathbb{C}(x),$$

$$\frac{d^m}{dx^m}(p\frac{d^k}{dx^k}) = \frac{d^{m-1}}{dx^{m-1}}\left\{\frac{d}{dx}\left(p\frac{d^k}{dx^k}\right)\right\},$$

where p' stands for the derivative dp/dx of p with respect to x. The vector space equipped with the multiplication has the structure of a ring; it is called the *ring of linear differential operators with coefficients in* $\mathbb{C}(x)$, and is denoted by

$$\mathcal{R} = \mathbb{C}(x)[\frac{d}{dx}].$$

We sometimes write $\left(\frac{d}{dx}\right)^k$ in place of $\frac{d^k}{dx^k}$. Note that the definition of multiplication is forced by the formula for the derivative of the product of two functions:

$$\frac{d}{dx}(uv) = \frac{du}{dx}v + u\frac{dv}{dx}.$$

We have the Leibniz rule:

$$\frac{d^k}{dx^k}p = \sum_{j=0}^{k}\binom{k}{j}p^{(j)}\frac{d^{k-j}}{dx^{k-j}},$$

where $p^{(j)}$ stands for the j-th derivative $\frac{d^j p}{dx^j}$ of p. This is proved inductively as follows:

$$\begin{aligned}
\frac{d^{k+1}}{dx^{k+1}}p &= \frac{d}{dx}\left(\frac{d^{k-j}}{dx^{k-j}}p\right)\\
&= \frac{d}{dx}\sum_{j=0}^{k}\binom{k}{j}p^{(j)}\frac{d^{k-j}}{dx^{k-j}}\\
&= \sum_{j=0}^{k}\binom{k}{j}\left\{p^{(j+1)}\frac{d^{k-j}}{dx^{k-j}}+p^{(j)}\frac{d^{k-j+1}}{dx^{k-j+1}}\right\}\\
&= \sum_{j=0}^{k+1}\left\{\binom{k}{j}+\binom{k}{j-1}\right\}p^{(j)}\frac{d^{k+1-j}}{dx^{k+1-j}}\\
&= \sum_{j=0}^{k+1}\binom{k+1}{j}p^{(j)}\frac{d^{k+1-j}}{dx^{k+1-j}}.
\end{aligned}$$

2.1 Contiguity relations

Let $\mathcal{S}(\alpha,\beta,\gamma)$ be the linear space of solutions of the hypergeometric equation $E(\alpha,\beta,\gamma)$ at a fixed point $\neq 0,1$. There are first order differential operators which send $\mathcal{S}(\alpha,\beta,\gamma)$ into $\mathcal{S}(\alpha+\varepsilon_1,\beta+\varepsilon_2,\gamma+\varepsilon_3)$, where $\varepsilon_j = 0,\pm 1$. Indeed, we have

THEOREM 2.1.1. *Let $H_j(\alpha,\beta,\gamma)$ and $B_j(\alpha,\beta,\gamma)$ be the first order differential operators defined by*

$$H_1(\alpha,\beta,\gamma) = x\frac{d}{dx}+\alpha,$$

$$H_2(\alpha,\beta,\gamma) = x\frac{d}{dx}+\beta,$$

$$H_3(\alpha,\beta,\gamma) = (1-x)\frac{d}{dx}+(\gamma-\alpha-\beta), \tag{2.1.1}$$

$$B_1(\alpha,\beta,\gamma) = x(1-x)\frac{d}{dx}+(\gamma-\alpha-\beta x),$$

$$B_2(\alpha,\beta,\gamma) = x(1-x)\frac{d}{dx}+(\gamma-\beta-\alpha x),$$

$$B_3(\alpha,\beta,\gamma) = x\frac{d}{dx}+(\gamma-1).$$

Then we have the following linear homomorphisms:

$$\begin{aligned}(1)\qquad H_1(\alpha,\beta,\gamma) &: \mathcal{S}(\alpha,\beta,\gamma)\to\mathcal{S}(\alpha+1,\beta,\gamma),\\ B_1(\alpha+1,\beta,\gamma) &: \mathcal{S}(\alpha+1,\beta,\gamma)\to\mathcal{S}(\alpha,\beta,\gamma),\end{aligned}$$

which are isomorphisms if and only if

$$c_1(\alpha,\beta,\gamma) := -\alpha(\alpha-\gamma+1)\neq 0.$$

$$\begin{aligned}(2)\qquad H_2(\alpha,\beta,\gamma) &: \mathcal{S}(\alpha,\beta,\gamma)\to\mathcal{S}(\alpha,\beta+1,\gamma),\\ B_2(\alpha,\beta+1,\gamma) &: \mathcal{S}(\alpha,\beta+1,\gamma)\to\mathcal{S}(\alpha,\beta,\gamma),\end{aligned}$$

which are isomorphisms if and only if

$$c_2(\alpha,\beta,\gamma) := -\beta(\beta-\gamma+1)\neq 0.$$

$$\begin{aligned}(3)\qquad H_3(\alpha,\beta,\gamma) &: \mathcal{S}(\alpha,\beta,\gamma)\to\mathcal{S}(\alpha,\beta,\gamma+1),\\ B_3(\alpha,\beta,\gamma+1) &: \mathcal{S}(\alpha,\beta,\gamma+1)\to\mathcal{S}(\alpha,\beta,\gamma),\end{aligned}$$

which are isomorphisms if and only if

$$c_3(\alpha,\beta,\gamma) := (\gamma-\alpha)(\gamma-\beta)\neq 0.$$

The operators $H_j(\alpha,\beta,\gamma),\ j=1,2$ and 3, increase the parameters α,β, and γ by 1, respectively, while the operators $B_j(\alpha,\beta,\gamma),\ j=1,2,$ and 3, decrease these parameters by 1, respectively. Thus $H_j(\alpha,\beta,\gamma)$ are called *step-up operators* and $B_j(\alpha,\beta,\gamma)$ are called *step-down operators.* The step-up operators $H_j(\alpha,\beta,\gamma),\ j=1,2,$ and the step-down operator $B_3(\alpha,\beta,\gamma)$ are obtained as follows: Let us define the hypergeometric differential operator $L(\alpha,\beta,\gamma)$ by

$$\begin{aligned}L(\alpha,\beta,\gamma) :&= \delta_x(\delta_x+\gamma-1)-x(\delta_x+\alpha)(\delta_x+\beta),\\ &= x\left[x(1-x)\frac{d^2}{dx^2}+\{\gamma-(\alpha+\beta+1)x\}\frac{d}{dx}-\alpha\beta\right]\end{aligned}\tag{2.1.2}$$

$$\delta_x = x\frac{d}{dx}$$

(see (1.1.7)). By the commutation relation $\delta_x x = x(\delta_x + 1)$, we have

$$(\delta_x + \alpha)L(\alpha,\beta,\gamma) = L(\alpha+1,\beta,\gamma)(\delta_x+\alpha),$$
$$(\delta_x + \gamma - 2)L(\alpha,\beta,\gamma) = L(\alpha,\beta,\gamma-1)(\delta_x+\gamma-1).$$

These formulae imply that $H_1(\alpha,\beta,\gamma) = \delta_x + \alpha$ is an step-up operator for α and $B_3(\alpha,\beta,\gamma) = \delta_x + \gamma - 1$ is a step-down operator for γ. The other step-up/step-down operators are slightly more difficult to find (see Remark 2.1.4). We can show

PROPOSITION 2.1.2.

(1) $$B_1(\alpha+1,\beta,\gamma)H_1(\alpha,\beta,\gamma) \in c_1(\alpha,\beta,\gamma) + \mathcal{R}L(\alpha,\beta,\gamma),$$
$$H_1(\alpha,\beta,\gamma)B_1(\alpha+1,\beta,\gamma) \in c_1(\alpha,\beta,\gamma) + \mathcal{R}L(\alpha+1,\beta,\gamma).$$

(2) $$B_2(\alpha,\beta+1,\gamma)H_2(\alpha,\beta,\gamma) \in c_2(\alpha,\beta,\gamma) + \mathcal{R}L(\alpha,\beta,\gamma),$$
$$H_2(\alpha,\beta,\gamma)B_2(\alpha,\beta+1,\gamma) \in c_2(\alpha,\beta,\gamma) + \mathcal{R}L(\alpha,\beta+1,\gamma).$$

(3) $$B_3(\alpha,\beta,\gamma+1)H_3(\alpha,\beta,\gamma) \in c_3(\alpha,\beta,\gamma) + \mathcal{R}L(\alpha,\beta,\gamma),$$
$$H_3(\alpha,\beta,\gamma)B_3(\alpha,\beta,\gamma+1) \in c_3(\alpha,\beta,\gamma) + \mathcal{R}L(\alpha,\beta,\gamma+1).$$

In fact, for example, we have

$$\begin{aligned}
&B_1(\alpha+1,\beta,\gamma)H_1(\alpha,\beta,\gamma)\\
&= \{(1-x)\delta + (\gamma-\alpha-1-\beta x)\}\{\delta+\alpha\}\\
&= (1-x)\delta^2 + \{\gamma - 1 - (\alpha+\beta)x\}\delta + \alpha(\gamma-\alpha-1-\beta x)\\
&= L(\alpha,\beta,\gamma) + \alpha(\gamma-\alpha-1).
\end{aligned}$$

Theorem 2.1.1 follows immediately from Proposition 2.1.2. The statement of Theorem 2.1.1 is referred to as *contiguity relations* for the hypergeometric equation. Moreover, we have *contiguity relations* for the hypergeometric series $F(\alpha,\beta,\gamma;x)$.

THEOREM 2.1.3.

(1)
$$H_1(\alpha,\beta,\gamma)F(\alpha,\beta,\gamma;x) = \alpha F(\alpha+1,\beta,\gamma;x),$$
$$B_1(\alpha+1,\beta,\gamma)F(\alpha+1,\beta,\gamma;x) = -(\alpha-\gamma+1)F(\alpha,\beta,\gamma;x).$$

(2)
$$H_2(\alpha,\beta,\gamma)F(\alpha,\beta,\gamma;x) = \beta F(\alpha,\beta+1,\gamma;x),$$
$$B_2(\alpha,\beta+1,\gamma)F(\alpha,\beta+1,\gamma;x) = -(\beta-\gamma+1)F(\alpha,\beta,\gamma:x).$$

(3)
$$H_3(\alpha,\beta,\gamma)F(\alpha,\beta,\gamma;x) = \frac{(\gamma-\alpha)(\gamma-\beta)}{\gamma}F(\alpha,\beta,\gamma+1;x),$$
$$B_3(\alpha,\beta,\gamma+1;x)F(\alpha,\beta,\gamma+1;x) = \gamma F(\alpha,\beta,\gamma;x).$$

The first formula of (1) is checked as follows:

$$\begin{aligned} H_1(\alpha,\beta,\gamma)F(\alpha,\beta,\gamma) &= (\delta+\alpha)\sum \frac{(\alpha)_m(\beta)_m}{(\gamma)_m(1)_m}x^m \\ &= \sum \frac{(\alpha)_m(\beta)_m}{(\gamma)_m(1)_m}(\alpha+m)x^m \\ &= \alpha\sum \frac{(\alpha+1)_m(\beta)_m}{(\gamma)_m(1)_m}x^m \\ &= \alpha F(\alpha+1,\beta,\gamma;x). \end{aligned}$$

The second formula of (1) is shown as follows: Since

$$L(\alpha,\beta,\gamma)F(\alpha,\beta,\gamma) = 0$$

by (1) of Proposition 2.1.2, we have

$$B_1(\alpha+1,\beta,\gamma)H_1(\alpha,\beta,\gamma)F(\alpha,\beta,\gamma;x) = c_1(\alpha,\beta,\gamma)F(\alpha,\beta,\gamma;x)$$

and, by using the first formula of (1),

$$\alpha B_1(\alpha+1,\beta,\gamma)F(\alpha+1,\beta,\gamma;x) = c_1(\alpha,\beta,\gamma)F(\alpha,\beta,\gamma;x),$$

which is the second formula of (1). The others are proved in a similar way.

REMARK 2.1.4. Consider the ring

$$\mathcal{R}_n = \mathbb{C}(x_1, \cdots, x_n)[\partial/\partial x_1, \cdots, \partial/\partial x_n]$$

of linear partial differential operators and its maximal left-ideals $\mathcal{I}_\lambda$ parametrized by a complex parameter λ. Denote by $\mathcal{S}_\lambda$ the collection of functions annihilated by $\mathcal{I}_\lambda$. If the step-up operators $H_\lambda : \mathcal{S}_\lambda \to \mathcal{S}_{\lambda+1}$ are known, we can find step-down operators $B_{\lambda+1} : \mathcal{S}_{\lambda+1} \to \mathcal{S}_\lambda$ by solving the equation

$$B_{\lambda+1} H_\lambda \equiv 1 \pmod{\mathcal{I}_\lambda} \tag{2.1.3}$$

for unknowns $B_{\lambda+1}$. Similarly, if step-down operators are known, the step-up operators are found by solving the equation

$$H_\lambda B_{\lambda+1} \equiv 1 \pmod{\mathcal{I}_{\lambda+1}} \tag{2.1.3$'$}$$

for unknowns H_λ. Let us explain this method for the hypergeometric differential operator $L(\alpha, \beta, \gamma)$, focusing our attention on the parameter α. Let $\mathcal{I}_\alpha$ be the ideal of $\mathcal{R} = \mathbb{C}(x)[d/dx]$ generated by $L(\alpha, \beta, \gamma)$, and let $H_\alpha := H_1(\alpha, \beta, \gamma)(= \delta_x + \alpha)$. We find the operator $B_{\alpha+1}$ so that

$$B_{\alpha+1} H_\alpha \equiv 1 \pmod{\mathcal{I}_\alpha}$$

as follows: Since

$$\begin{aligned}
L(\alpha, \beta, \gamma) &= \delta_x(\delta_x + \alpha + \gamma - 1 - \alpha) - x(\delta_x + \alpha)(\delta_x + \beta) \\
&= \delta_x(\gamma - 1 - \alpha) + \{\delta_x - x(\delta_x + \beta)\}(\delta_x + \alpha) \\
&= (\delta_x + \alpha - \alpha)(\gamma - 1 - \alpha) + \{\delta_x - x(\delta_x + \beta)\}(\delta_x + \alpha) \\
&= -\alpha(\gamma - 1 - \alpha) + \{\gamma - 1 - \alpha + \delta_x - x(\delta_x + \beta)\}(\delta_x + \alpha) \\
&= -\alpha(\gamma - 1 - \alpha) + \{(1 - x)\delta_x + \gamma - 1 - \alpha - x\beta)\}(\delta_x + \alpha),
\end{aligned}$$

we have

$$B_{\alpha+1} = \frac{1}{\alpha(\gamma - 1 - \alpha)} \{(1 - x)\delta_x + \gamma - 1 - \alpha - x\beta)\}(\delta_x + \alpha).$$

Many identities for special functions can be understood through the theory of rings of differential operators, where Göbner bases of an ideal play a crucial role to find explicit formulae. For more detail, see [Bern], [Zei], [Tky.1], [Tky.3], [Tky.5] and the references of the papers.

Considering the operator $H_1(\alpha, \beta, \gamma) H_2(\alpha, \beta, \gamma) H_3(\alpha, \beta, \gamma)$ modulo $L(\alpha, \beta, \gamma)$, we obtain

PROPOSITION 2.1.5. *If* $c_3(\alpha,\beta,\gamma) := (\gamma-\alpha)(\gamma-\beta) \neq 0$, *then*

$$\frac{d}{dx}\mathcal{S}(\alpha,\beta,\gamma) \subset \mathcal{S}(\alpha+1,\beta+1,\gamma+1).$$

Proof. For brevity, we write $H_j = H_j(\alpha,\beta,\gamma)$ and $L = L(\alpha,\beta,\gamma)$. By a straightforward calculation, one can check

$$H_1H_2H_3 = (x\frac{d}{dx} + \alpha + \beta - \gamma)L + c_3(\alpha,\beta,\gamma)\frac{d}{dx}. \tag{2.1.4}$$

Since Theorem 2.1.1 implies $H_1H_2H_3\mathcal{S}(\alpha,\beta,\gamma) \subset \mathcal{S}(\alpha+1,\beta+1,\gamma+1)$, (2.1.4) establishes the proposition. ∎

In Section 3.3, this proposition will provide a motivation to introduce the Euler transform D_a^α, which justifies the "α-times differentiation" $(d/dx)^\alpha$ for an arbitrary complex number α. The Euler transform will yield a clever way to derive integral representations of solutions of the hypergeometric equation.

We have also a contiguity relation for the hypergeometric series.

PROPOSITION 2.1.6.

$$\frac{d}{dx}F(\alpha,\beta,\gamma;x) = \frac{\alpha\beta}{\gamma}F(\alpha+1,\beta+1,\gamma+1;x)$$

Combining this proposition with Theorem 1.2.2, we obtain

PROPOSITION 2.1.7. *If* $\Re(\gamma-\alpha-\beta) > 1$, *then the power series* $\frac{d}{dx}F(\alpha,\beta,\gamma;x)$ *converges absolutely in the closed unit disc* $|x| \leq 1$.

In the next section, an application of the contiguity relations to the Toda equation will be given.

2.2 Contiguity relations and particular solutions of the Toda equation

The Toda equation, a famous equation in mathematical physics, is a system of non-linear differential-difference equations of the form

$$X^2 \log \phi_n = c_n \frac{\phi_{n-1}\phi_{n+1}}{\phi_n{}^2} \tag{2.2.1}$$

for unknowns $\phi_n(x)$ $(n \in \mathbb{Z})$, where X is a vector field, say $X = r(x)d/dx$, $r(x)$ a rational function, and c_n are constants. We shall give here solutions of (2.2.1) which are closely related to the contiguity relations for the hypergeometric equation. The method explained below can be developed not only for the hypergeometric equation but also for other differential equations (see [Kam.1], [Kam.2], [Okm.8] and [Okm.10]).

We first note that we can change the sequence c_n; a transformation $\phi_n = a_n\psi_n$, where a_n are non-zero constants, takes (2.2.1) into another Toda equation which is obtained by replacing ϕ_n by ψ_n and c_n by $c_n a_{n-1}a_{n+1}/a_n{}^2$. As we shall see soon, there is a convenient choice of the sequence c_n. Suppose that (2.2.1) has a solution of the form:

$$\phi_n(x) = \phi(x)^{\rho_n}, \quad (n \in \mathbb{Z}).$$

Substituting this into (2.2.1), we obtain

$$\rho_n X^2 \log \phi = c_n \phi^{\rho_{n-1}-2\rho_n+\rho_{n+1}},$$

so that

$$X^2 \log \phi = \phi$$

if the c_n satisfy

$$c_n = \rho_n, \quad \rho_{n-1} - 2\rho_n + \rho_{n+1} = 1.$$

Since the above difference equation is solved by

$$\rho_n = \frac{1}{2}n^2 + an + b \quad (a, b \in \mathbb{C}),$$

we have:

LEMMA 2.2.1. *If $\phi(x)$ is a solution of $X^2 \log \phi = \phi$ and if the sequence c_n is of the form: $c_n = \frac{1}{2}n^2 + an + b$ $(n \in \mathbb{Z})$, where a and b are fixed complex numbers, then the Toda equation* (2.2.1) *has a solution*

$$\phi_n(x) = \phi(x)^{c_n} \quad (n \in \mathbb{Z}). \tag{2.2.2}$$

In this section, we set a vector field X as

$$X = x(1-x)\frac{d}{dx}. \tag{2.2.3}$$

This choice is made because we are now considering the hypergeometric equation; when we consider other differential equations, suitable vector fields should be chosen. The following $\phi(x)$ and c_n satisfy the assumption made in Lemma 2.2.1.

$$\phi(x) = 2x(x-1), \quad c_n = \frac{1}{2}\{c_3(\alpha, \beta, \gamma+n-1) + \gamma + n - 1\}, \tag{2.2.4}$$

where $c_3(\alpha, \beta, \gamma) := (\gamma - \alpha)(\gamma - \beta)$ (see the statement of Theorem 2.1.1). Consider constant multiples of the step-up and step-down operators for γ:

$$\begin{aligned} U_n &= \frac{1}{\mu_n} H_3(\alpha, \beta, \gamma + n), \\ D_n &= \frac{1}{\nu_n} B_3(\alpha, \beta, \gamma + n), \end{aligned} \tag{2.2.5}$$

where μ_n and ν_n are non-zero constants such that

$$\mu_n \nu_n = c_3(\alpha, \beta, \gamma + n), \quad (n \in \mathbb{Z}). \tag{2.2.6}$$

Take a function $f_0 \in \mathcal{S}(\alpha, \beta, \gamma)$ and define a sequence f_n $(n \in \mathbb{Z})$ by

$$f_{n+1} = U_n f_n \quad (n \geq 0), \quad f_{n-1} = D_n f_n \quad (n \leq 0),$$

then assertion (3) of Theorem 2.1.1 implies that

$$\left.\begin{aligned} &f_n \in \mathcal{S}(\alpha, \beta, \gamma + n), \\ &f_{n+1} = U_n f_n, \quad f_{n-1} = D_n f_n. \end{aligned}\right. \quad (n \in \mathbb{Z}) \tag{2.2.7}$$

A sequence satisfying the condition (2.2.7) is called a *Laplace sequence* for the pair (U_n, D_n). By (2.2.5) and (2.2.7), we have

$$\{B_3(\alpha,\beta,\gamma+n)f_n\}\{H_3(\alpha,\beta,\gamma+n)f_n\} - 2\lambda_n f_{n-1}f_{n+1} = 0,$$

where

$$\lambda_n = \frac{1}{2}\mu_n\nu_n. \tag{2.2.8}$$

We carry out the differentiations:

$$\begin{aligned}
0 &= (1-x)f_n L(\alpha,\beta,\gamma+n)f_n \\
&\quad - x(1-x)[\{B_3(\alpha,\beta,\gamma+n)f_n\}\{H_3(\alpha,\beta,\gamma+n)f_n\} \\
&\qquad - 2\lambda_n f_{n-1}f_{n+1}] \\
&= (1-x)xf_n[x(1-x)f_n'' + \{\gamma+n-(\alpha+\beta+1)x\}f_n' - \alpha\beta f_n] \\
&\quad - x(1-x)\{xf_n' + (\gamma+n-1)f_n\}\{(1-x)f_n' + (\gamma+n-\alpha-\beta)f_n\} \\
&\quad + 2x(1-x)\lambda_n f_{n-1}f_{n+1} \\
&= x^2(1-x)^2\{f_n''f_n - (f_n')^2\} + x(1-x)(1-2x)f_nf_n' \\
&\quad - \lambda_n\phi f_{n-1}f_{n+1} + c_n\phi f_n^2 \\
&= f_n^2 x(1-x)\frac{d}{dx}\{x(1-x)\frac{f_n'}{f_n}\} - \lambda_n\phi f_{n-1}f_{n+1} + c_n\phi f_n^2
\end{aligned}$$

and find that $\{f_n\}$ satisfies

$$X^2 \log f_n = \lambda_n\phi(x)\frac{f_{n-1}f_{n+1}}{{f_n}^2} - c_n\phi(x), \tag{2.2.9}$$

where $\phi(x)$ and c_n are defined by (2.2.4). Let $\phi_n(x)$ be defined by (2.2.2) and (2.2.4), and put $\psi_n = \phi_n f_n$. Then, by (2.2.9), we see that $\psi_n (n \in \mathbb{Z})$ satisfy the Toda equation

$$X^2 \log \psi_n = \lambda_n \frac{\psi_{n-1}\psi_{n+1}}{{\psi_n}^2}. \tag{2.2.10}$$

THEOREM 2.2.2. *Suppose neither $\gamma-\alpha$ nor $\gamma-\beta$ is an integer. Let μ_n and ν_n be non-zero constants satisfying $\mu_n\nu_n = c_3(\alpha,\beta,\gamma+n)$ and put $\lambda_n = \mu_n\nu_n/2$. Then, for a Laplace sequence $f_n (n\in\mathbb{Z})$ for the pair (U_n, D_n), the sequence*

$$\psi_n = \phi_n f_n \quad (n\in\mathbb{Z}) \tag{2.2.11}$$

is a solution of the Toda equation (2.2.10), *where*

$$\phi_n(x) = [2x(x-1)]^{c_n}, \quad c_n = \frac{1}{2}[c_3(\alpha,\beta,\gamma+n)+\gamma+n-1]. \tag{2.2.12}$$

Notice that the assumption on α,β and γ made in Theorem 2.2.2 asserts the non-vanishing of $c_3(\alpha,\beta,\gamma+n)$ for all $n\in\mathbb{Z}$. Notice also that the theorem states that there is a class of solutions of the Toda equation of which elements can be decomposed as

(a very simple solution of the Toda equation)

×(a contiguous sequence of solutions of a linear

differential equation containing a parameter).

This phenomenon often occurs not only for the hypergeometric equation but also for many other differential equations. See [Okm.8] and [Okm.10].

3 Integral representations

3.1 Integral representations as a tool for global problems

One can know local properties of solutions of a linear differential equation in several ways (cf. Sections 1.2 and 1.3). However, one does not *a priori* know any global properties; for instance, to relate solutions at one point to those at another point, called the connection problem:

Connection problem. Let $f_j^{(1)}(1 \le j \le r)$ and $f_j^{(2)}(1 \le j \le r)$ be bases of local solutions of a differential equation of rank r at x_1 and x_2, respectively, let C be a path joining x_1 and x_2 and let $C_* f_j^{(1)}$ be the analytic continuation of $f_j^{(1)}$ along C. Find a linear relation between $C_* f_j^{(1)}(1 \le j \le r)$ and $f_j^{(2)}(1 \le j \le r)$.

If a solution $u(x)$ of the differential equation happens to admit an integral representation :

$$u(x) = \int_{\gamma(x)} K(x,t)dt,$$

it might be useful for attacking global problems; here the kernel $K(x,t)$ is expected, according to what one wants to do, to be a simpler function than $u(x)$, otherwise integral representations would give no information.

In this chapter two types of integrals
(i) Euler integral representation, and
(ii) Barnes integral representation
are treated in detail.

3.2 Euler integral representation derived from the power series

Hypergeometric functions admit the Euler integral representation. In this section we derive it from the hypergeometric series $F(\alpha, \beta, \gamma; x)$. Another clever derivation by using the Euler transform will be given in Sections 3.3 and 3.4.

THEOREM 3.2.1. *Suppose $\Re(\gamma) > \Re(\beta) > 0$. Then we have*

$$(3.2.1)\quad F(\alpha, \beta, \gamma; x) = \frac{\Gamma(\gamma)}{\Gamma(\beta)\Gamma(\gamma-\beta)} \int_0^1 t^{\beta-1}(1-t)^{\gamma-\beta-1}(1-xt)^{-\alpha} dt$$

for $|x| < 1$, *where the branch of the integrand is determined by the following assignment of the arguments*

$$\arg t = 0, \quad \arg(1-t) = 0, \quad |\arg(1-xt)| < \frac{\pi}{2}, \quad (0 < t < 1).$$

Proof. We use the well-known formula relating the Beta function with the Gamma function:

$$B(p,q) := \int_0^1 t^{p-1}(1-t)^{q-1}dt = \frac{\Gamma(p)\Gamma(q)}{\Gamma(p+q)}; \quad \Re(p),\ \Re(q) > 0. \tag{3.2.2}$$

The power series $F(\alpha, \beta, \gamma; x)$ can be rewritten as follows:

$$\begin{aligned}
F(\alpha,\beta,\gamma;x) &= \sum_{m=0}^{\infty} \frac{\Gamma(\alpha+m)\Gamma(\beta+m)\Gamma(\gamma)}{\Gamma(\alpha)\Gamma(\beta)\Gamma(\gamma+m)} \frac{x^m}{m!} \\
&= \frac{1}{B(\beta,\gamma-\beta)} \sum_{m=0}^{\infty} B(\beta+m,\gamma-\beta)\frac{\Gamma(\alpha+m)}{\Gamma(\alpha)m!}x^m \\
&= \frac{1}{B(\beta,\gamma-\beta)} \sum_{m=0}^{\infty} \int_0^1 t^{\beta+m-1}(1-t)^{\gamma-\beta-1}dt \cdot \binom{-\alpha}{m}(-x)^m \\
&= \frac{1}{B(\beta,\gamma-\beta)} \int_0^1 t^{\beta-1}(1-t)^{\gamma-\beta-1} \cdot \sum_{m=0}^{\infty} \binom{-\alpha}{m}(-tx)^m dt \\
&= \frac{\Gamma(\gamma)}{\Gamma(\beta)\Gamma(\gamma-\beta)} \int_0^1 t^{\beta-1}(1-t)^{\gamma-\beta-1}(1-xt)^{-\alpha}dt.
\end{aligned}$$

Here the third equality is obtained by applying the above formula for the Beta function, the fourth equality by exchanging the order of the summation and the integration, and the last equality by using the binomial theorem. Notice that the assumption $\Re(\gamma) > \Re(\beta) > 0$ is necessary to apply the formula for the Beta function. ∎

REMARK 3.2.2. The change of variable $t = 1/s$ takes the formula (3.2.1) into another formula (cf. Section 3.4)

$$\begin{aligned}
&F(\alpha,\beta,\gamma;x) \\
&\quad = \frac{\Gamma(\gamma)}{\Gamma(\beta)\Gamma(\gamma-\beta)} \int_1^{\infty} s^{\alpha-\gamma}(s-1)^{\gamma-\beta-1}(s-x)^{-\alpha}ds.
\end{aligned} \tag{3.2.3}$$

REMARK 3.2.3. The integrals on the right-hand sides of (3.2.1) and (3.2.3) make sense if x is in $\mathbb{C}\backslash[1,\infty)$. Hence $F(\alpha,\beta,\gamma;x)$ is analytically continuable to the domain $\mathbb{C}\backslash[1,\infty)$.

3.3 The Euler transform

Recall that the hypergeometric series admits an Euler integral representation (3.2.3). We shall consider why it admits such an integral representation. Using the contiguity relations given in Proposition 2.1.5 repeatedly, we obtain

$$\left(\frac{d}{dx}\right)^k \mathcal{S}(\alpha,\beta,\gamma) \subset \mathcal{S}(\alpha+k,\beta+k,\gamma+k),$$

where k is a positive integer. In particular, if α is a negative integer and $f(x) \in \mathcal{S}(\alpha,\beta,\gamma)$, then $g(x) := (d/dx)^{-\alpha} f(x) \in \mathcal{S}(0,\beta-\alpha,\gamma-\alpha)$. Since the hypergeometric differential operator $L(0,\beta-\alpha,\gamma-\alpha)$ is of the form

$$L(0,\beta-\alpha,\gamma-\alpha) = \{(\delta_x+\gamma-\alpha-1) - x(\delta_x+\beta-\alpha)\}\delta_x$$

(cf. (1.1.7)), $\mathcal{S}(0,\beta-\alpha,\gamma-\alpha)$ contains a function $g(x)$ such that

$$g'(x) = x^{\alpha-\gamma}(x-1)^{\gamma-\beta-1}. \tag{3.3.1}$$

Therefore, $(-\alpha+1)$-times integration of (3.3.1) yields a solution of the original hypergeometric equation. Since the k-th derivative of the integral

$$\frac{1}{\Gamma(k)}\int^x (x-t)^{k-1}\varphi(t)dt \tag{3.3.2}$$

(the Riemann-Liouville integral) is $\varphi(x)$, a function defined by

$$\int^x t^{\alpha-\gamma}(t-1)^{\gamma-\beta-1}(t-x)^{-\alpha}dt$$

is an element of $\mathcal{S}(\alpha,\beta,\gamma)$. This formula is the same as (3.2.3) except for the path of integration. We shall extend the above argument to the case where α is not necessarily a negative integer. To this end, we have to justify the operation $(d/dx)^{-k}$ for every complex number k. If k is a positive integer, then we may regard it as k-fold integration, whence $(d/dx)^{-k}\varphi$ is given by the Riemann-Liouville integral (3.3.2). Therefore, justification will be made by analytic continuation of (3.3.2) to the whole k-plane. The correspondence of $\varphi(x)$ to a function defined by (3.3.2) is called the *Euler transform* of $\varphi(x)$.

For the reason mentioned above, we shall consider the integral :

$$(D_a^{-\alpha}f)(x) := \frac{1}{\Gamma(\alpha)}\int_a^x (x-t)^{\alpha-1}f(t)dt, \tag{3.3.3}$$

where the path of integration is an arc C with initial point a and terminal point x and the integrand is of the form

$$f(t) = (t-a)^{\mu}g(t), \quad \mu\in\mathbb{C}, \tag{3.3.4}$$

where $g(t)$ is holomorphic in a neighborhood of C and $g(a)\neq 0$. For the present, we assume that the initial point a is finite. We shall consider the case $a=\infty$ later. We call μ the *exponent* of $f(t)$ at $t=a$. If we fix a branch of the multi-valued functions $\arg(t-x)$ and $\arg(t-a)$ along the arc C , then the integrand of (3.3.3) is well-defined. If $\Re(\alpha)>0$ and $\Re(\mu)>-1$, then the integral (3.3.3) converges. Now we want to generalize (3.3.3) in such a way that it makes sense for general α and μ. To this end, we introduce the notion of the finite part of a divergent integral.

Consider the integral

$$H(\nu) = \int_a^b (t-a)^{\nu-1}g(t)dt, \tag{3.3.5}$$

where the path of integration is an arc γ with initial point a and terminal point b, $g(t)$ is a holomorphic function in a neighborhood of the arc γ and a branch of the function $\arg(t-a)$ is fixed. We consider $H(\nu)$ as a function of ν. Clearly $H(\nu)$ is holomorphic in $\Re(\nu)>0$. Moreover we have the

LEMMA 3.3.1. *The function $H(\nu)$ can be continued analytically to a meromorphic function in $|\nu| < \infty$ which has simple poles at $\nu = 0, -1, -2, \cdots$. The residues at $\nu = -m$ are given by*

$$\operatorname*{Res}_{\nu=-m} H(\nu) = \frac{1}{m!} g^{(m)}(a).$$

Proof . Pick a point c on the path of integration in (3.3.5) within the circle of convergence of the Taylor series of g at a; then (initially for $\Re(\nu) > 0$)

$$H(\nu) = \int_a^c + \int_c^b = \sum_{m=0}^{\infty} \frac{g^{(m)}(a)}{m!} \int_a^c (t-a)^{\nu-1+m}\, dt + \int_c^b$$

$$= \sum_{m=0}^{\infty} \frac{g^{(m)}(a)}{m!} \frac{(c-a)^{\nu+m}}{\nu+m} + \text{holomorphic in } \nu,$$

from which the statement is clear. ∎

The integral (3.3.5) is, in general, divergent if $\Re(\nu) \le 0$. Even in such a case, if $\nu \neq 0, -1, -2, \cdots$, (3.3.5) has a meaning by Lemma 3.3.1, which is called *the finite part of a divergent integral.* The finite part of a divergent integral can be applied to the integral (3.3.3): one has only to divide the path of integration C into two arcs.

PROPOSITION 3.3.2. *The integral* (3.3.3) *makes sense, in the sense of finite part, for any value of α and defines an entire function in α, provided that the exponent μ in* (3.3.4) *satisfies*

(3.3.6) $$\mu \neq -1, -2, -3, \cdots.$$

Proof. Since the simple poles at $\alpha = 0, -1, -2, \cdots$ arising from the integral in (3.3.3) are canceled out by the simple poles of $\Gamma(\alpha)$ at those points, Lemma 3.3.1 implies the proposition. ∎

We shall next investigate some properties of the operator $D_a^{-\alpha}$. Our aim is
(i) to express the exponent of $D_a^{-\alpha} f$ at $x = a$ in terms of that of f,
(ii) to establish the composition rule : $D_a^{\alpha} \cdot D_a^{\beta} = D_a^{\alpha+\beta}$,
(iii) to make sure that $D_a^m = (d/dx)^m$, if m is a positive integer, and
(iv) to establish a kind of Leibniz rule for D_a^{α}.

PROPOSITION 3.3.3. *Let μ be the exponent of $f(x)$ at $x = a$ satisfying the condition* (3.3.6). *Then the exponent of $D_a^{-\alpha} f$ at that point is $\mu + \alpha$ unless $\mu + \alpha$ is a negative integer, in which case the exponent is 0 or a positive integer.*

Proof. We may assume, without loss of generality, that $a = 0$. If $f(x)$ is given by the power series

$$f(x) = x^{\mu} \sum_{m=0}^{\infty} c_m x^m,$$

then (3.3.3) for x within the circle of convergence of this series is given by the following expression

$$\begin{aligned}
(D_a^{-\alpha} f)(x) &= \frac{1}{\Gamma(\alpha)} \int_0^x (x-t)^{\alpha-1} t^{\mu} \sum_{m=0}^{\infty} c_m t^m dt \\
&= \sum_{m=0}^{\infty} c_m \frac{1}{\Gamma(\alpha)} \int_0^x (x-t)^{\alpha-1} t^{\mu+m} dt \\
&= \sum_{m=0}^{\infty} c_m \frac{x^{\alpha+m+\mu}}{\Gamma(\alpha)} \int_0^1 s^{\mu+m} (1-s)^{\alpha-1} ds \\
&= \sum_{m=0}^{\infty} c_m \frac{x^{\alpha+m+\mu}}{\Gamma(\alpha)} \frac{\Gamma(\mu+m+1)\Gamma(\alpha)}{\Gamma(\alpha+\mu+m+1)}, \\
&= x^{\mu+\alpha} \sum_{m=0}^{\infty} \frac{\Gamma(\mu+m+1)}{\Gamma(\alpha+\mu+m+1)} c_m x^m,
\end{aligned}$$

from which the assertion follows. ∎

PROPOSITION 3.3.4. *Let μ be the exponent of f at $x = a$, and suppose that μ is not a negative integer. Then we have*

$$D_a^{\beta} \cdot D_a^{\alpha} f(x) = D_a^{\beta+\alpha} f(x). \tag{3.3.7}$$

Proof. In view of the assumptions on μ, Propositions 3.3.2 and 3.3.3 imply that both sides of (3.3.7) make sense. Writing $f(x)$ in the form

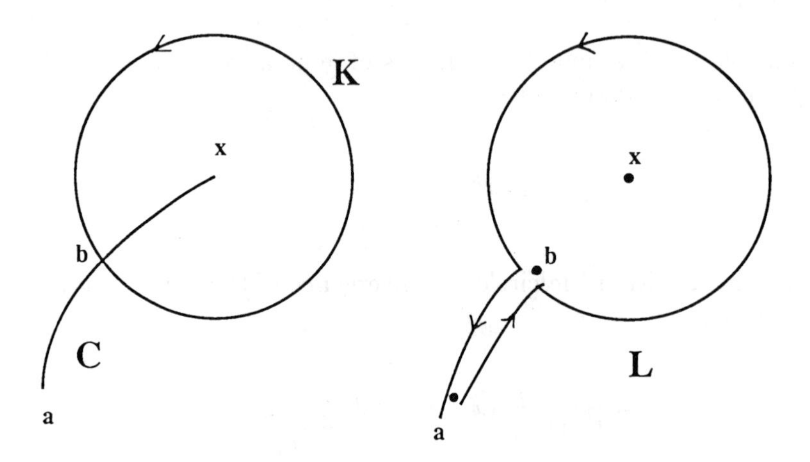

Figure 2.1.

(3.3.4), we can regard both sides of (3.3.7) as meromorphic functions in (α, β, μ) in $\mathbb{C}^3$. (Recall Lemma 3.3.1.) If $\Re(\alpha) < 0$, $\Re(\beta) < 0$ and $\Re(\mu) > -1$, then all three integrals in (3.3.7) are convergent and (3.3.7) is easily proved by exchanging the integral signs. By the use of the analytic continuation with respect to α and β, we see that (3.3.7) holds for all values of α and μ which satisfy the given condition. ∎

PROPOSITION 3.3.5. *If m is a positive integer, then*

$$D_a^m = \left(\frac{d}{dx}\right)^m.$$

Proof. This follows from Proposition 3.3.1, since $(-1)^\nu(D_a^{-\nu}f)(x)$ is the quotient of $H(\nu)$ (with a and b replaced by x and a in (3.3.5)) by $\Gamma(\nu)$, and these functions have simple poles at $\nu = -\mu$ with residues $g^{(m)}(x)/m!$ and $(-1)^m/m!$, respectively. ∎

PROPOSITION 3.3.6. *If $p(x)$ is a polynomial of degree n, then*

$$D_a^\alpha\{p(x)f(x)\} = \sum_{j=0}^{n} \binom{\alpha}{j} p^{(j)}(x)(D_a^{\alpha-j}f)(x),$$

where $p^{(j)}$ is the j-th derivative of p.

Proof. A polynomial $p(t)$ of degree n can be expressed by

$$p(t) = \sum_{j=0}^{n} \frac{p^{(j)}(x)}{j!}(t-x)^j.$$

Substituting this expression into (3.3.3) with $p(t)f(t)$ for $f(t)$, we obtain the desired formula. ∎

We shall next consider the Riemann-Liouville integral with initial point ∞ :

$$(D_\infty^{-\alpha}f)(x) = \frac{1}{\Gamma(\alpha)}\int_\infty^x (t-x)^{\alpha-1}f(t)dt, \tag{3.3.3$'$}$$

where $f(t)$ is assumed to be of the form

$$f(t) = t^{-\mu}g(t), \quad \mu \in \mathbb{C}, \tag{3.3.4$'$}$$

where $g(t)$ is holomorphic in a neighborhood of the path of integration such that $g(\infty) \neq 0$. The number μ is called the *exponent* of $f(t)$ at $t = \infty$. Changing the variable by $t = 1/s$ we have

$$(D_\infty^{-\alpha}f)(x) = \frac{-1}{\Gamma(\alpha)}\int_0^{1/x} s^{-\alpha+\mu-1}(1-sx)^{\alpha-1}g(\frac{1}{s})ds.$$

Then we obtain the following propositions similar to Propositions 3.3.2 - 3.3.4. Propositions 3.3.5 and 3.3.6 are valid for (3.3.3)$'$ without any modification, so we do not repeat them.

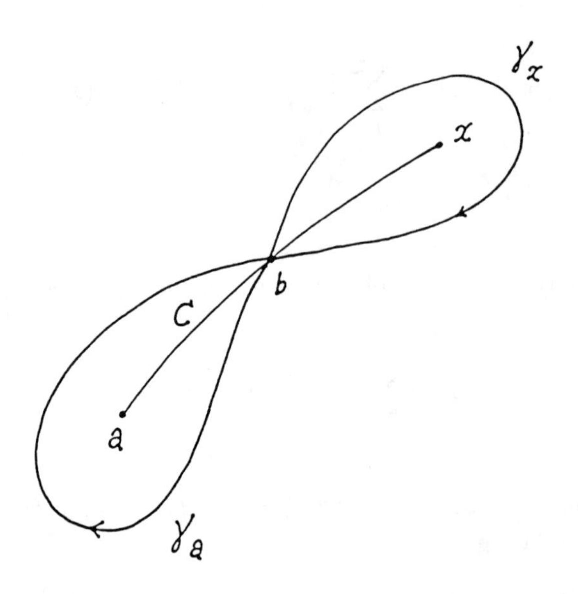

Figure 2.2.

PROPOSITION 3.3.2$'$. *In the sense of finite part, the integral* (3.3.3)$'$ *defines a meromorphic function in* α *in* $|\alpha| < \infty$ *with simple poles at* $\alpha = \mu, \mu + 1, \mu + 2, \cdots$.

PROPOSITION 3.3.3$'$. *If* $f(x)$ *has exponent* μ *at* $x = \infty$, *then* $D_\infty^{-\alpha} f$ *has exponent* $\mu - \alpha$ *(respectively, 0 or a positive integer if* $\mu - \alpha$ *is a negative integer) at* $x = \infty$.

PROPOSITION 3.3.4$'$. *Let* μ *be the exponent of* $f(x)$ *at* $x = \infty$. *If* μ *is not a negative integer, then*

$$D_\infty^\beta \cdot D_\infty^\alpha f = D_\infty^{\beta+\alpha} f.$$

The Euler transform is, by definition, expressed by an integral in a generalized sense over an arc (i.e., we take the finite part if the integral is divergent), but it can also be expressed as an integral over a double loop (the commutator of two loops with a fixed base point). Let b be a base point on C. Let γ_x be a loop starting and ending at b which encircles x once in the positive sense, and γ_a be a similar loop with respect to a. See Figure 2.2. The commutator $[\gamma_x, \gamma_a]$ of γ_x and γ_a is defined by

$$[\gamma_x, \gamma_a] = \gamma_x \cdot \gamma_a \cdot \gamma_x^{-1} \cdot \gamma_a^{-1}.$$

PROPOSITION 3.3.7. *Let $\mu \notin \mathbb{Z}$ be the exponent of $f(x)$ at $x = a$ and let $\alpha \notin \mathbb{N}$, then*

$$(D_a^{-\alpha} f)(x) = \frac{1}{A(\alpha, \mu)} \int_{[\gamma_x, \gamma_a]} (x - t)^{\alpha - 1} f(t) dt, \tag{3.3.9}$$

where

$$A(\alpha, \mu) = (1 - e^{2\pi i \alpha})(1 - e^{2\pi i \mu})\Gamma(\alpha).$$

Proof. One has only to trace the variation of branches of the integrand along the double loop. ∎

For $\alpha \in \mathbb{N}$ the right-hand side of (3.3.9) becomes $0/0$, but of course the limiting value of the right-hand side as α tends to a positive integer still equals the left-hand side, which is defined for any α.

In the next section, the Euler transform will be used for deriving the Euler integral representations for solutions of the hypergeometric equation.

3.4 The hypergeometric Euler transform

An n-th order differential operator

$$L = \sum_{j=0}^{n} p_j(x) D^j, \quad D = \frac{d}{dx}, \tag{3.4.1}$$

is said to be of *hypergeometric type* if the coefficients are polynomials such that $\deg p_j \leq j$ for all j and $\deg p_n = n$. We denote by $\mathcal{H}_n$ the set of n-th order differential operators of hypergeometric type. Let λ be a complex number and let $f(x)$ be a function holomorphic at $a \in \mathbb{C}$ with exponent μ such that $n - \mu \notin \mathbb{N}$ (see (3.3.4)). By using the calculus of the D_a^λ's established in the previous section, we have

$$\begin{aligned} D_a^\lambda L f &= \sum_{j=0}^{n} D_a^\lambda \{p_j(x) f^{(j)}\} \\ &= \sum_{j=0}^{n} \sum_{k=0}^{j} \binom{\lambda}{k} p_j^{(k)}(x) D_a^{\lambda-k} f^{(j)} \\ &= \sum_{j=0}^{n} \sum_{k=0}^{j} \binom{\lambda}{k} p_j^{(k)}(x) D_a^{\lambda+j-k} f \\ &= \sum_{j=0}^{n} \sum_{k=0}^{j} \binom{\lambda}{k} p_j^{(k)}(x) D^{j-k} D_a^\lambda f \\ &= \left(\sum_{k=0}^{n} \sum_{j=k}^{n} \binom{\lambda}{j-k} p_j^{(j-k)}(x) D^k \right) D_a^\lambda f, \end{aligned}$$

i.e.

$$D_a^\lambda L = (\mathcal{D}^\lambda L) D_a^\lambda, \tag{3.4.2}$$

where $\mathcal{D}^\lambda L$ is the differential operator

$$\mathcal{D}^\lambda L := \sum_{k=0}^{n} q_k(x) D^k \tag{3.4.3}$$

with $q_k(x)$ given by

$$q_k(x) = \sum_{j=k}^{n} \binom{\lambda}{j-k} p_j^{(j-k)}(x). \tag{3.4.4}$$

Therefore we have the following theorem.

THEOREM 3.4.1. *If the exponent μ of a function f at a satisfies $n - \mu \notin \mathbb{N}$, we have*

$$D_a^\lambda(Lf) = (\mathcal{D}^\lambda L)(D_a^\lambda f) \quad (\lambda \in \mathbb{C}).$$

Similarly, we have:

THEOREM 3.4.1′. *Let μ be the exponent of a function f at $x = \infty$ such that neither μ nor $\lambda + \mu$ is a negative integer . Then we have*

$$D_\infty^\lambda(Lf) = (\mathcal{D}^\lambda L)(D_\infty^\lambda f) \quad (\lambda \in \mathbb{C}).$$

The operator $\mathcal{D}^\lambda L$ is also of hypergeometric type, so we obtain the correspondence $\mathcal{D}^\lambda : \mathcal{H}_n \to \mathcal{H}_n$. We call it the *hypergeometric Euler transform.* We have the composition rule:

$$\mathcal{D}^\nu \mathcal{D}^\lambda = \mathcal{D}^{\nu+\lambda} = \mathcal{D}^\lambda \mathcal{D}^\nu \quad (\lambda, \nu \in \mathbb{C}). \tag{3.4.5}$$

This follows formally from Theorem 3.4.1 and the corresponding property of D_a^λ (Proposition 3.3.4), or from the symbolic formula $\mathcal{D}^\lambda = \left(1 + \frac{d}{dx} D^{-1}\right)^\lambda$, or, of course, by direct computation using (3.4.3) and (3.4.4).

Theorems 3.4.1 and 3.4.1′ give us a method of obtaining solutions of $(\mathcal{D}^\lambda L)f = 0$ from those of $Lf = 0$. Indeed we derive now, as an application, integral representations of solutions of the so-called Jordan-Pochhammer equation. For two vectors $a = (a_1, \cdots, a_n) \in \mathbb{C}^n$, $a_i \neq$

a_j $(i \neq j)$ and $\mu = (\mu_1, \cdots, \mu_n) \in \mathbb{C}^n$, we define two polynomials p_n and p_{n-1} by

$$\text{(3.4.6)} \qquad \begin{aligned} p_n(x;a) &:= \prod_{j=1}^{n}(x-a_j), \\ p_{n-1}(x;a,\mu) &:= \sum_{j=1}^{n} \mu_j \prod_{k \neq j}(x-a_k) \end{aligned}$$

and define the differential operator L of hypergeometric type by

$$\text{(3.4.7)} \qquad L = p_n(x;a)D^n - p_{n-1}(x;a,\mu)D^{n-1}.$$

Since

$$\begin{aligned} D_a^{n-1}(Lf) &= D^{n-1}\{p_n D - p_{n-1}\}D^{n-1}f \\ &= D^{n-1}\{p_n D - p_{n-1}\}D_a^{n-1}f, \end{aligned}$$

we have

$$\mathcal{D}^{n-1}L = D^{n-1}\{p_n(x;a)D - p_{n-1}(x;a,\mu)\},$$

so the differential equation $(\mathcal{D}^{n-1}L)g = 0$ has a solution

$$g(x) = \prod_{j=1}^{n}(x-a_j)^{\mu_j}.$$

If $n - \mu_j \notin \mathrm{N}$, then Theorem 3.4.1 implies that $D_{a_j}^{\lambda} g(x)$ is a solution of $(\mathcal{D}^{\lambda+n-1}L)f = 0$. Similarly, Theorem 3.4.1′ implies that if neither $\mu_1 + \cdots + \mu_n$ nor $\mu_1 + \cdots + \mu + \lambda$ is a positive integer, then $D_{\infty}^{\lambda} g(x)$ is a solution of $(\mathcal{D}^{\lambda+n-1}L)f = 0$. By (3.4.4) and (3.4.7), the differential equation $(\mathcal{D}^{\lambda+n-1}L)f = 0$ can be written as

$$\text{(3.4.8)} \qquad \sum_{k=0}^{n} q_k(x;a,\mu,\lambda)D^k f = 0,$$

where

$$\text{(3.4.9)} \qquad \begin{aligned} q_k(x;a,\mu,\lambda) :&= \binom{\lambda+n-1}{n-k} p_n^{(n-k)}(x;a) \\ &\quad - \binom{\lambda+n-1}{n-k-1} p_{n-1}^{(n-k-1)}(x;a,\mu). \end{aligned}$$

The differential equation (3.4.8) with (3.4.6) and (3.4.9) is called the *Jordan-Pochhammer equation*. Thus we have proved:

THEOREM 3.4.2. *Let $\mu = (\mu_1, \cdots, \mu_n) \in \mathbb{C}^n$ be constants such that $n - \mu_j, \mu_1 + \cdots + \mu_n, \mu_1 + \cdots + \mu_n + \lambda \notin \mathbb{N}$. The Jordan-Pochhammer equation (3.4.8) admits solutions $f_j(x) = f_j(x; a, \mu, \lambda)$ $(j = 0, 1, \cdots, n)$ defined by the integral*

$$f_j(x) = \frac{1}{\Gamma(-\lambda)} \int_{a_j}^{x} (x-t)^{-\lambda-1} \prod_{k=1}^{n} (t - a_k)^{\mu_k} dt,$$

where $a_0 = \infty$. The right-hand side makes sense as the finite part of a divergent integral, even if it is divergent in the usual sense.

The hypergeometric equation is a special case of the Jordan-Pochhammer equation: Indeed, in case

$$n = 2, \quad a = (0,1), \quad \mu = (\alpha - \gamma, \gamma - \beta - 1) \text{ and } \lambda = \alpha - 1,$$

(3.4.8) becomes the hypergeometric equation with the parameters α, β and γ. Thus we obtain the Euler integral representation of solutions of the hypergeometric equation.

THEOREM 3.4.3. *Let α, β and γ be constants such that*

$$2 - (\alpha - \gamma), 2 - (\gamma - \beta - 1), \alpha - \beta - 1, -\beta - 1, \alpha \notin \mathbb{N}.$$

Then the hypergeometric equation admits solutions $F_{pq}(x)$ defined by

$$F_{pq}(x) = \int_p^q t^{\alpha-\gamma}(1-t)^{\gamma-\beta-1}(t-x)^{-\alpha} dt, \tag{3.4.10}$$

where $p = 0$, 1, ∞, or x. If the integral is divergent, then it is regarded as the finite part of a divergent integral.

This theorem gives us six solutions $F_{pq}(x)$ where $p, q \in \{0, 1, \infty, x\}$, $p \neq q$, which will be used in Section 4.4 to find the monodromy of the hypergeometric equation. Each function $F_{pq}(x)$ can also be expressed by the integral over a double loop around p and q (cf. Proposition 3.3.7).

3.5 Barnes integral representation: interpolation method

As we announced in Section 3.1, solutions of the hypergeometric equation admit two kinds of integral representations. One is the Euler integral representation and was disussed in Sections 3.2 - 3.4. Here, we shall discuss the other, i.e., the Barnes integral representation. We have, at least, three different approaches to it:

(1) to derive it from the power series representation of the solutions,
(2) to derive it by transforming the differential equation into a difference equation by the Mellin transform, and
(3) to derive it by using the representation theory of Lie groups.

In this section, we take Approach (1). Approach (2) will be explained in Section 3.6. Approach (3) is not treated in this book; readers interested in this method can consult [Vil]. The idea of finding the Barnes integral representation described in this section is based on an interpolation of an infinite sequence of numbers by a meromorphic function defined in the entire plane.

Consider a function $f(x)$ defined by the power series

$$f(x) = \sum_{m=0}^{\infty} a_m x^m.$$

Suppose that there is a function $g(t)$ satisfying the conditions

(i) $g(t)$ is meromorphic in $|t| < \infty$,
(ii) $g(t)$ is holomorphic at $t = 0, 1, 2, \cdots$,
(iii) $g(t)$ interpolates the sequence $\{a_m\}$, i.e., $g(m) = a_m$.

Then the function

$$h(t) := -g(t)\frac{\pi}{\sin(\pi t)}(-x)^t,$$

has simple poles at $t = m$ $(= 0,\ 1,\ 2,\ \ldots)$ with the residue $-a_m x^m$. Therefore, if C_N is a loop which encircles the points $t = 0, 1, \cdots, N$ in the negative sense and leaves the other poles of $h(t)$ outside, then we have

$$\frac{1}{2\pi i}\int_{C_N} h(t)dt = \sum_{m=0}^{N} a_m x^m. \tag{3.5.1}$$

If $h(t)$ has a suitable asymptotic property, then letting $N \to \infty$, we may obtain an integral representation for $f(x)$ of the form

$$f(x) = \frac{1}{2\pi i} \int_C h(t)dt,$$

where C is a certain path of integration.

Now we apply this idea when a_m are given by the coefficients

$$a_m = \frac{(\alpha)_m(\beta)_m}{(\gamma)_m(1)_m} = \frac{\Gamma(\gamma)}{\Gamma(\alpha)\Gamma(\beta)} \frac{\Gamma(\alpha+m)\Gamma(\beta+m)}{\Gamma(\gamma+m)\Gamma(1+m)}, \tag{3.5.2}$$

of the power series $F(\alpha, \beta, \gamma; x)$. The right-hand side makes sense only when

$$\alpha, \beta, \gamma \neq 0, -1, -2, \cdots. \tag{3.5.3}$$

Then the function

$$g(t) = \frac{\Gamma(\gamma)}{\Gamma(\alpha)\Gamma(\beta)} \frac{\Gamma(\alpha+t)\Gamma(\beta+t)}{\Gamma(\gamma+t)\Gamma(1+t)}$$

satisfies the above conditions (i)-(iii). By using the well-known formula

$$\Gamma(t)\Gamma(1-t) = \frac{\pi}{\sin(\pi t)}, \tag{3.5.4}$$

we find that

$$h(t) = \frac{\Gamma(\gamma)}{\Gamma(\alpha)\Gamma(\beta)} \frac{\Gamma(\alpha+t)\Gamma(\beta+t)\Gamma(-t)}{\Gamma(\gamma+t)}(-x)^t. \tag{3.5.5}$$

We have the following theorem.

THEOREM 3.5.1. *Suppose that α, β and γ satisfy* (3.5.3). *Then*

$$F(\alpha, \beta, \gamma; x) = \frac{1}{2\pi i} \frac{\Gamma(\gamma)}{\Gamma(\alpha)\Gamma(\beta)} \int_C \frac{\Gamma(\alpha+t)\Gamma(\beta+t)\Gamma(-t)}{\Gamma(\gamma+t)}(-x)^t dt$$

holds for $|x| < 1$ *and* $|\arg(-x)| < \pi$. *Here the path of integration* C *runs from* $-i\infty$ *to* $+i\infty$ *in such a way that, if* $|t|$ *is sufficiently large, then* $t \in C$ *lies on the imaginary axis, the poles of* $\Gamma(\alpha+t)\Gamma(\beta+t)$ *lie to the left of* C *and those of* $\Gamma(-t)$ *lie to the right of* C *(see Figure 2.3.).*

We shall prove this theorem along the lines explained above. Let $C_N = C_N^{(1)} + C_N^{(2)}$ be a loop indicated in Figure 2.4. Then we obtain the equality (3.5.1), where a_m and $h(t)$ are defined by (3.5.2) and (3.5.5) respectively. Therefore, in order to prove Theorem 3.5.1, it suffices to prove:

(3.5.6) $$\int_{C_N^{(2)}} h(t)dt \text{ converges to zero as } N \text{ tends to } \infty.$$

By using (3.5.4), $h(t)$ can be rewritten as

(3.5.7) $$h(t) = -\frac{\Gamma(\gamma)}{\Gamma(\alpha)\Gamma(\beta)} \frac{\Gamma(\alpha+t)\Gamma(\beta+t)}{\Gamma(\gamma+t)\Gamma(1+t)} \frac{\pi}{\sin(\pi t)}(-x)^t.$$

LEMMA 3.5.2. *Suppose* $|x| < 1$ *and* $|\arg(-x)| < \pi$, *then we have*

$$h(t) = \mathcal{O}\left(t^{(\alpha+\beta-\gamma-1)} e^{-\{\pi|\Im(t)| + \arg(-x)\cdot\Im(t)\}}\right)$$

as t *tends to* ∞ *in such a way that*

$$\delta < |\arg t| \le \frac{\pi}{2} \quad \text{or} \quad \Re(t) \in \mathrm{N} + \frac{1}{2}$$

holds, δ *being an arbitrary positive number. (Here* $\mathcal{O}$ *is Landau's symbol.)*

Proof. Recall the Stirling formula

$$\log\Gamma(t) = t\log t - t - \frac{1}{2}\log t + \mathcal{O}(1) \quad (t \to \infty, |\arg t| < \pi - \delta),$$

where δ is an arbitrary positive number. Apply this formula to the expression (3.5.5) for h for $\delta < |\arg t| \le \pi/2$, and apply it to (3.5.7) for $\Re(t) \in \mathrm{N} + 1/2$. ∎

Lemma 3.5.2 shows that, if $|x| < 1$ and $|\arg(-x)| < \pi$, then $h(t)$ is exponentially small on all of $C_N^{(2)}$ as N tends to ∞. This proves the assertion (3.5.6) and hence also Theorem 3.5.1.

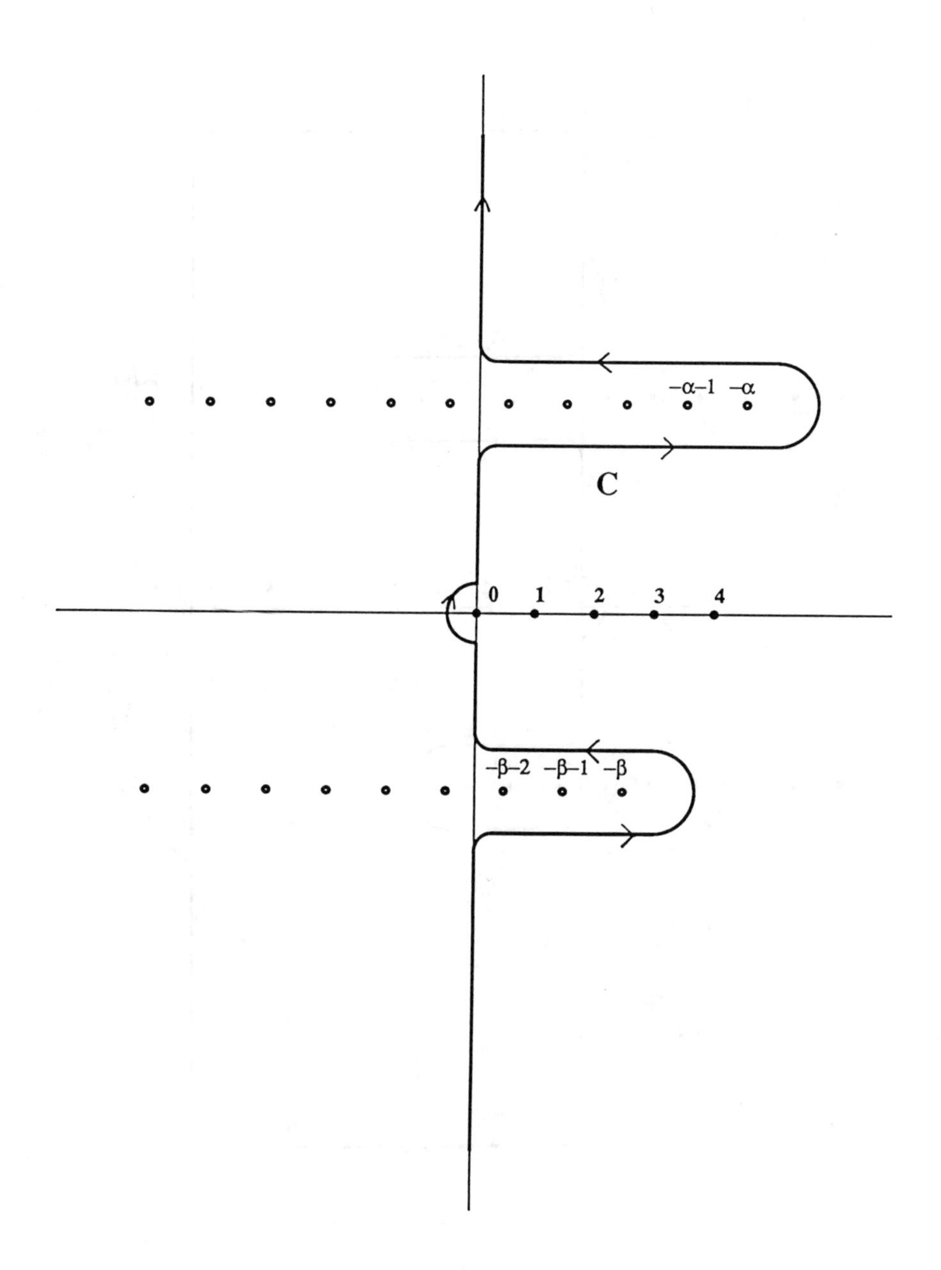

Figure 2.3.

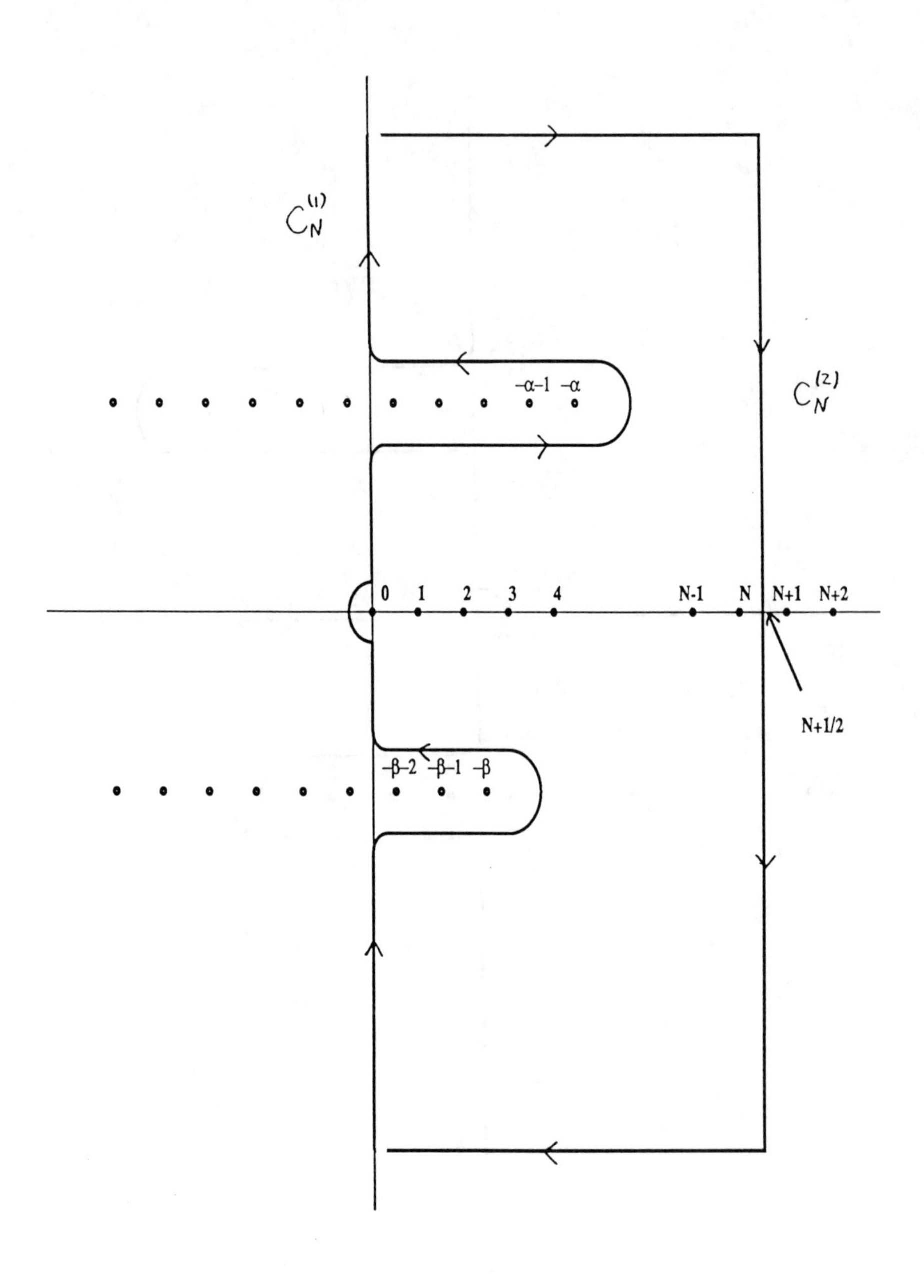

Figure 2.4.

3.6 Barnes integral representation: difference equation method

In this section, we are concerned with Approach (2) to the Barnes integral representation introduced in Section 3.5.

The *inverse Mellin transform* $F(x)$ of a function $G(t)$ is defined *formally* by the integral

$$F(x) = \frac{1}{2\pi i}\int G(t)\,(-x)^t\,dt,$$

where "formally" means that no path of integration has yet been specified and that no convergence has been discussed.

It happens that the Mellin transform converts a differential equation into a difference equation, when the differential equation is in a special form. Then, if the resulting difference equation is simple and can be solved explicitly, the inverse Mellin transform yields an integral representation of a solution of the original differential equation. The hypergeometric equation is a typical example where this observation really works.

The crucial fact used in this section is that the hypergeometric equation can be written in the form

$$\{\delta_x(\delta_x + \gamma - 1) - x(\delta_x + \alpha)(\delta_x + \beta)\}f = 0, \tag{3.6.1}$$

where $\delta_x = x(d/dx)$ (see (1.1.7)). We note that the method explained below is applicable in general to any differential equation with polynomial coefficients.

We shall find a condition on a function $g(t)$ under which the inverse Mellin transform

$$f(x) = \frac{1}{2\pi i}\int_C g(t)(-x)^t dt \tag{3.6.2}$$

converges and gives a solution of the hypergeometric equation, where C is a vertical line with possible deviation to avoid singularities of the integrand. From (3.6.1) and (3.6.2) we would have

$$\int_C t(t+\gamma-1)g(t)(-x)^t dt + \int_C (t+\alpha)(t+\beta)g(t)(-x)^{t+1} dt = 0. \tag{3.6.3}$$

If the path of integration in the first integral can be shifted to the left by 1, then we would obtain

$$\int_C \{(t+1)(t+\gamma)g(t+1)+(t+\alpha)(t+\beta)g(t)\}(-x)^{t+1}dt = 0.$$

Therefore $f(x)$ would be a solution of the hypergeometric equation if $g(t)$ satisfied the difference equation

$$g(t+1) = -\frac{(t+\alpha)(t+\beta)}{(t+\gamma)(t+1)}g(t). \tag{3.6.4}$$

Take a path of integration C as mentioned above. Then if the condition

$$\lim_{\tau\to\infty} t^2 g(t)(-x)^t = 0 \quad (t=\sigma+i\tau) \tag{3.6.5}$$

holds uniformly as t tends to ∞ in any finite vertical strip, then the integrals (3.6.2) and (3.6.3) converge uniformly with respect to x and so the above formal calculation can be legitimated. For instance, take the function

$$g(t) = \frac{\Gamma(t+\alpha)\Gamma(t+\beta)\Gamma(-t)}{\Gamma(t+\gamma)} \tag{3.6.6}$$

which is a solution of (3.6.4) satisfying the condition (3.6.5) (see Lemma 3.5.2). Then, substituting (3.6.6) into (3.6.2), we obtain a solution

$$f(x) = \frac{1}{2\pi i}\int_C \frac{\Gamma(t+\alpha)\Gamma(t+\beta)\Gamma(-t)}{\Gamma(t+\gamma)}(-x)^t\,dt \tag{3.6.7}$$

of the hypergeometric equation, where the path of integration C is taken in the manner indicated in Figure 2.3.

3.7 The Gauss-Kummer identity

We know that the hypergeometric series $F(\alpha, \beta, \gamma; x)$ is convergent on the circle $|x| = 1$ if $\Re(\gamma - \alpha - \beta) > 0$, (Assertion (ii) of Theorem 1.2.2). In this section, we shall find its explicit value at $x = 1$.

THEOREM 3.7.1. *If* $\Re(\gamma - \alpha - \beta) > 0$, *then*

$$F(\alpha, \beta, \gamma; 1) = \frac{\Gamma(\gamma)\Gamma(\gamma - \alpha - \beta)}{\Gamma(\gamma - \alpha)\Gamma(\gamma - \beta)}. \tag{3.7.1}$$

This formula is called the *Gauss-Kummer identity.* It provides a clue to calculate the connection matrices (see Section 4.7). A generalization of the Gauss-Kummer identity was made by [Okub.1] and [Okub.2], which plays an important role in computing monodromy of the so-called hypergeometric type (or Okubo type) equations. We shall give two proofs of the theorem:

Proof 1 is based on the Euler integral representation;
Proof 2 is based on the contiguity relation for the parameter γ.

Proof 1. Suppose $\Re(\alpha) > 0$ and $\Re(\beta) > 0$. Under the condition $\Re(\gamma - \alpha - \beta) > 0$, we have $\Re(\gamma) > \Re(\beta) > 0$. Hence the Euler integral representation (3.2.1) can be applied. Putting $x = 1$ in (3.2.1), we obtain

$$\begin{aligned} F(\alpha, \beta, \gamma; 1) &= \frac{1}{B(\beta, \gamma - \beta)} \int_0^1 t^{\beta-1}(1-t)^{\gamma-\alpha-\beta-1} dt \\ &= \frac{B(\beta, \gamma - \alpha - \beta)}{B(\beta, \gamma - \beta)} \\ &= \frac{\Gamma(\gamma)\Gamma(\gamma - \alpha - \beta)}{\Gamma(\gamma - \alpha)\Gamma(\gamma - \beta)}. \end{aligned}$$

Now analytic continuation gives the result in general. ∎

Proof 2. We use the following contiguity relation (Theorem 2.1.3, (3)) for the parameter γ:

$$\begin{aligned} (\gamma - \alpha)(\gamma - \beta)F(\alpha, \beta, \gamma + 1; x) &= \gamma(\gamma - \alpha - \beta)F(\alpha, \beta, \gamma; x) \\ &\quad + \gamma(1 - x)\frac{d}{dx}F(\alpha, \beta, \gamma; x). \end{aligned} \tag{3.7.2}$$

If $\Re(\gamma - \alpha - \beta) > 1$, then the power series $F(\alpha, \beta, \gamma; x), F(\alpha, \beta, \gamma + 1; x), \frac{d}{dx}F(\alpha, \beta, \gamma; x)$ converge on $|x| \leq 1$ (Theorem 1.2.2, Proposition 2.1.7). So letting x tend to 1 in (3.7.2), we obtain

$$F(\alpha, \beta, \gamma; 1) = \frac{(\gamma - \alpha)(\gamma - \beta)}{\gamma(\gamma - \alpha - \beta)} F(\alpha, \beta, \gamma + 1; 1).$$

By the analyticity of $F(\alpha, \beta, \gamma; 1)$ in (α, β, γ), this formula is valid for $\Re(\gamma - \alpha - \beta) > 0$. Using this formula repeatedly, we obtain

$$\begin{aligned} &F(\alpha, \beta, \gamma; 1) \\ &= \frac{\Gamma(\gamma - \alpha + m)\Gamma(\gamma - \beta + m)}{\Gamma(\gamma + m)\Gamma(\gamma - \alpha - \beta + m)} \cdot \frac{\Gamma(\gamma)\Gamma(\gamma - \alpha - \beta)}{\Gamma(\gamma - \alpha)\Gamma(\gamma - \beta)} \cdot F(\alpha, \beta, \gamma + m; 1). \end{aligned}$$

To show (3.7.1), we must show that the first and third factors on the right-hand side tend to 1 as m tends to infinity. For the first factor this follows immediately from the formula

$$\frac{\Gamma(m + c)}{\Gamma(m)} \sim m^c \qquad (c \text{ fixed}, m \to \infty).$$

For the second, we verify that all the terms except the constant term 1 in the hypergeometric series representation of $F(\alpha, \beta, \gamma + m; 1)$ tend to 0 in a sufficiently uniform way as $m \to \infty$; details are left to the reader.∎

4 Monodromy of the hypergeometric equation

4.1 Fundamental group of $\mathbb{P}^1 \backslash \{0, 1, \infty\}$ and the monodromy of the Riemann equation

Let D be a domain in the complex plane. Consider an n-th order differential equation

$$\frac{d^n f}{dx^n} + a_1(x) \frac{d^{n-1} f}{dx^{n-1}} + \cdots + a_n(x) f = 0, \tag{4.1.1}$$

where the $a_j(x)$'s are holomorphic functions in D. If D is not simply connected, then a solution of (4.1.1) may be a multi-valued holomorphic function in D. To describe the multi-valuedness of solutions of (4.1.1), we can associate to (4.1.1) the conjugacy class of a subgroup of $GL(n, \mathbb{C})$, which will be called the *monodromy* of (4.1.1).

We introduce the *fundamental group* $\pi_1(D, b)$ of D with a base point $b \in D$. A *loop* γ *in* D *with the base point* b is a curve

$$\gamma : I = [0, 1] \longrightarrow D$$

starting and ending at b. Let $L(D, b)$ be the set of loops in D with the base point b. Two loops $\gamma_0, \gamma_1 \in L(D, b)$ are said to be *homotopy-equivalent* and denoted by $\gamma_0 \simeq \gamma_1$ if and only if γ_0 can be deformed continuously to γ_1 keeping the base point b fixed. One can check that $\simeq$ is actually an equivalence relation. Let $\pi_1(D, b)$ be the set of all equivalence classes of loops in $L(D, b)$, i.e., $\pi_1(D, b) = L(D, b)/\simeq$. The class, which is called the *homotopy class*, containing $\gamma \in L(D, b)$ will be denoted by $[\gamma]$. The product $\gamma_1 \cdot \gamma_0$ of two loops γ_0 and $\gamma_1 \in L(D, b)$ is defined to be a loop going firstly along γ_0 and then γ_1 in the obvious sense. The product is compatible with the equivalence relation $\simeq$. Namely, if $\gamma_j \simeq \gamma_j'$ $(\gamma_j, \gamma_j' \in L(D, b), j = 0, 1)$, then $\gamma_1 \cdot \gamma_0 \simeq \gamma_1' \cdot \gamma_0'$. Thus the product is naturally defined on $\pi_1(D, b)$, which in this way becomes a group. In fact, the unit element $e \in \pi_1(D, b)$ is given by $[c]$, where $c \in L(D, b)$ is the constant map $c : I \to D$, $c(t) \equiv b$. The inverse element α^{-1} of $\alpha = [\gamma] \in \pi_1(D, b)$ is given by $[\gamma^{-1}]$, where γ^{-1} is a loop defined by $\gamma^{-1}(t) := \gamma(1 - t)$. Proof of these assertions and the associativity of the product is left to the reader. The group $\pi_1(D, b)$ is called the *fundamental group of* D *with base point* b.

Let U be a simply connected neighborhood of b in D and let $\mathcal{F} = (f_1, \cdots, f_n)$ be a fundamental system of solutions of (4.1.1) in U. Given any element $\alpha \in \pi_1(D, b)$, choose a representative loop $\gamma \in L(D, b)$ of α. Let $\gamma_* \mathcal{F}$ be the analytic continuation of $\mathcal{F}$ along the loop γ; then the monodromy theorem for analytic continuation implies that $\gamma_* \mathcal{F}$ depends only on the homotopy class α containing the loop γ. Thus we write $\alpha_* \mathcal{F}$ for $\gamma_* \mathcal{F}$. Since (4.1.1) is a linear equation, $\alpha_* \mathcal{F}$ is also a fundamental system of solutions of (4.1.1) in U and there is a unique non-singular matrix $M(\alpha; \mathcal{F}) \in GL(n, \mathbb{C})$ such that

$$\alpha_* \mathcal{F} = \mathcal{F} M(\alpha; \mathcal{F}). \tag{4.1.2}$$

Since $e_* \mathcal{F} = \mathcal{F}$ and $(\alpha\beta)_* \mathcal{F} = \alpha_*(\beta_* \mathcal{F})$ for $\alpha, \beta \in \pi_1(D, b)$, we have

$$M(e; \mathcal{F}) = I, \quad M(\alpha\beta; \mathcal{F}) = M(\alpha; \mathcal{F}) M(\beta; \mathcal{F}),$$

where I is the unit matrix in $GL(n, \mathbb{C})$. These imply that the map

$$\rho_{\mathcal{F}} : \pi_1(D, b) \to GL(n, \mathbb{C}), \quad \alpha \mapsto M(\alpha; \mathcal{F})$$

is a group homomorphism. A group homomorphism $\rho : \pi_1(D, b) \to GL(n, \mathbb{C})$ is also called a (*linear*) *representation* of rank n. We call $\rho_{\mathcal{F}}$ the *monodromy representation* and the image group $\rho_{\mathcal{F}}(\pi_1(D, b)) \subset GL(n, \mathbb{C})$ the *monodromy group* of (4.1.1) with respect to the fundamental system of solutions $\mathcal{F}$. Let $\mathcal{G}$ be another fundamental system of solutions at another point, say a. If we denote also by $\mathcal{G}$ the analytic continuation of $\mathcal{G}$ along a curve joining a and b, then there is a matrix $C \in GL(n, \mathbb{C})$ such that $\mathcal{G} = \mathcal{F} C$. Hence we have

$$\mathcal{G} M(\alpha; \mathcal{G}) = \alpha_* \mathcal{G} = (\alpha_* \mathcal{F}) C = \mathcal{F} M(\alpha; \mathcal{F}) C = \mathcal{G} C^{-1} M(\alpha; \mathcal{F}) C,$$

i.e., $M(\alpha; \mathcal{G}) = C^{-1} M(\alpha; \mathcal{F}) C$. In other words, we have

$$\rho_{\mathcal{G}}(\alpha) = C^{-1} \rho_{\mathcal{F}}(\alpha) C \quad \text{for} \quad \alpha \in \pi_1(D, b). \tag{4.1.3}$$

In general, two representations $\rho, \rho' : \pi_1(D, b) \to GL(n, \mathbb{C})$ are said to be *conjugate* if and only if there is a matrix $C \in GL(n, \mathbb{C})$ such that $\rho'(\alpha) = C^{-1} \rho(\alpha) C$ for all $\alpha \in \pi_1(D, b)$. Conjugacy is clearly an equivalence relation. The monodromy representation depends not only on the differential equation (4.1.1), but also on its fundamental system of solutions. Note that, however, (4.1.3) implies that every two monodromy representations of (4.1.1) are conjugate, so that the conjugacy class of the monodromy representation is uniquely determined by the differential equation (4.1.1). We call this conjugacy class the *monodromy* of (4.1.1). Two subgroups G_1, G_2 of $GL(n,\mathbb{C})$ are said to be *conjugate* if and only if there exists a matrix $C \in GL(n,\mathbb{C})$ such that $G_2 = C^{-1}G_1C$. It is clear from the above argument that the monodromy group $\rho_{\mathcal{F}}(\pi_1(D,b))$ of (4.1.1) with respect to any $\mathcal{F}$ belongs to the same conjugacy class which is also called the *monodromy* of (4.1.1).

PROBLEM 4.1.1. (Monodromy problem) *For a given linear differential equation, find an explicit expression of its monodromy; or find generators of the monodromy group with respect to a fundamental system of solutions.*

When (4.1.1) is a Fuchsian differential equation on the Riemann sphere with regular singular points at $p_1, \cdots, p_m, p_{m+1} = \infty$, then we take $D = \mathbb{C}\backslash\{p_1, \cdots, p_m\}$. For each $j = 1, \cdots, m$, let U_j be an open disc in $D \cup \{p_j\}$ centered at p_j, and let ℓ_j be a loop in $U_j\backslash\{p_j\}$ with a base point $q_j \in U_j\backslash\{p_j\}$ which encircles p_j once counterclockwise, and let $\mathcal{F}_j$ be a fundamental system of solutions of (4.1.1) in $U_j\backslash\{p_j\}$. Let b be a point in D, let $\mathcal{F}$ be a fundamental system of solutions of (4.1.1) in a simply connected neighborhood of b and γ_j $(j = 1, \cdots, m)$ be arcs starting at b and terminating at q_j. The *connection matrices* $C_j \in GL(n,\mathbb{C})$ $(j = 1, \cdots, m)$ are defined by

$$\gamma_{j*}\mathcal{F} = \mathcal{F}_jC_j. \tag{4.1.4}$$

The connection problem mentioned in section 3.1 can be stated as follows:

PROBLEM 4.1.2. (Connection problem) *For a given linear differential equation, let $\mathcal{F}, \mathcal{F}_j$ and C_j be as above. Find an explicit expression of the C_j's.*

The *circuit matrices M_j around p_j* can be defined by

$$\ell_{j*}\mathcal{F}_j = \mathcal{F}_jM_j. \tag{4.1.5}$$

Since generators of the monodromy group with respect to $\mathcal{F}$ are given by

$$C_j^{-1}M_jC_j,$$

we see

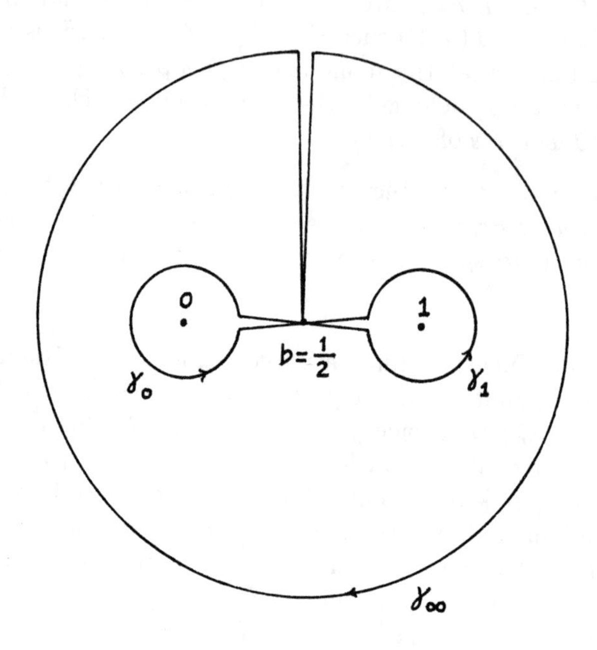

Figure 2.5.

REMARK 4.1.3. The monodromy problem is a part of the connection problem.

Unfortunately, we know only a very restricted number of equations whose monodromy problem we can solve. To each of such equations, one applies an appropriate method, according to the property of the equation, which stems from a method used for the hypergeometric differential equation. So, we shall present in Sections 4.3-4.7 several known methods to find generators of the monodromy group of the hypergeometric differential equation.

We briefly recall the structure of the fundamental group of $D = \mathbb{P}^1\backslash\{0, 1, \infty\}$. Take a point $b \in D$ and three loops γ_0, γ_1 and γ_∞ with the base point $b\,(= 1/2)$ in a manner indicated in Figure 2.5. For brevity, we also denote by γ_j the homotopy class to which γ_j belongs. We can easily show that the fundamental group $G = \pi_1(D, b)$ is a *free group* generated by the two elements γ_0 and γ_1; γ_∞ is given by

$$\gamma_\infty{}^{-1} = \gamma_1\gamma_0. \tag{4.1.6}$$

Notice that $\gamma_1\gamma_0$ and $\gamma_0\gamma_1$ are conjugate ; in fact

$$\begin{aligned}\gamma_0\gamma_1 &= \gamma_0(\gamma_1\gamma_0)\gamma_0{}^{-1}\\ &= \gamma_1{}^{-1}(\gamma_1\gamma_0)\gamma_1.\end{aligned}$$

4.2 Classification of 2-dimensional representations of the free group with 2 generators

The monodromy representation of a Riemann equation is a 2-dimensional representation of a free group with 2 generators. So we classify 2-dimensional representations of the free group with 2 generators in advance. Let

$$\begin{aligned} G = \langle u, v\rangle &: \text{free group generated by } u \text{ and } v,\\ V &: 2 \text{ dimensional complex vector space},\\ \rho : G \to GL(V) &: \text{a 2 dimensional representation}, \end{aligned}$$

and

$$\begin{array}{lll} \{\lambda_1,\lambda_2\} & : \text{set of eigenvalues of} & \rho(u),\\ \{\mu_1,\mu_2\} & : \text{set of eigenvalues of} & \rho(v),\\ \{\nu_1,\nu_2\} & : \text{set of eigenvalues of} & \rho(uv). \end{array} \tag{4.2.1}$$

Notice that the eigenvalues (4.2.1) depend only on the conjugacy class of ρ , and that the equality $\det\rho(u)\cdot\det\rho(v) = \det\rho(uv)$ leads to the following relation among these eigenvalues :

$$\lambda_1\lambda_2\mu_1\mu_2 = \nu_1\nu_2. \tag{4.2.2}$$

One may think that a class of the representation is determined by the eigenvalues (4.2.1) but it is true only when the representation is irreducible. A subspace $W \subset V$ is called *ρ-invariant* if and only if $\rho(G)W \subset W$ and W is called *proper* if it is neither $\{0\}$ nor V. A representation ρ is called *irreducible* if there is no ρ-invariant proper subspace.

THEOREM 4.2.1. (i) *ρ is irreducible if and only if*

$$\lambda_i \mu_j \neq \nu_k \quad \text{for all} \quad i, j, k = 1, 2. \tag{4.2.3}$$

(ii) *If ρ is irreducible, then there is a basis of V such that $\rho(u)$ and $\rho(v)$ are represented by the following matrices*

$$\begin{aligned} \rho(u) &\leftrightarrow \begin{pmatrix} \lambda_1 & 1 \\ 0 & \lambda_2 \end{pmatrix}, \\ \rho(v) &\leftrightarrow \begin{pmatrix} \mu_1 & 0 \\ (\nu_1 + \nu_2) - (\lambda_1 \mu_1 + \lambda_2 \mu_2) & \mu_2 \end{pmatrix}. \end{aligned} \tag{4.2.4}$$

All representations (4.2.4) given by exchanging λ_1 and λ_2 and/or μ_1 and μ_2 are mutually conjugate. In particular, if a conjugacy class is irreducible, it is determined uniquely by the eigenvalues $\{\lambda_1, \lambda_2\}, \{\mu_1, \mu_2\}$ and $\{\nu_1, \nu_2\}$.

(iii) *If ρ is reducible, then the conjugacy class of ρ is not determined only by the eigenvalues; there are several cases.*

(a) *In case $\lambda_1 \neq \lambda_2, \mu_1 \neq \mu_2$ and $\nu_1 \neq \nu_2$, we label (there are two ways) indices so that*

$$\lambda_1 \mu_1 = \nu_1 \quad (\Leftrightarrow \lambda_2 \mu_2 = \nu_2).$$

There are three conjugacy classes given by

$$\rho(u) \leftrightarrow \begin{pmatrix} \lambda_1 & 0 \\ 0 & \lambda_2 \end{pmatrix}, \quad \rho(v) \leftrightarrow \begin{pmatrix} \mu_1 & 0 \\ 0 & \mu_2 \end{pmatrix};$$

$$\rho(u) \leftrightarrow \begin{pmatrix} \lambda_1 & 0 \\ 0 & \lambda_2 \end{pmatrix}, \quad \rho(v) \leftrightarrow \begin{pmatrix} \mu_1 & 1 \\ 0 & \mu_2 \end{pmatrix};$$

$$\rho(u) \leftrightarrow \begin{pmatrix} \lambda_2 & 0 \\ 0 & \lambda_1 \end{pmatrix}, \quad \rho(v) \leftrightarrow \begin{pmatrix} \mu_2 & 1 \\ 0 & \mu_1 \end{pmatrix}.$$

(b) *In case $\lambda_1 = \lambda_2 (=: \lambda), \mu_1 \neq \mu_2$ and $\nu_1 \neq \nu_2$, there are three conjugacy classes:*

$$\rho(u) \leftrightarrow \begin{pmatrix} \lambda & 1 \\ 0 & \lambda \end{pmatrix}, \quad \rho(v) \leftrightarrow \begin{pmatrix} \mu_1 & 0 \\ 0 & \mu_2 \end{pmatrix};$$

$$\rho(u) \leftrightarrow \begin{pmatrix} \lambda & 1 \\ 0 & \lambda \end{pmatrix}, \quad \rho(v) \leftrightarrow \begin{pmatrix} \mu_2 & 0 \\ 0 & \mu_1 \end{pmatrix};$$

$$\rho(u) \leftrightarrow \begin{pmatrix} \lambda & 0 \\ 0 & \lambda \end{pmatrix}, \quad \rho(v) \leftrightarrow \begin{pmatrix} \mu_1 & 0 \\ 0 & \mu_2 \end{pmatrix}.$$

(b)′ *In case* $\lambda_1 \neq \lambda_2, \mu_1 = \mu_2 (=: \mu)$ *and* $\nu_1 \neq \nu_2$, *there are three conjugacy classes:*

$$\rho(u) \leftrightarrow \begin{pmatrix} \lambda_1 & 0 \\ 0 & \lambda_2 \end{pmatrix}, \quad \rho(v) \leftrightarrow \begin{pmatrix} \mu & 1 \\ 0 & \mu \end{pmatrix};$$

$$\rho(u) \leftrightarrow \begin{pmatrix} \lambda_2 & 0 \\ 0 & \lambda_1 \end{pmatrix}, \quad \rho(v) \leftrightarrow \begin{pmatrix} \mu & 1 \\ 0 & \mu \end{pmatrix};$$

$$\rho(u) \leftrightarrow \begin{pmatrix} \lambda_1 & 0 \\ 0 & \lambda_2 \end{pmatrix}, \quad \rho(v) \leftrightarrow \begin{pmatrix} \mu & 0 \\ 0 & \mu \end{pmatrix}.$$

(b)″ *In case* $\lambda_1 \neq \lambda_2, \mu_1 \neq \mu_2$ *and* $\nu_1 = \nu_2 (=: \nu)$, *there are three conjugacy classes:*

$$\rho(u) \leftrightarrow \begin{pmatrix} \lambda_1 & 0 \\ 0 & \lambda_2 \end{pmatrix}, \quad \rho(v) \leftrightarrow \begin{pmatrix} \mu_1 & \lambda_1^{-1} \\ 0 & \mu_2 \end{pmatrix}, \quad \rho(uv) \leftrightarrow \begin{pmatrix} \nu & 1 \\ 0 & \nu \end{pmatrix};$$

$$\rho(u) \leftrightarrow \begin{pmatrix} \lambda_2 & 0 \\ 0 & \lambda_1 \end{pmatrix}, \quad \rho(v) \leftrightarrow \begin{pmatrix} \mu_2 & \lambda_2^{-1} \\ 0 & \mu_1 \end{pmatrix}, \quad \rho(uv) \leftrightarrow \begin{pmatrix} \nu & 1 \\ 0 & \nu \end{pmatrix};$$

$$\rho(u) \leftrightarrow \begin{pmatrix} \lambda_1 & 0 \\ 0 & \lambda_2 \end{pmatrix}, \quad \rho(v) \leftrightarrow \begin{pmatrix} \mu_1 & 0 \\ 0 & \mu_2 \end{pmatrix}, \quad \rho(uv) \leftrightarrow \begin{pmatrix} \nu & 0 \\ 0 & \nu \end{pmatrix}.$$

(c) *In case* $\lambda_1 = \lambda_2 (=: \lambda), \mu_1 = \mu_2 (=: \mu)$ *and* $\nu_1 = \nu_2 (=: \nu)$, *there are a* 1*-parameter family of conjugacy classes:*

$$\rho(u) \leftrightarrow \begin{pmatrix} \lambda & 1 \\ 0 & \lambda \end{pmatrix}, \quad \rho(v) \leftrightarrow \begin{pmatrix} \mu & b \\ 0 & \mu \end{pmatrix} \quad (b \in \mathbb{C}),$$

and two classes:

$$\rho(u) \leftrightarrow \begin{pmatrix} \lambda & 0 \\ 0 & \lambda \end{pmatrix}, \quad \rho(v) \leftrightarrow \begin{pmatrix} \mu & \delta \\ 0 & \mu \end{pmatrix} \quad (\delta = 0, 1).$$

Proof. Proof of the assertions (i) and (ii): Suppose that ρ is irreducible. Let e_1 be an eigenvector of $\rho(u)$ corresponding to the eigenvalue λ_1, and let e_2 be an eigenvector of $\rho(v)$ corresponding to the eigenvalue μ_2. By the irreducibility of ρ, the two vectors e_1 and e_2 are linearly independent, and the line spanned by e_2 is not $\rho(u)$-invariant. Hence e_2 can be normalized so that $\rho(u)e_2 = e_1 + \text{const}\cdot e_2$. Then, in terms of the basis $\{e_1, e_2\}$, $\rho(u)$ and $\rho(v)$ are represented by the following matrices,

$$(4.2.5) \qquad \rho(u) \leftrightarrow A = \begin{pmatrix} \lambda_1 & 1 \\ 0 & \lambda_2 \end{pmatrix}, \quad \rho(v) \leftrightarrow B = \begin{pmatrix} \mu_1 & 0 \\ b & \mu_2 \end{pmatrix};$$

and so $\rho(uv)$ is represented by

$$\rho(uv) \leftrightarrow AB = \begin{pmatrix} b + \lambda_1\mu_1 & \mu_2 \\ \lambda_2 b & \lambda_2\mu_2 \end{pmatrix}.$$

Computing the trace of $\rho(uv)$ in two ways, we obtain

$$(4.2.6) \qquad b + \lambda_1\mu_1 + \lambda_2\mu_2 = \nu_1 + \nu_2.$$

Since ρ is assumed to be irreducible, (4.2.5) implies that

$$(4.2.7) \qquad b \neq 0,$$

and that the line spanned by e_1 is not $\rho(v)$-invariant. Now let us consider a $\rho(u)$-invariant line. In case $\lambda_1 = \lambda_2$, such a line is spanned by e_1. Hence in this case (4.2.7) is also sufficient for ρ to be irreducible. In case $\lambda_1 \neq \lambda_2$, the vector $e' = e_1/(\lambda_2 - \lambda_1) + e_2$ spans another $\rho(u)$-invariant line. For ρ to be irreducible, it is sufficient that the line spanned by e' is not $\rho(v)$-invariant and this is equivalent to the condition that e' is not an eigenvector of $\rho(v)$ corresponding to the eigenvalue μ_1. This condition is given by

$$(4.2.8) \qquad b + (\lambda_2 - \lambda_1)(\mu_2 - \mu_1) \neq 0.$$

If $\lambda_1 = \lambda_2$, then (4.2.8) clearly reduces to (4.2.7). Hence (4.2.8) covers the first case. Therefore (4.2.7) together with (4.2.8) is a necessary and sufficient condition for ρ to be irreducible. Using (4.2.6), we find that this condition is nothing but

$$\lambda_1\mu_i + \lambda_2\mu_j \neq \nu_1 + \nu_2 \quad (i, j = 1, 2, i \neq j). \tag{4.2.9}$$

Since we have (4.2.2), the relation between roots and coefficients of a quadratic equation (e.g. $(x - \lambda_1\mu_2)(x - \lambda_2\mu_1) = 0$ etc.) implies that (4.2.9) is equivalent to (4.2.3). Hence the assertion (i) is proved. The assertion (ii) is already established by (4.2.5) and (4.2.6).

Proof of the assertion (iii): Suppose ρ is reducible. Then there is a 1-dimensional $\rho(G)$-invariant subspace. Let e_1 be its basis vector, which is a simultaneous eigenvector of $\rho(u)$ and $\rho(v)$. We assume that e_1 corresponds to the eigenvalues λ_1 of $\rho(u)$ and μ_1 of $\rho(v)$. If we extend e_1 to a basis $\{e_1, e_2\}$ of V, then $\rho(u)$ and $\rho(v)$ are expressed by the matrices of the form

$$\rho(u) \leftrightarrow \begin{pmatrix} \lambda_1 & \varepsilon \\ 0 & \lambda_2 \end{pmatrix}, \quad \rho(v) \leftrightarrow \begin{pmatrix} \mu_1 & \delta \\ 0 & \mu_2 \end{pmatrix}, \quad (\varepsilon, \delta \in \mathbb{C}).$$

Since any 2 by 2 matrix with distinct eigenvalues is diagonalizable, the following elementary lemma leads to the assertion (iii).

LEMMA 4.2.2. *Let P be a 2 by 2 non-singular matrix.*
(i) *If P commutes with*

$$\begin{pmatrix} \lambda_1 & 0 \\ 0 & \lambda_2 \end{pmatrix}, \quad (\lambda_1 \neq \lambda_2)$$

then P is of the form

$$\begin{pmatrix} p & 0 \\ 0 & q \end{pmatrix},$$

and we have

$$P \begin{pmatrix} \mu_1 & \delta \\ 0 & \mu_2 \end{pmatrix} P^{-1} = \begin{pmatrix} \mu_1 & pq^{-1}\delta \\ 0 & \mu_2 \end{pmatrix}.$$

(ii) *If P commutes with*

$$\begin{pmatrix} \lambda & 1 \\ 0 & \lambda \end{pmatrix},$$

then P is of the form

$$\begin{pmatrix} p & q \\ 0 & p \end{pmatrix},$$

and we have

$$P\begin{pmatrix} \mu_1 & \delta \\ 0 & \mu_2 \end{pmatrix}P^{-1} = \begin{pmatrix} \mu_1 & p^{-1}q(\mu_2 - \mu_1) + \delta \\ 0 & \mu_2 \end{pmatrix}.$$

4.3 Finding the monodromy by local properties and the Fuchs relation

Let us consider the Riemann equation $RE(\rho, \sigma, \tau)$ with a Riemann scheme

$$\begin{pmatrix} 0 & 1 & \infty \\ \rho_1 & \sigma_1 & \tau_1 \\ \rho_2 & \sigma_2 & \tau_2 \end{pmatrix}. \tag{4.3.1}$$

In this section we shall find the *monodromy* of $RE(\rho, \sigma, \tau)$ by using local properties and the Fuchs relation. On the other hand, in Sections 4.4 - 4.7 we shall find the monodromy representation of $RE(\rho, \sigma, \tau)$ with respect to a certain fundamental system of solutions by using explicit expressions of solutions, (e.g., by using integral representations and by solving the connection problem). In the former method, although the monodromy representation is not given, one can know the monodromy for arbitrary values of the characteristic exponents ρ_i, σ_j, τ_k $(i, j, k = 1, 2)$. In particular, one can know which classes of 2-dimensional representations of the free group generated by two elements can be realized as the monodromy of a Riemann equation (see Corollary 4.3.4). On the other hand, in the latter method, one can get more detailed information about the monodromy representation; however, these methods cannot be applied when the characteristic exponents ρ_i, σ_j, τ_k $(i, j, k = 1, 2)$ take certain special values. Notice that the monodromy depends only on the Riemann equation and the Riemann equation depends only on its characteristic exponents, so its monodromy can be described only in terms of its characteristic exponents.

Let $G = \pi_1(D, b)$ and $\gamma_j \in G$ $(j = 0, 1, \infty)$ be as in Section 4.1. Let $\rho : G \to GL(2, \mathbb{C})$ be a monodromy representation of the Riemann equation $RE(\rho,\sigma,\tau)$. Since $\rho(\gamma_1 \cdot \gamma_0) = \rho(\gamma_\infty)^{-1}$ by (4.1.6) and $\gamma_0\gamma_1$ is conjugated to $\gamma_1\gamma_0$, we have

$$
\begin{array}{llll}
 & \{\varepsilon(\rho_1),\varepsilon(\rho_2)\} & : \text{set of eigenvalues of} & \rho(\gamma_0), \\
(4.3.2) & \{\varepsilon(\sigma_1),\varepsilon(\sigma_2)\} & : \text{set of eigenvalues of} & \rho(\gamma_1), \\
 & \{\varepsilon(-\tau_1),\varepsilon(-\tau_2)\} & : \text{set of eigenvalues of} & \rho(\gamma_0\gamma_1),
\end{array}
$$

where $\varepsilon(\cdot) = \exp(2\pi i\cdot)$. Thus, putting $\lambda_j = \varepsilon(\rho_j), \mu_j = \varepsilon(\sigma_j)$ and $\nu_j = \varepsilon(-\tau_j)$, we arrive at the situation in Section 4.2. As was seen in Section 4.2, the conjugacy class of ρ , i.e., the monodromy of $RE(\rho,\sigma,\tau)$, is almost determined by the eigenvalues (4.3.2); and is completely determined, if the monodromy is irreducible.

DEFINITION 4.3.1. The Riemann equation $RE(\rho,\sigma,\tau)$ is said to be *(ir)reducible* if its monodromy is (ir)reducible.

We first treat the irreducible case. As an immediate consequence of Theorem 4.2.1 (i) and (ii), we obtain the

THEOREM 4.3.2 (Irreducible case). *The Riemann equation $RE(\rho,\sigma,\tau)$ is irreducible if and only if*

$$\rho_i + \sigma_j + \tau_k \notin \mathbb{Z} \quad (i,j,k = 1,2). \tag{4.3.3}$$

Under this condition, the representation ρ is expressed, up to conjugacy, by the following matrices,

$$\rho(\gamma_0) \leftrightarrow \begin{pmatrix} \varepsilon(\rho_1) & 1 \\ 0 & \varepsilon(\rho_2) \end{pmatrix}, \quad \rho(\gamma_1) \leftrightarrow \begin{pmatrix} \varepsilon(\sigma_1) & 0 \\ b & \varepsilon(\sigma_2) \end{pmatrix}. \tag{4.3.4}$$

Here the number b is given by

$$b = \varepsilon(-\tau_1) + \varepsilon(-\tau_2) - \varepsilon(\rho_1+\sigma_1) - \varepsilon(\rho_2+\sigma_2), \tag{4.3.5}$$

where $\varepsilon(\cdot) = \exp(2\pi i\cdot)$. *All representations* (4.3.4) *obtained by exchanging ρ_1 and ρ_2 and/or σ_1 and σ_2 are mutually conjugate.*

We shall next consider reducible cases; which are more complicated than the irreducible case. Recall that a reducible representation class cannot be determined only by the eigenvalues (4.2.1). So the monodromy of $RE(\rho, \sigma, \tau)$ cannot be determined only by the exponentials (4.3.2) of the characteristic exponents $\rho_i, \sigma_j, \tau_k (i, j, k = 1, 2)$. Namely, there still remain some informations to be taken account of. Of course, the monodromy is determined by the characteristic exponents ρ_i, σ_j, τ_k, because the Riemann equation is uniquely determined by its characteristic exponents. The key is to know which singular points are (non-) logarithmic. Keeping this point in mind, we state the result, by using the following notation:

$\mathbb{N}$: the set of positive integers,
$\mathbb{N}_0$: the set of non-negative integers,

$$\langle m \rangle = \begin{cases} \phi : \text{the empty set} & (m = 0) \\ \{1, 2, \cdots, m\} & (m \in \mathbb{N}), \end{cases}$$

$$\varepsilon(\cdot) = \exp(2\pi i \cdot).$$

THEOREM 4.3.3 (Reducible cases). *Suppose that the Riemann equation $RE(\rho, \sigma, \tau)$ is reducible, namely, the following condition holds*

$$\rho_i + \sigma_j + \tau_k \in \mathbb{Z} \quad \textit{for some} \quad i, j, k \in \{1, 2\}. \tag{4.3.6}$$

Then $\rho(\gamma_0)$ and $\rho(\gamma_1)$ are uniquely represented up to conjugacy by the following matrices (the result is divided into several cases):

(A) *Case $\rho_1 - \rho_2, \sigma_1 - \sigma_2, \tau_1 - \tau_2 \notin \mathbb{Z}$. We can uniquely label the exponents so that*

$$\rho_1 + \sigma_1 + \tau_1 \in -\mathbb{N}_0$$

which is equivalent, by the Fuchs relation, to

$$\rho_2 + \sigma_2 + \tau_2 \in \mathbb{N}.$$

The representation ρ is given by

$$\rho(\gamma_0) \leftrightarrow \begin{pmatrix} \varepsilon(\rho_1) & 0 \\ 0 & \varepsilon(\rho_2) \end{pmatrix}, \quad \rho(\gamma_1) \leftrightarrow \begin{pmatrix} \varepsilon(\sigma_1) & 1 \\ 0 & \varepsilon(\sigma_2) \end{pmatrix}. \tag{4.3.7}$$

(B) *Case* $\rho_1 - \rho_2 \in \mathbb{Z}, \sigma_1 - \sigma_2 \notin \mathbb{Z}, \tau_1 - \tau_2 \notin \mathbb{Z}$. *We label the exponents at* $x = 0$ *so that*

$$\rho_1 - \rho_2 \in \mathbb{N}_0.$$

(B.1) *Case when*

$$\rho_1 + \sigma_i + \tau_j \notin \langle \rho_1 - \rho_2 \rangle \quad (i, j = 1, 2) \tag{4.3.8}$$

holds (i.e., $x = 0$ *is logarithmic). We can uniquely label the exponents at* $x = 1$ *and* ∞ *so that*

$$\rho_1 + \sigma_1 + \tau_1 \in -\mathbb{N}_0.$$

The representation is given by

$$\rho(\gamma_0) \leftrightarrow \begin{pmatrix} \varepsilon(\rho) & 1 \\ 0 & \varepsilon(\rho) \end{pmatrix}, \quad \rho(\gamma_1) \leftrightarrow \begin{pmatrix} \varepsilon(\sigma_1) & 0 \\ 0 & \varepsilon(\sigma_2) \end{pmatrix} \tag{4.3.9}$$

where $\varepsilon(\rho) := \varepsilon(\rho_1) = \varepsilon(\rho_2)$.
(B.2) *Case* (4.3.8) *does not hold (i.e.,* $x = 0$ *is non-logarithmic). The representation is given by*

$$\rho(\gamma_0) \leftrightarrow \begin{pmatrix} \varepsilon(\rho) & 0 \\ 0 & \varepsilon(\rho) \end{pmatrix}, \quad \rho(\gamma_1) \leftrightarrow \begin{pmatrix} \varepsilon(\sigma_1) & 0 \\ 0 & \varepsilon(\sigma_2) \end{pmatrix} \tag{4.3.10}$$

(B)$'$ *Case* $\rho_1 - \rho_2 \notin \mathbb{Z}, \sigma_1 - \sigma_2 \in \mathbb{Z}, \tau_1 - \tau_2 \notin \mathbb{Z}$. *We label the exponents at* $x = 1$ *so that*

$$\sigma_1 - \sigma_2 \in \mathbb{N}_0.$$

(B.1)$'$ *Case when*

$$\rho_i + \sigma_1 + \tau_j \notin \langle \sigma_1 - \sigma_2 \rangle \quad (i, j = 1, 2) \tag{4.3.8$'$}$$

holds (i.e., $x = 1$ *is logarithmic). We can uniquely label the exponents at* $x = 0$ *and* ∞ *so that*

$$\rho_1 + \sigma_1 + \tau_1 \in -\mathbb{N}_0.$$

The representation is given by

$$\rho(\gamma_0) \leftrightarrow \begin{pmatrix} \varepsilon(\rho_1) & 0 \\ 0 & \varepsilon(\rho_2) \end{pmatrix}, \quad \rho(\gamma_1) \leftrightarrow \begin{pmatrix} \varepsilon(\sigma) & 1 \\ 0 & \varepsilon(\sigma) \end{pmatrix} \tag{4.3.9$'$}$$

where $\varepsilon(\sigma) := \varepsilon(\sigma_1) = \varepsilon(\sigma_2)$.
(B.2)′ *Case* (4.3.8)′ *does not hold (i.e.,* $x = 1$ *is non-logarithmic). The representation is given by*

$$(4.3.10)' \qquad \rho(\gamma_0) \leftrightarrow \begin{pmatrix} \varepsilon(\rho_1) & 0 \\ 0 & \varepsilon(\rho_2) \end{pmatrix}, \quad \rho(\gamma_1) \leftrightarrow \begin{pmatrix} \varepsilon(\sigma) & 0 \\ 0 & \varepsilon(\sigma) \end{pmatrix}$$

(B)″ *Case where* $\rho_1 - \rho_2 \notin \mathbb{Z}, \sigma_1 - \sigma_2 \notin \mathbb{Z}, \tau_1 - \tau_2 \in \mathbb{Z}$. *We label the exponents at* $x = \infty$ *so that*

$$\tau_1 - \tau_2 \in \mathbf{N}_0.$$

(B.1)″ *Case when*

$$(4.3.8)'' \qquad \rho_i + \sigma_j + \tau_1 \notin \langle \tau_1 - \tau_2 \rangle \quad (i, j = 1, 2)$$

holds (i.e., $x = \infty$ *is logarithmic). We can uniquely label the exponents at* $x = 0$ *and* 1 *so that*

$$\rho_1 + \sigma_1 + \tau_1 \in -\mathbf{N}_0.$$

The representation is given by

$$(4.3.9)'' \qquad \begin{aligned} &\rho(\gamma_0) \leftrightarrow \begin{pmatrix} \varepsilon(\rho_1) & 0 \\ 0 & \varepsilon(\rho_2) \end{pmatrix}, \quad \rho(\gamma_1) \leftrightarrow \begin{pmatrix} \varepsilon(\sigma_1) & \varepsilon(\rho_1)^{-1} \\ 0 & \varepsilon(\sigma_2) \end{pmatrix}, \\ &\rho(\gamma_0\gamma_1) \leftrightarrow \begin{pmatrix} \varepsilon(\tau) & 1 \\ 0 & \varepsilon(\tau) \end{pmatrix}, \end{aligned}$$

where $\varepsilon(\tau) := \varepsilon(\tau_1) = \varepsilon(\tau_2)$.
(B.2)″ *Case* (4.3.8)″ *does not hold (i.e.,* $x = \infty$ *is non-logarithmic). The representation is given by*

$$(4.3.10)'' \qquad \begin{aligned} &\rho(\gamma_0) \leftrightarrow \begin{pmatrix} \varepsilon(\rho_1) & 0 \\ 0 & \varepsilon(\rho_2) \end{pmatrix}, \quad \rho(\gamma_1) \leftrightarrow \begin{pmatrix} \varepsilon(\sigma_1) & 0 \\ 0 & \varepsilon(\sigma_2) \end{pmatrix}, \\ &\rho(\gamma_0\gamma_1) \leftrightarrow \begin{pmatrix} \varepsilon(\tau) & 0 \\ 0 & \varepsilon(\tau) \end{pmatrix}. \end{aligned}$$

(C) *Case* $\rho_1 - \rho_2, \sigma_1 - \sigma_2, \tau_1 - \tau_2 \in \mathbb{Z}$. *We label the exponents so that*

$$\rho_1 - \rho_2, \sigma_1 - \sigma_2 \in \mathbf{N}_0,$$

and set $\varepsilon(\rho) := \varepsilon(\rho_1) = \varepsilon(\rho_2)$. *and* $\varepsilon(\sigma) := \varepsilon(\sigma_1) = \varepsilon(\sigma_2)$.
(C.1) *Case* (4.3.8) *and* (4.3.8)′ *hold (i.e.,* $x = 0$ *and* $x = 1$ *are logarithmic).*

$$\rho(\gamma_0) \leftrightarrow \begin{pmatrix} \varepsilon(\rho) & 1 \\ 0 & \varepsilon(\rho) \end{pmatrix}, \quad \rho(\gamma_1) \leftrightarrow \begin{pmatrix} \varepsilon(\sigma) & -\varepsilon(\sigma-\rho) \\ 0 & \varepsilon(\sigma) \end{pmatrix}, \tag{4.3.11}$$

(C.1)′ *Case* (4.3.8) *holds and* (4.3.8)′ *does not hold (i.e.,* $x = 0$ *is logarithmic and* $x = 1$ *is non-logarithmic).*

$$\rho(\gamma_0) \leftrightarrow \begin{pmatrix} \varepsilon(\rho) & 1 \\ 0 & \varepsilon(\rho) \end{pmatrix}, \quad \rho(\gamma_1) \leftrightarrow \begin{pmatrix} \varepsilon(\sigma) & 0 \\ 0 & \varepsilon(\sigma) \end{pmatrix}, \tag{4.3.11$'$}$$

(C.1)″ *Case* (4.3.8) *does not hold and* (4.3.8)′ *holds (i.e.,* $x = 0$ *is non-logarithmic and* $x = 1$ *is logarithmic).*

$$\rho(\gamma_0) \leftrightarrow \begin{pmatrix} \varepsilon(\rho) & 0 \\ 0 & \varepsilon(\rho) \end{pmatrix}, \quad \rho(\gamma_1) \leftrightarrow \begin{pmatrix} \varepsilon(\sigma) & 1 \\ 0 & \varepsilon(\sigma) \end{pmatrix}, \tag{4.3.11$''$}$$

(C.2) *Case neither* (4.3.8) *nor* (4.3.8)′ *holds (i.e.,* $x = 0$ *and* $x = 1$ *are non-logarithmic).*

$$\rho(\gamma_0) \leftrightarrow \begin{pmatrix} \varepsilon(\rho) & 0 \\ 0 & \varepsilon(\rho) \end{pmatrix}, \quad \rho(\gamma_1) \leftrightarrow \begin{pmatrix} \varepsilon(\sigma) & 0 \\ 0 & \varepsilon(\sigma) \end{pmatrix}, \tag{4.3.12}$$

The above cases (A), (B), (B)′, (B)″ *and* (C) *cover all possibilities.*

COROLLARY 4.3.4. *Except for the cases mentioned below, any conjugacy class of* 2*-dimensional representations of the free group with two generators can be realized as the monodromy of a Riemann equation. The exceptions are classes determined by the representations:*

$$\rho(u) \leftrightarrow \begin{pmatrix} \lambda_1 & 0 \\ 0 & \lambda_2 \end{pmatrix}, \quad \rho(v) \leftrightarrow \begin{pmatrix} \mu_1 & 0 \\ 0 & \mu_2 \end{pmatrix}$$

where

$$\lambda_i, \mu_j \in \mathbb{C}^\times, \lambda_1 \neq \lambda_2, \mu_1 \neq \mu_2, \lambda_1\mu_1 \neq \lambda_2\mu_2;$$

and

$$\rho(u) \leftrightarrow \begin{pmatrix} \lambda & 1 \\ 0 & \lambda \end{pmatrix}, \quad \rho(v) \leftrightarrow \begin{pmatrix} \mu & b \\ 0 & \mu \end{pmatrix}$$

where

$$\lambda, \mu \in \mathbb{C}^\times, \ b \in \mathbb{C} \text{ and } b \neq -\mu/\lambda.$$

Acknowledgement. This corollary was first proved in [KS]. (c.f. [Bieb] and [KimT.6]). The authors are grateful to Professors T. Kimura and K. Shima for having informed us of their results.

LEMMA 4.3.5. *If the generators $\rho(\gamma_0)$ and $\rho(\gamma_1)$ of a monodromy representation of a Riemann equation are simultaneously diagonalizable then at least one of $\rho(\gamma_0), \rho(\gamma_1)$ and $\rho(\gamma_0\gamma_1)$ is a scalar matrix.*

Proof. If $\rho(\gamma_0)$ and $\rho(\gamma_1)$ are diagonal and neither of them is a scalar matrix, the Riemann equation admits two linearly independent solutions:

$$x^\rho(1-x)^\sigma p(x) \quad \text{and } x^{\rho'}(1-x)^{\sigma'}p'(x)$$

where $\{\rho, \rho'\} = \{\rho_1, \rho_2\}, \{\sigma, \sigma'\} = \{\sigma_1, \sigma_2\}$, and p and p' are polynomials of degree n and n', respectively, since p and p' are single-valued and $x = \infty$ is a regular singular point of the equation. If $\rho(\gamma_0\gamma_1)$ is not a scalar matrix, we conclude that $\{-\rho-\sigma-n, -\rho'-\sigma'-n'\} = \{\tau_1, \tau_2\}$. The Fuchs relation tells us

$$1 = \rho + \rho' + \sigma + \sigma' + \tau + \tau' = -n - n',$$

which is a contradiction. ∎

LEMMA 4.3.6. *If*

$$\rho_1 + \sigma_1 + \tau_1 \in -\mathbb{N}_0$$

and if at least one of $\rho_1 - \rho_2, \sigma_1 - \sigma_2$ and $\tau_1 - \tau_2$ is not an integer then there is a solution of the Riemann equation $RE(\rho, \sigma, \tau)$ of the form

$$x^{\rho_1}(1-x)^{\sigma_1}p(x)$$

where p is a polynomial of degree at most $-(\rho_1 + \sigma_1 + \tau_1)$.

Proof. We assume, for example, $\rho_1 - \rho_2 \notin \mathbb{Z}$. Put $n := -(\rho_1 + \sigma_1 + \tau_1) \geq 0$. Since we have

$$P\begin{pmatrix} 0 & 1 & \infty & \\ \rho_1 & \sigma_1 & \tau_1 & ;x \\ \rho_2 & \sigma_2 & \tau_2 & \end{pmatrix}, \qquad \sum_{i=1}^{2}(\rho_i + \sigma_i + \tau_i) = 1$$

$$= x^{\rho_1}(1-x)^{\sigma_1} P\begin{pmatrix} 0 & 1 & \infty & \\ 0 & 0 & \rho_1+\sigma_1+\tau_1 & ;x \\ \rho_2-\rho_1 & \sigma_2-\sigma_1 & \rho_1+\sigma_1+\tau_2 & \end{pmatrix}$$

$$= x^{\rho_1}(1-x)^{\sigma_1} P\begin{pmatrix} 0 & 1 & \infty & \\ 0 & 0 & \alpha & ;x \\ 1-\gamma & \gamma-\alpha-\beta & \beta & \end{pmatrix}$$

where

$$\alpha = \rho_1 + \sigma_1 + \tau_1 = -n,$$

$$\beta = \rho_1 + \sigma_1 + \tau_2,$$

$$\gamma = 1 - \rho_2 + \rho_1 \notin \mathbb{Z},$$

the Riemann equation $E(\rho, \sigma, \tau)$ admits a solution

$$x^{\rho_1}(1-x)^{\sigma_1} \sum_{i=0}^{\infty} \frac{(\alpha)_i(\beta)_i}{(\gamma)_i(1)_i} x^i$$

$$= x^{\rho_1}(1-x)^{\sigma_1} \sum_{i=0}^{n} \frac{(-n)_i(\beta)_i}{(\gamma)_i(1)_i} x^i. \blacksquare$$

LEMMA 4.3.7.
(i) *Suppose* $\Re(\rho_1) \geq \Re(\rho_2)$. *Then the Riemann equation* $RE(\rho, \sigma, \tau)$ *has a logarithmic singularity at* $x = 0$ *if and only if the following conditions hold:*

$$\rho_1 - \rho_2 \in \mathbb{N}_0, \quad \rho_1 + \sigma_i + \tau_j \notin \langle \rho_1 - \rho_2 \rangle \quad (i, j = 1, 2).$$

(ii) *Suppose* $\Re(\sigma_1) \geq \Re(\sigma_2)$. *Then the Riemann equation* $RE(\rho, \sigma, \tau)$ *has a logarithmic singularity at* $x = 1$ *if and only if the following conditions hold:*

$$\sigma_1 - \sigma_2 \in \mathbb{N}_0, \quad \rho_i + \sigma_1 + \tau_j \notin \langle \sigma_1 - \sigma_2 \rangle \quad (i, j = 1, 2).$$

SUBLEMMA 4.3.8. *When $\gamma \in -\mathbb{N}_0$, the hypergeometric equation with parameters (α, β, γ) has a non-logarithmic singularity at $x = 0$ if and only if*

$$\alpha \text{ or } \beta \in \{\gamma, \ldots, -2, -1, 0\}.$$

Proof. Since $2 - \gamma \neq 0, -1, -2, \ldots$, the equation has the Kummer's solution

$$f_0(x; 1-\gamma) = x^{1-\gamma} F(\alpha - \gamma + 1, \beta - \gamma + 1, 2 - \gamma; x)$$

corresponding to the exponent $1 - \gamma$. Let us consider the possibility of the solution of the form

$$u = \sum_{n=0}^{\infty} a_n x^n \quad (a_0 = 1)$$

corresponding to the exponent 0. Substituting the series u into the equation we have the recurrence formula (c.f. Chapter 1, Section 1.3, Frobenius's method):

$$(\gamma + n)(1 + n)a_{n+1} = (\alpha + n)(\beta + n)a_n \quad (n \geq 0)$$

Thus if the condition in the sublemma holds, we can take

$$a_n = 0 \quad n \geq 1 \qquad \text{if } \gamma = 0$$

$$a_n = 0 \quad n \geq -\gamma \qquad \text{if } \gamma \in -\mathbb{N}$$

so that the finite series u is another solution. Conversely, if the condition does not hold, there is no series $\{a_n\}$ $(a_0 = 1)$ satisfying the above recurrence formula for $n = -\gamma$. ∎

Proof of Lemma 4.3.7. We prove only (i). For a Riemann equation $RE(\rho, \sigma, \tau)$, let us assume $\rho_1 - \rho_2 \in \mathbb{N}$. (If $\rho_1 = \rho_2$, the singularity at $x = 0$ is logarithmic (c.f. Chapter 1, Section 1.3).) We have

$$P\begin{pmatrix} 0 & 1 & \infty & \\ \rho_1 & \sigma & \tau & ;x \\ \rho_2 & \sigma' & \tau' & \end{pmatrix} \qquad \{\sigma,\sigma'\}=\{\sigma_1,\sigma_2\},\ \{\tau,\tau'\}=\{\tau_1,\tau_2\}$$

$$= x^{\rho_2}(1-x)^{\sigma}P\begin{pmatrix} 0 & 1 & \infty & \\ 0 & 0 & \rho_2+\sigma+\tau & ;x \\ \rho_1-\rho_2 & \sigma'-\sigma & \rho_2+\sigma+\tau' & \end{pmatrix}$$

Put

$$1-\gamma=\rho_1-\rho_2\in\mathbb{N},$$

$$\alpha=\rho_2+\sigma+\tau,$$

$$\beta=\rho_2+\sigma+\tau'.$$

Since $\gamma\in-\mathbb{N}_0$, by the sublemma, $x=0$ is logarithmic if and only if

$$\alpha \text{ or } \beta\in\{\gamma,\dots,-2,-1,0\}.$$

Since

$$\rho_1+\sigma+\tau=\rho_1-\rho_2+\alpha\ (=1-\gamma+\alpha)$$

$$\rho_1+\sigma+\tau'=\rho_1-\rho_2+\beta\ (=1-\gamma+\beta),$$

we have only to recall the definition

$$\langle\rho_1-\rho_2\rangle=\{1,2,\dots,\rho_1-\rho_2\}. \blacksquare$$

LEMMA 4.3.9. *If the Riemann equation $E(\rho,\sigma,\tau)$ is reducible, then it cannot happen that all three singular points are simultaneously logarithmic.*

Proof. Suppose that the reducible Riemann equation $RE(\rho,\sigma,\tau)$ has three logarithmic singular points. We assume that

$$\Re(\xi_1)\geq\Re(\xi_2)\quad(\xi_j=\rho_j,\sigma_j,\tau_j). \tag{4.3.14}$$

Since $RE(\rho,\sigma,\tau)$ is reducible, there is a solution $f(x)$ of $RE(\rho,\sigma,\tau)$ invariant under the action of $\pi_1(\mathbb{P}^1\backslash\{0,1,\infty\})$ up to multiplication by constants; it must be of the form

$$f(x)=x^{\xi}(x-1)^{\eta}p(x),\quad p(x):\ \text{a polynamial of degree } n;$$

we can assume that $p(x)$ is not divisible by x nor $x-1$, so by the definition of characteristic exponents, we conclude that ξ, η and $-\xi-\eta-n$ are characteristic exponents at $x=0,1$ and ∞, respectively. Since all singular points are now logarithmic and (4.3.14) is assumed, the exponents ρ_2, σ_2 and τ_2 correspond to logarithmic solutions. Hence we have

$$\xi=\rho_1, \quad \eta=\sigma_1, \quad -\xi-\eta-n=\tau_1.$$

Summing up the both sides, we obtain $\rho_1+\sigma_1+\tau_1=-n\leq 0$. The assumption (4.3.14) and this lead to the inequality

$$\Re(\rho_1+\rho_2+\sigma_1+\sigma_2+\tau_1+\tau_2)\leq 0.$$

This contradicts the Fuchs relation. ∎

Proof of Theorem 4.3.3. Since the representation is reducible, we assume $\rho(\gamma_0)$ and $\rho(\gamma_1)$ are upper triangular. Let us consider three numbers

$$\rho_1-\rho_2, \sigma_1-\sigma_2, \tau_1-\tau_2.$$

By the Fuchs relation and the reducibility condition (4.3.6), we see that only three cases occur: (A) none of the three is an integer, (B) exactly one of them is an integer, (C) all of them are integers.
(A) In this case, we can uniquely label the exponents so that

$$\rho_1+\sigma_1+\tau_1\in -\mathrm{N}_0.$$

We take $\rho(\gamma_0)$ diagonal. By Lemma 4.3.5, $\rho(\gamma_1)$ is not diagonal. Since, by Lemma 4.3.6, the equation admits a solution of the form $x^{\rho_1}(1-x)^{\sigma_1}$ (a polynomial), the representation is given by (4.3.7).
(B) We take $\rho(\gamma_1)$ diagonal. Let us label the exponents at $x=0$ so that $\rho_1-\rho_2\in \mathrm{N}_0$, and consider the four numbers:

$$\rho_1+\sigma+\tau$$

where σ (resp.τ) is one of the exponents at $x=1$ (resp. ∞). Since the representation is reducible, Theorem 4.3.2 and the Fuchs relation tell us that at least one of these numbers is an integer. Since $\sigma_1-\sigma_2, \tau_1-\tau_2\notin \mathbb{Z}$, exactly two of the four numbers are integers which can be written as

$$\varepsilon=\rho_1+\sigma+\tau$$

$$\varepsilon'=\rho_1+\sigma'+\tau' \quad \sigma\neq\sigma', \tau\neq\tau'.$$

By the Fuchs relation we have

$$\varepsilon + \varepsilon' = \rho_1 - \rho_2 + 1.$$

If $x = 0$ is a logarithmic singular point, then by Lemma 4.3.7, we have

$$\rho_1 + \sigma + \tau,\ \rho_1 + \sigma' + \tau' \notin \langle \rho_1 - \rho_2 \rangle.$$

By this condition together with the above equality, we conclude that ε or $\varepsilon' \in -\mathrm{N}_0$. Thus we can uniquely label the exponents at $x = 1$ and ∞ so that

$$\rho_1 + \sigma_1 + \tau_1 \in -\mathrm{N}_0;$$

so, by Lemma 4.3.6, the representation is given by (4.3.9).
If $x = 0$ is non-logarithmic, there is no need to label the exponents at $x = 1$, because if we exchange σ_1 and σ_2 in the representation (4.3.10) it gives a conjugate one.
(B)$'$, (B)$''$: Same as (B).
(C) When $x = 0$ and 1 are both logarithmic, Lemma 4.3.9 tells us that $x = \infty$ is non-logarithmic, i.e., not of type

$$\begin{pmatrix} \varepsilon(\tau) & \delta \\ 0 & \varepsilon(\tau) \end{pmatrix} \quad \delta \neq 0,$$

thus we have (4.3.11). When $x = 0, 1$ and ∞ are all non-logarithmic, we have (4.3.12).
Theorem 4.3.3 is completely proved.

4.4 Finding the monodromy by Euler integrals over arcs

In this section, we shall solve the connection problem with respect to solutions expressed by Euler integrals over arcs starting and ending at one of the four points: $0, 1, \infty$ and x. As is mentioned in Remark 4.1.3, the monodromy group with respect to such solutions is obtained from the solution of the connection problem.

Consider the integral

$$F_{pq}(x) = \int_p^q \varphi(t, x)dt, \tag{4.4.1}$$

where

$$\varphi(t, x) = \varphi(t, x; \lambda, \mu, \nu) = t^\lambda (1-t)^\mu (x-t)^\nu, \tag{4.4.2}$$

$$\lambda = \alpha - \gamma, \quad \mu = \gamma - \beta - 1, \quad \nu = -\alpha \tag{4.4.3}$$

and

$$p, q \in \{0, 1, \infty, x\}, \quad p \neq q.$$

We assume that x lies in the upper-half plane: $\Im(x) > 0$. If $\Re(\beta) < \Re(\gamma) < \Re(\alpha) + 1 < 2$, then the integral (4.4.1) converges; anyhow, this integral makes sense as the finite part of a divergent integral (cf. Section 3.3), provided that

(4.4.4) none of $\alpha, 1-\beta, \gamma-\alpha$ and $\beta - \gamma + 1$ is a positive integer.

Under this condition, the integral (4.4.1) gives a solution of the hypergeometric equation (Section 3.4). There are $\binom{4}{2} = 6$ integrals to be considered:

$$F_{01}, F_{1\infty}, F_{\infty 0}, F_{0x}, F_{1x}, F_{x\infty}. \tag{4.4.5}$$

The paths of integration for F_{pq} are denoted by $\overline{pq}$ and indicated in Figure 2.6. The branches of integrand of F_{pq} along $\overline{pq}$ are determined by the assignment of $\arg(t), \arg(1-t)$ and $\arg(x-t)$ tabulated in Table 1.

We shall solve the connection problem for the hypergeometric equation with respect to the 6 solutions (4.4.5). Since the solution space is 2-dimensional, the solution of the connection problem is given by four homogeneous linear equations for (4.4.5); any four solutions are linear combinations of the remaining two.

Assignment of branchs in case $\Im x > 0$

	$\arg t$	$\arg(1-t)$	$\arg(x-t)$
$\overline{01}$	0	0	$[\xi, \eta]^*$
$\overline{1\infty}$	0	$-\pi$	$[\eta, \pi]$
$\overline{\infty 0}$	π	0	$[0, \xi]$
$\overline{0x}$	ξ	$[\eta - \pi, 0]$	ξ
$\overline{1x}$	$[0, \xi]$	$\eta - \pi$	η
$\overline{x\infty}$	ξ	$[\xi - \pi, \eta - \pi]^*$	$\xi + \pi$

$\xi := \arg x,\ \eta := \arg(x-1),\ 0 < \xi, \eta < \pi,$

$[a, b]^* := [\min(a, b), \max(a, b)].$

Table 1

THEOREM 4.4.1. *The solutions $F_{pq}(x)$ listed in (4.4.5) satisfy the following linear relations:*

$$
\begin{array}{lllllll}
(1) & F_{01} + & F_{1\infty} + & F_{\infty 0} & & & = 0, \\
(2) & F_{01} & & & - F_{0x} + F_{1x} & & = 0, \\
(3) & & \varepsilon(-\mu) F_{1\infty} & & & - F_{1x} - \varepsilon(\nu) F_{x\infty} & = 0, \\
(4) & & & \varepsilon(\lambda) F_{\infty 0} + F_{0x} & & + \quad F_{x\infty} & = 0,
\end{array}
$$

where $\varepsilon(\cdot) = \exp(2\pi i \cdot)$.

Proof. Let $D_j (j = 0, \dots, 4)$ be domains indicated in Figure 2.6 and denote by ∂D_j their boundaries with the natural orientation. If (λ, μ, ν) lies in a suitable open subset of $\mathbb{C}^3$, then the integration of $\varphi(\cdot, x; \lambda, \mu, \nu)$ over D_j makes sense and Cauchy's theorem is applicable:

$$\int_{\partial D_j} \varphi(t, x; \lambda, \mu, \nu) dt = 0. \tag{4.4.6}$$

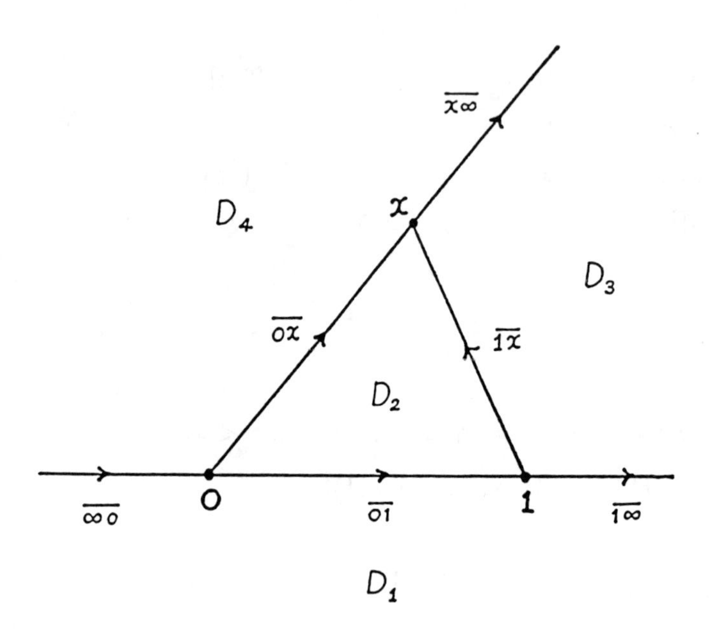

Figure 2.6.

Divide ∂D_j into three arcs of the form $\overline{pq}$; the integral (4.4.6) is the sum of the corresponding three. Comparing branches of these integrands with those of F_{pq} (see Table 1), we can easily show that (4.4.6) leads to the linear relation (j) of the theorem. For a general (λ, μ, ν), the theorem is proved by making an analytic continuation with respect to (λ, μ, ν). ∎

To find generators of the monodromy group with respect to two solutions in (4.4.5), we establish relations between F_{pq} and local solutions $f_a(x; \xi)$ which has characteristic exponent ξ at $x = a$ defined in Section 1.3.

THEOREM 4.4.2. *We have*

$$
(4.4.7)\qquad \begin{cases} F_{1\infty}(x) = c_{1\infty} f_0(x;0), \\ F_{0x}(x) = c_{0x} f_0(x; 1-\gamma), \\ c_{1\infty} = -e^{\pi i(\gamma-\alpha-\beta)} \dfrac{\Gamma(\beta)\Gamma(\gamma-\beta)}{\Gamma(\gamma)}, \\ c_{0x} = \dfrac{\Gamma(\alpha-\gamma+1)\Gamma(1-\alpha)}{\Gamma(2-\gamma)}, \end{cases}
$$

$$
(4.4.8)\qquad \begin{cases} F_{\infty 0}(x) = c_{\infty 0} f_1(x;0), \\ F_{1x}(x) = c_{1x} f_1(x; \gamma-\alpha-\beta), \\ c_{\infty 0} = e^{\pi i(\gamma-\alpha)} \dfrac{\Gamma(\beta)\Gamma(\alpha-\gamma+1)}{\Gamma(\alpha+\beta-\gamma+1)}, \\ c_{1x} = -e^{-\pi i\alpha} \dfrac{\Gamma(\gamma-\beta)\Gamma(1-\alpha)}{\Gamma(\gamma-\alpha-\beta+1)}, \end{cases}
$$

$$
(4.4.9)\qquad \begin{cases} F_{01}(x) = c_{01} f_\infty(x;\alpha), \\ F_{x\infty}(x) = c_{x\infty} f_\infty(x;\beta), \\ c_{01} = \dfrac{\Gamma(\alpha-\gamma+1)\Gamma(\gamma-\beta)}{\Gamma(\alpha-\beta+1)}, \\ c_{x0} = -e^{-\pi i(\gamma-\alpha-\beta)} \dfrac{\Gamma(\beta)\Gamma(1-\alpha)}{\Gamma(\beta-\alpha+1)}. \end{cases}
$$

Proof. The first formula of (4.4.7) follows immediately from (3.2.2) since $f_0(x;0) = F(\alpha,\beta,\gamma;x)$. To prove the second formula of (4.4.7), we make a change of variable $t = x/s$ in (4.4.1) for F_{0x} to obtain

$$
F_{0x}(x) = x^{1-\gamma} \int_1^\infty s^{\beta-1}(s-1)^{-\alpha}(s-x)^{\gamma-\beta-1} ds.
$$

The integral on the right-hand side is holomorphic at $x = 0$, whence $F_{0x}(x)$ has the exponent $1-\gamma$ at $x = 0$. So $F_{0x}(x)$ is a constant multiple of $f_0(x;1-\gamma)$. The constant c_{0x} is found by evaluating the above integral at $x=0$. For the first formula of (4.4.8), we make a change of variable $t=1-s$ in (4.4.1) for $F_{\infty 0}(x)$ to obtain

$$F_{\infty 0}(x)=\int_1^\infty s^{\gamma-\beta-1}(1-s)^{\alpha-\gamma}\{s-(1-x)\}^{-\alpha}ds.$$

Hence $F_{\infty 0}(x)$ is holomorphic at $x=1$, and therefore $F_{\infty 0}(x)=c_{\infty 0}$ $f_1(x;0)$ with $c_{\infty 0}=F_{\infty 0}(1)$. Evaluating $F_{\infty 0}(1)$, we obtain the first formula of (4.4.8). To show the second formula of (4.4.8), we make a change of variable $t=1-(1-x)s$ in (4.4.1) for F_{1x} to obtain

$$F_{1x}(x)=-e^{-\pi i\alpha}(1-x)^{\gamma-\alpha-\beta}\int_0^1 s^{\gamma-\beta-1}(1-s)^{-\alpha}\{s-(1-x)\}^{\alpha-\gamma}ds.$$

The integral on the right-hand side is holomorphic at $x=1$. Hence $F_{1x}(x)$ has the exponent $\gamma-\alpha-\beta$ at $x=1$. Evaluating the integral at $x=1$, we obtain the second formula of (4.4.8). For the first formula of (4.4.9), we rewrite $F_{01}(x)$ in the form:

$$F_{01}(x)=x^{-\alpha}\int_0^1 t^\lambda(1-t)^\mu(1-\frac{t}{x})^\nu dt.$$

The integral on the right-hand side is holomorphic at $x=\infty$. Evaluating it at $x=\infty$, we obtain the first formula of (4.4.9). Finally, to show the second formula of (4.4.9), we make a change of variable $t=x/s$ in (4.4.1) for $F_{x\infty}$ to obtain

$$F_{x\infty}(x)=x^{-\beta}\int_0^1 s^{\beta-1}(s-1)^{-\alpha}(\frac{s}{x}-1)^{\gamma-\beta-1}ds.$$

The integral on the right-hand side is holomorphic at $x=\infty$. Evaluating it at $x=\infty$, we obtain the second formula of (4.4.9). ∎

Now we turn to the monodromy group. Solving the equations (1) - (4) of Theorem 4.4.1 with respect to $F_{\infty 0}$ and F_{1x}, we obtain

$$(F_{1\infty},F_{0x})=(F_{\infty 0},F_{1x})\frac{P}{\varepsilon(\nu)-\varepsilon(-\mu)}, \tag{4.4.10}$$

where

$$P = \begin{pmatrix} \varepsilon(\lambda+\nu)-\varepsilon(\nu) & \varepsilon(-\mu)-\varepsilon(\lambda+\nu) \\ \varepsilon(\nu)-1 & 1-\varepsilon(-\mu) \end{pmatrix}. \tag{4.4.11}$$

We finally have generators of the monodromy group of the hypergeometric equation with respect to the fundamental system of solutions $(F_{1\infty}, F_{0x})$.

THEOREM 4.4.3. *Let $\gamma_j (j = 0, 1)$ be the loops defined in Figure 2.5 with base point $b = 1/2$. Suppose α, β and γ satisfy (4.4.4). Then the analytic continuation ${\gamma_j}_* \mathcal{F}$ of $\mathcal{F} := (F_{1\infty}, F_{0x})$ along γ_j are given by*

$$\gamma_{0*}\mathcal{F} = \mathcal{F}A_0, \quad \gamma_{1*}\mathcal{F} = \mathcal{F}A_1, \tag{4.4.12}$$

where

$$A_0 = \begin{pmatrix} 1 & 0 \\ 0 & \varepsilon(-\gamma) \end{pmatrix}, \quad A_1 = P^{-1}\begin{pmatrix} 1 & 0 \\ 0 & \varepsilon(\gamma-\alpha-\beta) \end{pmatrix} P. \tag{4.4.13}$$

The matrix P is given by (4.4.11). The monodromy group with respect to the fundamental system $\mathcal{F}$ of solutions is generated by A_0 and A_1.

4.5 Finding the monodromy by Euler integrals over double loops

Euler integrals over double loops also provide us with solutions of the hypergeometric equation (Section 3.4). We shall compute the monodromy group with respect to two such integrals. This method is more elegant than that of using Euler integrals over arcs; we need not worry about divergence of integrals. Moreover, we can know the monodromy directly, i.e., not by solving the connection problem. Let T be the equilateral triangular domain with vertices $0, 1$ and $p := (1 + i\sqrt{3})/2$. For $x \in T$, we set

$$Y_x = \mathbb{P}^1 \backslash \{0, 1, \infty, x\}$$

and take p as base point of the fundamental group of Y_x.

We consider an Euler integral over a double loop:

$$\langle \xi, \eta \rangle (x) := \int_{[\xi,\eta]} \varphi(t, x)dt \quad (x \in T), \tag{4.5.1}$$

where

$$(4.5.2) \qquad \varphi(t,x) := \varphi(t,x;\alpha,\beta,\gamma) = t^{\alpha-\gamma}(t-1)^{\gamma-\beta-1}(t-x)^{-\alpha},$$

$$(4.5.3) \qquad [\xi,\eta] := \xi\eta\xi^{-1}\eta^{-1} : \text{commutator of } \xi,\eta \in \pi_1(Y_x,p).$$

The branch of $\varphi(\cdot,x)$ at the starting point $t=p$ can be determined by the assignment: $\arg t = \pi/3, \arg(t-1) = 2\pi/3, \pi/3 < \arg(t-x) < 2\pi/3$. The integral $\langle\xi,\eta\rangle$ represents a solution of the hypergeometric equation (Section 3.4). The integrand $\varphi(\cdot,x)$ enjoys a very simple monodromy property, which is expressed by the 1-dimensional representation:

$$(4.5.4) \qquad c : \pi_1(Y_x,p) \to \mathbb{C}^\times, \quad \xi \mapsto c(\xi),$$

where $c(\xi)$ is a constant caused by the analytic continuation of $\varphi(\cdot,x)$ along ξ. Let us denote by $\xi_*\varphi$ the analytic continuation of φ along ξ. Notice that $[\eta,\zeta]_*\varphi = \varphi$ and so that $c([\eta,\zeta]) = 1$. Notice also that

$$\begin{aligned}\int_{\xi\eta}\varphi dt &= \int_\xi \eta_*\varphi dt + \int_\eta \varphi dt\\ &= c(\eta)\int_\xi \varphi dt + \int_\eta \varphi dt.\end{aligned}$$

The monodromy of the hypergeometric equation can be derived by the 1-dimensional representation (4.5.4) and the distributive law (4.5.5) mentioned below.

LEMMA 4.5.1. *For $\xi,\eta,\zeta \in \pi_1(Y_x,p)$, we have*

$$(4.5.5) \qquad \langle\xi\eta,\zeta\rangle = \langle\xi,\zeta\rangle + c(\xi)^{-1}\langle\eta,\zeta\rangle.$$

Proof. Since

$$\begin{aligned}[\xi\eta,\zeta] &= \xi\eta\zeta\eta^{-1}\xi^{-1}\zeta^{-1}\\ &= \xi\eta\zeta\eta^{-1}\zeta^{-1}\xi^{-1}\xi\zeta\xi^{-1}\zeta^{-1}\\ &= \xi[\eta,\zeta]\xi^{-1}[\xi,\zeta],\end{aligned}$$

we have

$$\begin{aligned}
\langle \xi\eta, \zeta \rangle &= \int_{\xi[\eta,\zeta]\xi^{-1}[\xi,\zeta]} \varphi dt \\
&= \int_{\xi[\eta,\zeta]\xi^{-1}} [\xi,\zeta]_* \varphi dt + \int_{[\xi,\zeta]} \varphi dt \\
&= \int_{\xi[\eta,\zeta]\xi^{-1}} \varphi dt + \langle \xi, \zeta \rangle \\
&= \int_{\xi[\eta,\zeta]} \xi_*^{-1} \varphi dt + \int_{\xi^{-1}} \varphi dt + \langle \xi, \zeta \rangle \\
&= c(\xi^{-1}) \{ \int_{\xi} [\eta,\zeta]_* \varphi dt + \int_{[\eta,\zeta]} \varphi dt \} + \int_{\xi^{-1}} \varphi dt + \langle \xi, \zeta \rangle \\
&= c(\xi^{-1}) \int_{\xi} \varphi dt + \int_{\xi^{-1}} \varphi dt + c(\xi)^{-1} \langle \eta, \zeta \rangle + \langle \xi, \zeta \rangle \\
&= \int_{\xi\xi^{-1}} \varphi dt + c(\xi)^{-1} \langle \eta, \zeta \rangle + \langle \xi, \zeta \rangle \\
&= c(\xi)^{-1} \langle \eta, \zeta \rangle + \langle \xi, \zeta \rangle. \quad \blacksquare
\end{aligned}$$

COROLLARY 4.5.2. *For $\xi, \eta, \zeta \in \pi_1(Y_x, p)$, we have*

$$\langle \xi^{-1}, \zeta \rangle = -c(\xi) \langle \xi, \zeta \rangle, \tag{4.5.6}$$

$$\langle \xi^{-1}\eta\xi, \zeta \rangle = c(\xi)\{c(\eta)^{-1} - 1\} \langle \xi, \zeta \rangle + c(\xi) \langle \eta, \zeta \rangle. \tag{4.5.7}$$

Proof. They follow from

$$0 = \langle \xi\xi^{-1}, \zeta \rangle = \langle \xi, \zeta \rangle + c(\xi)^{-1} \langle \xi^{-1}, \zeta \rangle$$

and

$$\begin{aligned}
\langle \xi^{-1}\eta\xi, \zeta \rangle &= \langle \xi^{-1}, \zeta \rangle + c(\xi^{-1})^{-1} \langle \eta\xi, \zeta \rangle \\
&= -c(\xi) \langle \xi, \zeta \rangle + c(\xi) \{ \langle \eta, \zeta \rangle + c(\eta)^{-1} \langle \xi, \zeta \rangle \}. \quad \blacksquare
\end{aligned}$$

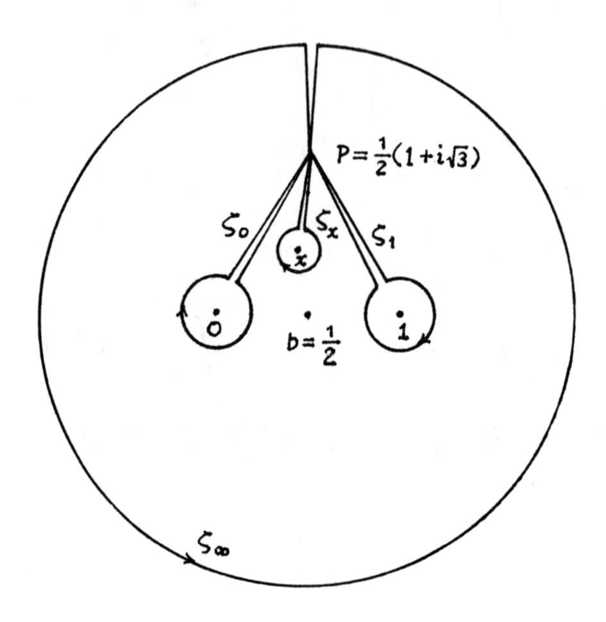

Figure 2.7.

Let $\zeta_j (j = 0, 1, \infty$ and $x)$ be the loops with base point p indicated in Figure 2.7. We note that ζ_j encircles j once in the negative direction. The fundamental group $\pi_1(Y_x, p)$ is a free group generated by ζ_0, ζ_1 and ζ_x; we have a relation

$$\zeta_0 \zeta_x \zeta_1 = \zeta_\infty^{-1}. \tag{4.5.8}$$

Operating $\langle \cdot, \zeta_\infty \rangle$ on (4.5.8) and using (4.5.5), we have

$$\begin{aligned} 0 = \langle \zeta_\infty^{-1}, \zeta_\infty \rangle &= \langle \zeta_0 \zeta_x \zeta_1, \zeta_\infty \rangle \\ &= \langle \zeta_0, \zeta_\infty \rangle + c(\zeta_0)^{-1} \langle \zeta_x \zeta_1, \zeta_\infty \rangle \\ &= \langle \zeta_0, \zeta_\infty \rangle + c(\zeta_0)^{-1} \{ \langle \zeta_x, \zeta_\infty \rangle + c(\zeta_x)^{-1} \langle \zeta_1, \zeta_\infty \rangle \}, \end{aligned}$$

and so

$$\langle\zeta_0,\zeta_\infty\rangle + c(\zeta_0)^{-1}\langle\zeta_x,\zeta_\infty\rangle + c(\zeta_0\zeta_x)^{-1}\langle\zeta_1,\zeta_\infty\rangle = 0. \tag{4.5.9}$$

Let $\gamma_\nu(\nu = 0,1)$ be the loops with base $b(= 1/2)$ defined in Figure 2.5. Recall that the fundamental group $\pi_1(X,b)$ of $X = \mathbb{P}^1\backslash\{0,1,\infty\}$ is generated by γ_0 and γ_1. We see that $\langle\zeta_j,\zeta_\infty\rangle(j = 0,1,\infty)$ are holomorphic functions of x in the triangle T. Their analytic continuations along $\gamma_\nu(\nu = 0,1)$ are denoted by $\gamma_{\nu*}\langle\zeta_j,\zeta_\infty\rangle$. Notice that $\langle\zeta_0,\zeta_\infty\rangle$ (resp. $\langle\zeta_1,\zeta_\infty\rangle$) is holomorphic at $x = 1$ (resp. $x = 0$). Let us analytically continue the functions $\langle\zeta_1,\zeta_\infty\rangle$ and $\langle\zeta_x,\zeta_\infty\rangle$ along the loop γ_0. The analytic continuation is found by deforming the paths of integration ζ_1,ζ_x and ζ_∞ in $Y_x = \mathbb{P}^1 - \{0,1,\infty,x\}$ as $x \in \mathbb{P}^1 - \{0,1,\infty\}$ travels along γ_0. The paths ζ_1 and ζ_∞ have nothing to do with the moving point x so they can remain as they were. Thus we have

$$\gamma_{0*}\langle\zeta_1,\zeta_\infty\rangle = \langle\zeta_1,\zeta_\infty\rangle, \tag{4.5.10}$$

and by exchanging the role of 0 and 1, we have

$$\gamma_{1*}\langle\zeta_0,\zeta_\infty\rangle = \langle\zeta_0,\zeta_\infty\rangle. \tag{4.5.10$'$}$$

The following illustration (Figure 2.8) shows the deformation of ζ_x.

The last picture (Figure 2.8 (7)) shows the result when the point x has ended its journey and the path ζ_x has been deformed into $\zeta_0^{-1}\zeta_x\zeta_0$. Therefor we obtain

$$\begin{aligned}\gamma_{0*}\langle\zeta_x,\zeta_\infty\rangle &= \langle\zeta_0^{-1}\zeta_x\zeta_0,\zeta_\infty\rangle \\ &= c(\zeta_0)\{c(\zeta_x)^{-1} - 1\}\langle\zeta_0,\zeta_\infty\rangle + c(\zeta_0)\langle\zeta_x,\zeta_\infty\rangle,\end{aligned} \tag{4.5.11}$$

where (4.5.7) is used to derive the second equality. Making a similar consideration also for ζ_1, we have

$$\begin{aligned}\gamma_{1*}\langle\zeta_x,\zeta_\infty\rangle &= \langle\zeta_1^{-1}\zeta_x\zeta_1,\zeta_\infty\rangle \\ &= c(\zeta_1)\{c(\zeta_x)^{-1} - 1\}\langle\zeta_1,\zeta_\infty\rangle + c(\zeta_1)\langle\zeta_x,\zeta_\infty\rangle.\end{aligned} \tag{4.5.11$'$}$$

Perform on (4.5.9) the operation γ_{1*} and make use of (4.5.10) and (4.5.11); then we have

$$\begin{aligned}0 &= \gamma_{1*}\langle\zeta_0,\zeta_\infty\rangle + c(\zeta_0)^{-1}\gamma_{1*}\langle\zeta_x,\zeta_\infty\rangle + c(\zeta_0\zeta_x)^{-1}\gamma_{1*}\langle\zeta_1,\zeta_\infty\rangle \\ &= \langle\zeta_0,\zeta_\infty\rangle + c(\zeta_0)^{-1}[c(\zeta_1)\{c(\zeta_x)^{-1} - 1\}\langle\zeta_1,\zeta_\infty\rangle + c(\zeta_1)\langle\zeta_x,\zeta_\infty\rangle] \\ &\quad + c(\zeta_0\zeta_x)^{-1}\gamma_{1*}\langle\zeta_1,\zeta_\infty\rangle,\end{aligned}$$

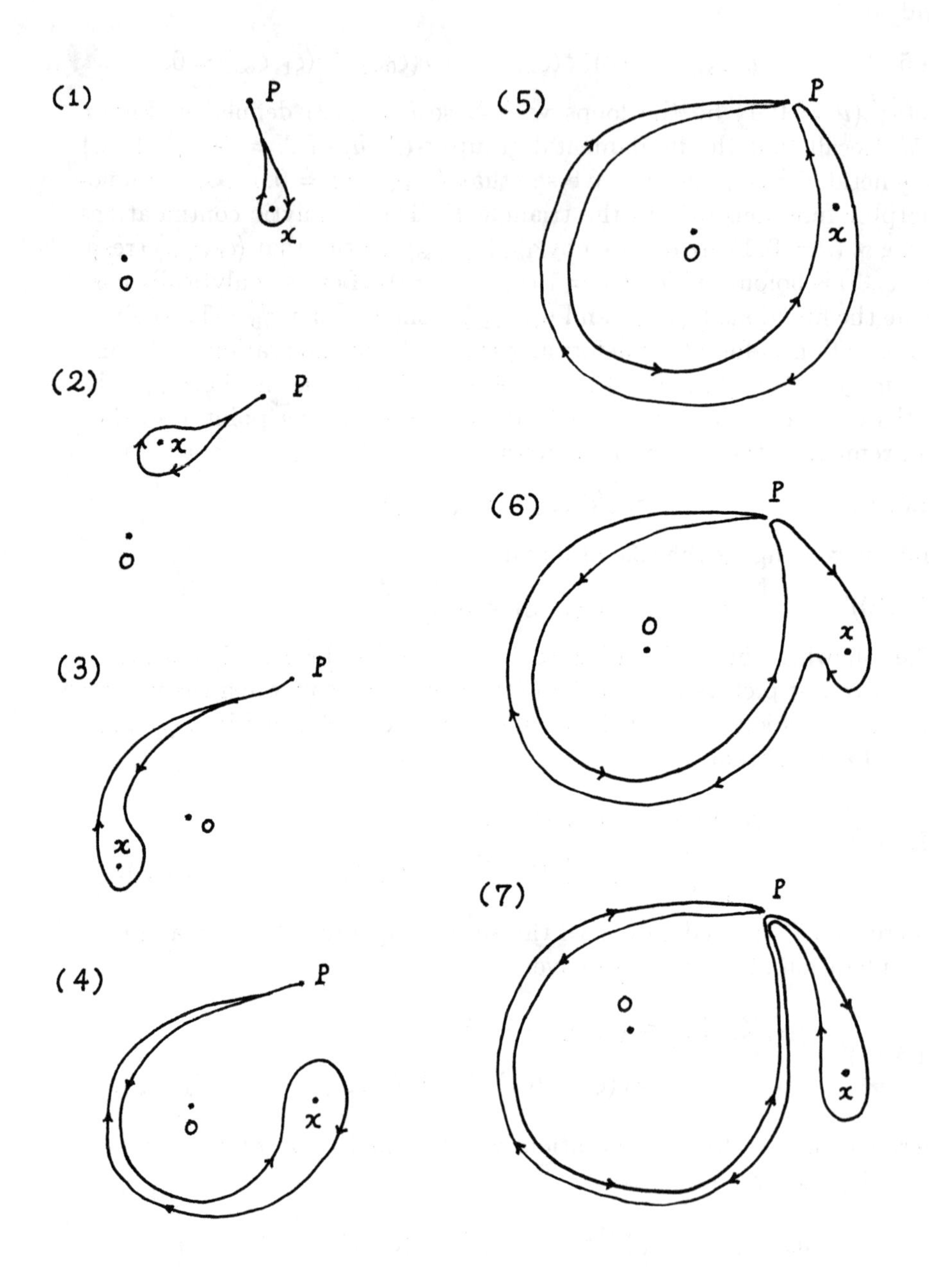

Figure 2.8.

and so

$$\gamma_{1*}\langle\zeta_1,\zeta_\infty\rangle = -c(\zeta_0\zeta_x)\langle\zeta_0,\zeta_\infty\rangle - c(\zeta_1)\{1-c(\zeta_x)\}\langle\zeta_1,\zeta_\infty\rangle$$
$$- c(\zeta_1\zeta_x)\langle\zeta_x,\zeta_\infty\rangle.$$

Performing on (4.5.9) the operation γ_{0*}, we have similarly,

$$\gamma_{0*}\langle\zeta_0,\zeta_\infty\rangle = \{1-c(\zeta_x)^{-1}\}\langle\zeta_0,\zeta_\infty\rangle - c(\zeta_0\zeta_x)^{-1}\langle\zeta_1,\zeta_\infty\rangle$$
$$- \langle\zeta_x,\zeta_\infty\rangle.$$

Elimination of $\langle\zeta_x,\zeta_\infty\rangle$ from these formulae by using (4.5.9) yields

$$\gamma_{1*}\langle\zeta_1,\zeta_\infty\rangle = c(\zeta_1\zeta_x)\langle\zeta_1,\zeta_\infty\rangle + c(\zeta_0\zeta_x)\{c(\zeta_1)-1\}\langle\zeta_0,\zeta_\infty\rangle, \tag{4.5.12}$$

$$\gamma_{0*}\langle\zeta_0,\zeta_\infty\rangle = c(\zeta_x)^{-1}\{c(1-\zeta_0)^{-1}\}\langle\zeta_1,\zeta_\infty\rangle + \{1+c(\zeta_0)-c(\zeta_x)^{-1}\}\langle\zeta_0,\zeta_\infty\rangle. \tag{4.5.13}$$

On the other hand, by the definition of the loops ζ_0, ζ_1 and ζ_∞, we see

$$c(\zeta_0) = \varepsilon(\gamma-\alpha), \quad c(\zeta_1) = \varepsilon(\beta-\gamma), \quad c(\zeta_x) = \varepsilon(\alpha), \tag{4.5.14}$$

where $\varepsilon(\cdot) = \exp(2\pi i \cdot)$. The following lemma tells us under which conditions $\langle\zeta_0,\zeta_\infty\rangle$ and $\langle\zeta_1,\zeta_\infty\rangle$ are linearly independent.

LEMMA 4.5.3. *Suppose that α, β, and γ satisfy*

$$\begin{cases} \alpha \neq 0, -1, -2, -3, \cdots, \\ \alpha - \beta \notin \mathbb{Z}, \ \textit{and} \\ \gamma - \beta \neq 1, 2, 3, \cdots \ \textit{or} \ \alpha - \gamma \neq 0, 1, 2, \cdots. \end{cases} \tag{4.5.15}$$

Then, $\langle\zeta_0,\zeta_\infty\rangle$ and $\langle\zeta_1,\zeta_\infty\rangle$ are linearly independent.

Proof. Since $\langle\zeta_0,\zeta_\infty\rangle$ is holomorphic near $x = 1$, and $\langle\zeta_1,\zeta_\infty\rangle$ is holomorphic near $x = 0$, if they are linearly dependent, one of them, say $\langle\zeta_0, \zeta_\infty\rangle$, is holomorphic also at $x = 0$, and hence holomorphic in the entire plane $\mathbb{C}$. Since it has polynomial growth order, we conclude that it is a polynomial. We shall show that, under the condition (4.5.15), neither $\langle\zeta_0, \zeta_\infty\rangle$ nor $\langle\zeta_1, \zeta_\infty\rangle$ is a polynomial, by expanding them into power series at 1 and 0, respectively. To carry out the computation, we use *Pochhammer's integral representation of the Beta function*:

$$\begin{aligned} B(p, q) &= \frac{1}{\{1 - \varepsilon(p)\}\{1 - \varepsilon(q)\}} \int_{[\gamma_0^{-1}, \gamma_1^{-1}]} t^{p-1}(1-t)^{q-1} dt \\ &= \frac{\varepsilon(p+q)}{\{1 - \varepsilon(p)\}\{1 - \varepsilon(q)\}} \int_{[\gamma_0, \gamma_1]} t^{p-1}(1-t)^{q-1} dt, \end{aligned}$$

where the argument of the integrand is assigned at the starting point $b = 1/2$ by $\arg t = \arg(1-t) = 0$. This formula is a direct consequence of the definition (3.2.2) and Proposition 3.3.7.

Expanding $\langle\zeta_0, \zeta_\infty\rangle$ and $\langle\zeta_1, \zeta_\infty\rangle$ around $x = 1$ and 0, respectively, we have

$$\langle\zeta_0, \zeta_\infty\rangle = \sum_{m=0}^{\infty} c_m^{(0)} (x-1)^m,$$

$$\langle\zeta_1, \zeta_\infty\rangle = \sum_{m=0}^{\infty} c_m^{(1)} x^m,$$

where

$$c_m^{(0)} = c_0(\alpha)_m \int_{[\zeta_0, \zeta_\infty]} t^{\alpha-\gamma}(t-1)^{-\gamma-m} dt,$$

$$c_m^{(1)} = c_1(\alpha)_m \int_{[\zeta_1, \zeta_\infty]} t^{-\gamma-m}(t-1)^{\gamma-\beta-1} dt,$$

and c_0 and c_1 are constants independent of α, β, γ, m. By changing the variable of integration, we have

$$\int_{[\zeta_1,\zeta_\infty]} t^{-\gamma-m}(t-1)^{\gamma-\beta-1}dt$$

$$= \int_{[\gamma_0,\gamma_1]} s^{(m+\beta)-1}(1-s)^{(\gamma-\beta)-1}ds$$

$$= \frac{[1-\varepsilon(m+\beta)][1-\varepsilon(\gamma-\beta)]}{\varepsilon(m+\gamma)}B(m+\beta,\gamma-\beta)$$

$$= \frac{[1-\varepsilon(\beta)][1-\varepsilon(\gamma-\beta)]}{\varepsilon(\gamma)}\frac{\Gamma(m+\beta)\Gamma(\gamma-\beta)}{\Gamma(m+\gamma)}.$$

In this way, we have

$$c_m^{(0)} = c_0(\alpha)_m\{1-\varepsilon(\alpha-\gamma)\}\{1-\varepsilon(\beta-\alpha)\}\frac{\Gamma(\alpha-\gamma+1)\Gamma(\beta-\alpha+m)}{\Gamma(\beta-\gamma+m+1)},$$

$$c_m^{(1)} = c_1(\alpha)_m\{1-\varepsilon(\beta-\gamma)\}\{1-\varepsilon(\gamma-\beta)\}\frac{\Gamma(\beta-\alpha+m)\Gamma(\gamma-\beta)}{\Gamma(\gamma-\alpha+m)}.$$

So, under the condition (4.5.15), $c_m^{(0)}, c_m^{(1)} \neq 0$ $(m \geq 0)$. Hence $\langle\zeta_0,\zeta_\infty\rangle$ and $\langle\zeta_1,\zeta_\infty\rangle$ are not polynomials. ∎

THEOREM 4.5.4. *Suppose that α, β and γ satisfy (4.5.15), then $\mathcal{F} = (\langle\zeta_1,\zeta_\infty\rangle, \langle\zeta_0,\zeta_\infty\rangle)$ forms a fundamental system of solutions of the hypergeometric equation. Let γ_ν $(\nu = 0,1)$ be the loops with the base point $b = 1/2$ defined in Figure 2.5. Analytic continuation of $\mathcal{F}$ along $\gamma_\nu(\nu = 0,1)$ is given by*

$$\gamma_{\nu*}\mathcal{F} = \mathcal{F}A_\nu \quad (\nu = 0,1), \tag{4.5.16}$$

where the matrices A_ν are given by

$$A_0 = \begin{pmatrix} 1 & \varepsilon(-\alpha)-\varepsilon(-\gamma) \\ 0 & 1+\varepsilon(\gamma-\alpha)-\varepsilon(-\alpha) \end{pmatrix},$$

$$A_1 = \begin{pmatrix} \varepsilon(\alpha+\beta-\gamma) & 0 \\ \varepsilon(\beta)-\varepsilon(\gamma) & 1 \end{pmatrix},$$

where $\varepsilon(\cdot) = \varepsilon(2\pi i \cdot)$. The monodromy group with respect to the fundamental system of solutions $\mathcal{F}$ is generated by A_0 and A_1.

Proof. By (4.5.10) and (4.5.13), we have

$$\begin{aligned}
\gamma_{0*}\mathcal{F} &= (\gamma_{0*}\langle\zeta_1,\zeta_\infty\rangle, \gamma_{0*}\langle\zeta_0,\zeta_\infty\rangle) \\
&= (\langle\zeta_1,\zeta_\infty\rangle, \langle\zeta_0,\zeta_\infty\rangle)\begin{pmatrix} 1 & c(\zeta_0\zeta_x)^{-1}\{c(\zeta_0)-1\} \\ 0 & 1+c(\zeta_0)-c(\zeta_x)^{-1} \end{pmatrix} \\
&= \mathcal{F}\begin{pmatrix} 1 & \varepsilon(\alpha-\gamma)\varepsilon(-\alpha)\{\varepsilon(\gamma-\alpha)-1\} \\ 0 & 1+\varepsilon(\gamma-\alpha)-\epsilon(-\alpha) \end{pmatrix} \\
&= \mathcal{F}A_0.
\end{aligned}$$

Analogously, we have

$$\begin{aligned}
\gamma_{1*}\mathcal{F} &= (\gamma_{1*}\langle\zeta_1,\zeta_\infty\rangle, \gamma_{1*}\langle\zeta_0,\zeta_\infty\rangle) \\
&= (\langle\zeta_1,\zeta_\infty\rangle, \langle\zeta_0,\zeta_\infty\rangle)\begin{pmatrix} c(\zeta_1\zeta_x) & 0 \\ c(\zeta_0\zeta_x)\{c(\zeta_1)-1\} & 1 \end{pmatrix} \\
&= \mathcal{F}\begin{pmatrix} \varepsilon(\beta-\alpha)\varepsilon(\alpha) & 0 \\ \varepsilon(\gamma-\alpha)\varepsilon(\alpha)\{\varepsilon(\beta-\gamma)-1\} & 1 \end{pmatrix} \\
&= \mathcal{F}A_1. \quad \blacksquare
\end{aligned}$$

Note. A generalization of the method used in sections 4.4 and 4.5 is made in [Aom.5], [HK] and [KN].

4.6 Finding the monodromy by Barnes integrals

The Barnes integral representation (Section 3.5) for $F(\alpha,\beta,\gamma;x)$ takes the form

$$F(\alpha,\beta,\gamma;x) = \frac{1}{2\pi i}\int_C h(t,x)dt, \tag{4.6.1}$$

where the integrand is given by

$$h(t,x) = \frac{\Gamma(\gamma)}{\Gamma(\alpha)\Gamma(\beta)} \frac{\Gamma(\alpha+t)\Gamma(\beta+t)\Gamma(-t)}{\Gamma(t+\gamma)}(-x)^t. \tag{4.6.2}$$

All poles of $h(\cdot, x)$ are on the union of $\mathcal{P}_+$ and $\mathcal{P}_-$, where

$$\mathcal{P}_+ = \{0, 1, 2, 3, \cdots\},$$

$$\mathcal{P}_- = \{-\alpha, -\alpha-1, -\alpha-2, \cdots\} \cup \{-\beta, -\beta-1, -\beta-2, \cdots\}.$$

The path of integration C is a vertical line along the imaginary axis with deviation so that $\mathcal{P}_+$ lies to the right of C and $\mathcal{P}_-$ lies to the left of C (Figure 2.3). Let C_N be the path indicated in Figure 2.9, where N is an integer. We change the path of integration C into C_N and let $N \to \infty$. Following this line, we obtain an answer to the connection problem, which leads us to the monodromy group of the hypergeometric equation.

THEOREM 4.6.1. *Suppose none of $\alpha, \beta, \gamma, \gamma-\alpha$ and $\gamma-\beta$ is an integer. Then we have the connection formula*

$$(f_0(x;0), f_0(x;1-\gamma)) = (f_\infty(x;\alpha), f_\infty(x;\beta))P, \tag{4.6.3}$$

where P is a matrix given by

$$P = \begin{pmatrix} c(\alpha,\beta,\gamma) & c(\alpha-\gamma+1, \beta-\gamma+1, 2-\gamma) \\ c(\beta,\alpha,\gamma) & c(\beta-\gamma+1, \alpha-\gamma+1, 2-\gamma) \end{pmatrix}, \tag{4.6.4}$$

and $c(\alpha,\beta,\gamma)$ is defined by

$$c(\lambda,\mu,\nu) = e^{-\pi i\lambda} \frac{\Gamma(\nu)\Gamma(\mu-\lambda)}{\Gamma(\mu)\Gamma(\nu-\lambda)}, \tag{4.6.5}$$

and the function $f_a(x;\lambda)$ is a Kummer's solution with the exponent λ at $x=a$, given in Section 1.3.

Proof. The integrand $h(\cdot, x)$ of (4.6.1) has simple poles in $\mathcal{P}_+ \cup \mathcal{P}_-$, whose residues at $\mathcal{P}_-$ are given by

$$\begin{aligned} &\operatorname*{Res}_{t=-\nu-n} h(t,x) \\ &= \frac{\Gamma(\gamma)}{\Gamma(\alpha)\Gamma(\beta)} \frac{\Gamma(\alpha+\beta-2\nu-n)\Gamma(\nu+n)}{\Gamma(\gamma-\nu-n)\Gamma(1+n)}(-x)^{-\nu}x^{-n} \end{aligned} \tag{4.6.6}$$

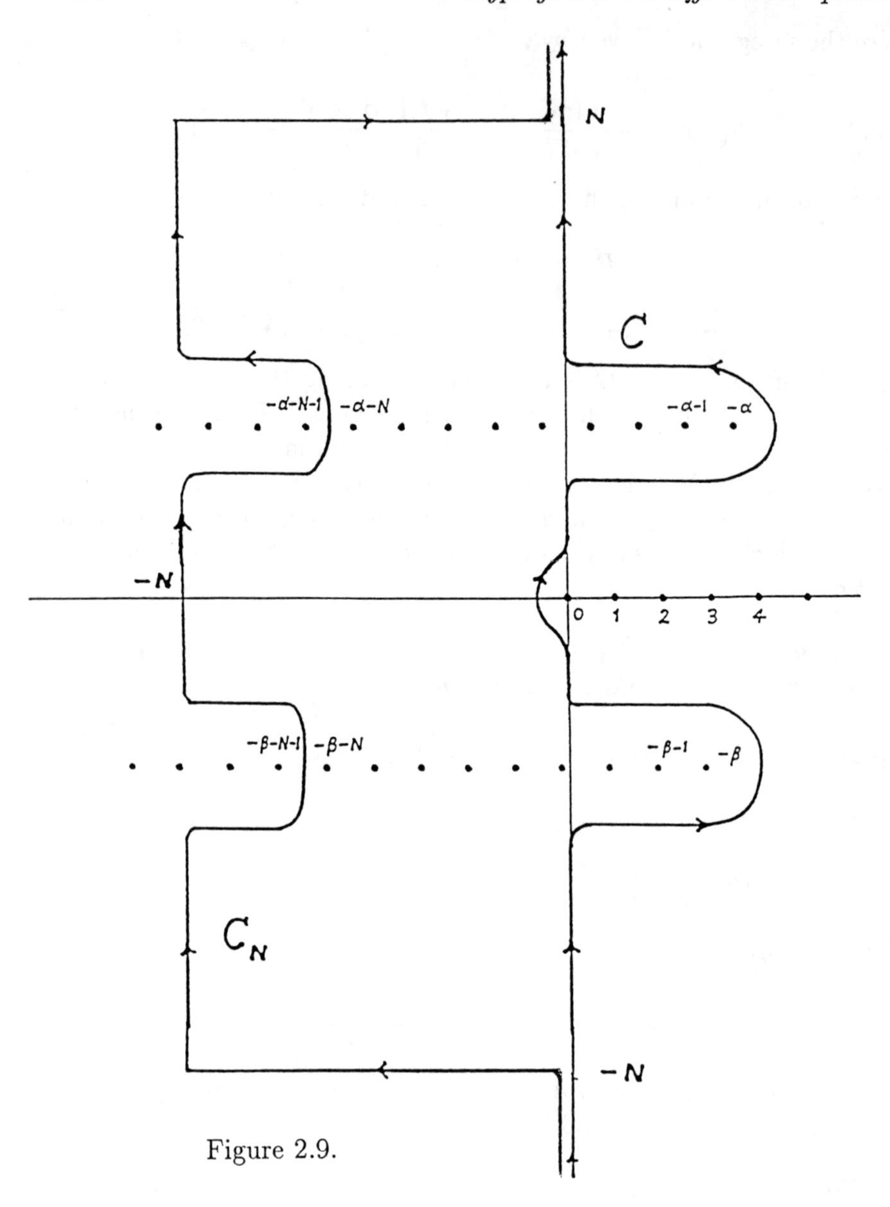

Figure 2.9.

for $\nu = \alpha, \beta$, where the identity $\mathrm{Res}_{t=-n}\Gamma(t) = (-1)^n/n!$ is used. On the other hand, we have

$$\begin{aligned} \Gamma(\alpha+\beta-2\nu-n) &= (-1)^n \frac{\Gamma(\alpha+\beta-2\nu)}{(2\nu-\alpha-\beta+1)_n}, \quad (\nu=\alpha,\beta), \\ \Gamma(\gamma-\nu-n) &= (-1)^n \frac{\Gamma(\gamma-\nu)}{(\nu-\gamma+1)_n}, \end{aligned} \tag{4.6.7}$$

which are easily verified by using

$$\Gamma(t+1) = t\Gamma(t).$$

Substituting (4.6.7) into (4.6.6), we obtain

$$\begin{aligned}(4.6.8)\qquad &\operatorname*{Res}_{t=-\nu-n} h(t,x)\\ &=c(\nu,\alpha+\beta-\nu,\gamma)\frac{(\nu-\gamma+1)_n(\nu)_n}{(2\nu-\alpha-\beta+1)_n n!}x^{-\nu-n}\end{aligned}$$

for $\nu=\alpha,\beta$. Hence, (4.6.1) and the residue theorem imply

$$f_0(x;0) = c(\alpha,\beta,\gamma)f_{\infty,N}(x;\alpha) + c(\beta,\alpha,\gamma)f_{\infty,N}(x;\beta) + I_N(x),$$

where

$$f_{\infty,N}(x,\nu) = \sum_{n=0}^{N} \frac{(\nu-\gamma+1)_n(\nu)_n}{(2\nu-\alpha-\beta+1)_n n!}x^{-\nu-n},$$

$$I_N(x) = \frac{1}{2\pi i}\int_{C_N} h(t,x)dt.$$

We can show that $I_N(x)$ converges to 0 as N tends to ∞ in a manner similar to the proof of Lemma 3.5.2. Moreover we have $f_{\infty,N}(x;\nu) \to f_\infty(x;\nu)$ as $N\to\infty$ for $\nu=\alpha,\beta$. Hence, we obtain

$$(4.6.9)\qquad f_0(x;0) = c(\alpha,\beta,\gamma)f_\infty(x;\alpha) + c(\beta,\alpha,\gamma)f_\infty(x;\beta).$$

If we replace (α,β,γ) by $(\alpha',\beta',\gamma') = (\alpha-\gamma+1,\beta-\gamma+1,2-\gamma)$ in (4.6.9) and multiply it by $x^{1-\gamma}$, we see

$$\begin{aligned}(4.6.10)\qquad &f_0(x;1-\gamma)\\ &= c(\alpha-\gamma+1,\beta-\gamma+1,2-\gamma)f_\infty(x;\alpha)\\ &+ c(\beta-\gamma+1,\alpha-\gamma+1,2-\gamma)f_\infty(x;\beta).\end{aligned}$$

In fact, putting superscripts $(\alpha\beta\gamma)$ on the symbols of Kummer's solutions in order to indicate the parameters, we have

$$x^{1-\gamma}f_0^{(\alpha'\beta'\gamma')}(x;0) = f_0^{(\alpha\beta\gamma)}(x;1-\gamma),$$

$$x^{1-\gamma}f_\infty^{(\alpha'\beta'\gamma')}(x;\alpha') = f_\infty^{(\alpha\beta\gamma)}(x;\alpha),$$

$$x^{1-\gamma}f_\infty^{(\alpha'\beta'\gamma')}(x;\beta') = f_\infty^{(\alpha\beta\gamma)}(x;\beta).$$

Equation (4.6.3) follows readily from (4.6.9) and (4.6.10). ∎

THEOREM 4.6.2. *Let $\gamma_\nu(\nu = 0, \infty)$ be the loops with base point $b = 1/2$ defined in Figure 2.5. Suppose that none of $\alpha, \beta, \gamma, \gamma - \alpha$ and $\gamma - \beta$ is an integer. Then the analytic continuations of $\mathcal{F} = (f_0(x; 0), f_0(x; 1-\gamma))$ along $\gamma_\nu(\nu = 0, \infty)$ are given by*

$$\gamma_{\nu *}\mathcal{F} = \mathcal{F}A_\nu \quad (\nu = 0, \infty), \tag{4.6.11}$$

where A_ν are the matrices

$$A_0 = \begin{pmatrix} 1 & 0 \\ 0 & \varepsilon(-\gamma) \end{pmatrix}, \quad A_\infty = P^{-1} \begin{pmatrix} \varepsilon(-\alpha) & 0 \\ 0 & \varepsilon(-\beta) \end{pmatrix} P, \tag{4.6.12}$$

$\varepsilon(\cdot) = \exp(2\pi i \cdot)$, and P is the matrix given by (4.6.4). The monodromy group with respect to the fundamental system of solutions $\mathcal{F}$ is generated by A_0 and A_∞.

4.7 Finding the monodromy by Gauss-Kummer's identity

We shall solve the connection problem by using Gauss-Kummer's identity and then find generators of the monodromy group. We consider two fundamental systems of solutions $(f_0(x; 0), f_0(x; 1 - \gamma))$ and $(f_1(x; 0), f_1(x; \gamma - \alpha - \beta))$, and find a relation between these systems, where the functions $f_a(x; \nu)$ are defined in Section 1.3.

THEOREM 4.7.1. *If neither γ nor $\gamma - \alpha - \beta$ is an integer, then*

$$(f_0(x; 0), f_0(x; 1 - \gamma)) = (f_1(x; 0), f_1(x; \gamma - \alpha - \beta))P, \tag{4.7.1}$$

where P is the matrix defined by

$$P = \begin{pmatrix} \dfrac{\Gamma(\gamma)\Gamma(\gamma - \alpha - \beta)}{\Gamma(\gamma - \alpha)\Gamma(\gamma - \beta)} & \dfrac{\Gamma(2 - \gamma)\Gamma(\gamma - \alpha - \beta)}{\Gamma(1 - \alpha)\Gamma(1 - \beta)} \\[2ex] \dfrac{\Gamma(\gamma)\Gamma(\gamma - \alpha - \beta)}{\Gamma(\alpha)\Gamma(\beta)} & \dfrac{\Gamma(\gamma - \alpha - \beta)\Gamma(2 - \gamma)}{\Gamma(\alpha - \gamma + 1)\Gamma(\beta - \gamma + 1)} \end{pmatrix}. \tag{4.7.2}$$

Proof. Let us put

$$f_0(x; 0) = b(\alpha, \beta, \gamma)f_1(x; 0) + c(\alpha, \beta, \gamma)f_1(x; \gamma - \alpha - \beta), \tag{4.7.3}$$

and find the constants $b = b(\alpha, \beta, \gamma)$ and $c = c(\alpha, \beta, \gamma)$, which are meromorphic functions of α, β and γ. To find b, we put $x = 1$ in (4.7.3) under the condition $\Re(\gamma - \alpha - \beta) > 0$. Taking Gauss-Kummer's identity (3.7.1) into account, we obtain

$$b(\alpha, \beta, \gamma) = f_0(1; 0) = \frac{\Gamma(\gamma)\Gamma(\gamma - \alpha - \beta)}{\Gamma(\gamma - \alpha)\Gamma(\gamma - \beta)}. \tag{4.7.4}$$

On the other hand, applying Gauss-Kummer's identity to

$$f_1(x; 0) = F(\alpha, \beta, \alpha + \beta - \gamma + 1; 1 - x)$$

and

$$f_1(x; \gamma - \alpha - \beta) = (1 - x)^{\gamma - \alpha - \beta} F(\gamma - \alpha, \gamma - \beta, \gamma - \alpha - \beta; 1 - x),$$

we obtain

$$\begin{aligned} f_1(0; 0) &= \frac{\Gamma(\alpha + \beta - \gamma + 1)\Gamma(1 - \gamma)}{\Gamma(\beta - \gamma + 1)\Gamma(\alpha - \gamma + 1)}, \\ f_1(0; \gamma - \alpha - \beta) &= \frac{\Gamma(\gamma - \alpha - \beta + 1)\Gamma(1 - \gamma)}{\Gamma(1 - \alpha)\Gamma(1 - \beta)}. \end{aligned} \qquad (\Re(1 - \gamma) > 0) \tag{4.7.5}$$

Using $\Gamma(t)\Gamma(1 - t) = \pi / \sin \pi t$, we obtain from (4.7.4) and (4.7.5)

$$\begin{aligned} b f_1(0; 0) &= \frac{\sin \pi(\gamma - \alpha) \cdot \sin \pi(\gamma - \beta)}{\sin \pi\gamma \cdot \sin \pi(\gamma - \alpha - \beta)}, \\ f_1(0; \gamma - \alpha - \beta) &= \frac{\sin \pi\alpha \cdot \sin \pi\beta}{\sin \pi\gamma \cdot \sin \pi(\alpha + \beta - \gamma)} \\ &\quad \times \frac{\Gamma(\alpha)\Gamma(\beta)}{\Gamma(\gamma)\Gamma(\alpha + \beta - \gamma)}. \end{aligned} \tag{4.7.6}$$

Assuming $\Re(1 - \gamma) > 0$ and putting $x = 0$ in (4.7.3) , we obtain

$$1 = b f_1(0; 0) + c f_1(0; \gamma - \alpha - \beta). \tag{4.7.7}$$

Substituting (4.7.6) into (4.7.7), we find that

$$c(\alpha,\beta,\gamma)=\frac{\Gamma(\gamma)\Gamma(\gamma-\alpha-\beta)}{\Gamma(\alpha)\Gamma(\beta)}. \tag{4.7.8}$$

If we replace (α,β,γ) by $(\alpha-\gamma+1,\beta-\gamma+1,2-\gamma)$ in (4.7.3) and make use of identities in the table of Kummer's 24 solutions (Section 1.3), we see

$$\begin{aligned} & f_0(x;1-\gamma) \\ =& b(\alpha-\gamma+1,\beta-\gamma+1,2-\gamma)f_1(x;0) \\ &+c(\alpha-\gamma+1,\beta-\gamma+1,2-\gamma)f_1(x;\gamma-\alpha-\beta). \end{aligned} \tag{4.7.9}$$

Now (4.7.1) is readily proved by (4.7.3) and (4.7.9) together with (4.7.4) and (4.7.8). ∎

THEOREM 4.7.2. *Let $\gamma_\nu(\nu=0,1)$ be the loops with base point $b=1/2$ defined in Figure 2.5. Suppose that neither γ nor $\gamma-\alpha-\beta$ is an integer. Analytic continuation of the fundamental system of solutions $\mathcal{F}=(f_0(x;0),f_0(x;1-\gamma))$ along γ_ν is given by*

$$\gamma_{\nu *}\mathcal{F}=\mathcal{F}A_\nu \quad (\nu=0,1), \tag{4.7.10}$$

where A_ν are matrices defined by

$$A_0=\begin{pmatrix}1 & 0\\ 0 & \varepsilon(-\gamma)\end{pmatrix}, \quad A_1=P^{-1}\begin{pmatrix}1 & 0\\ 0 & \varepsilon(\gamma-\alpha-\beta)\end{pmatrix}P, \tag{4.7.11}$$

where P is given by (4.7.2), and $\varepsilon(\cdot)=\exp(2\pi i\cdot)$. The monodromy group with respect to the fundamental system $\mathcal{F}$ is generated by A_0 and A_1.

3 Monodromy Preserving Deformation, Painlevé Equations and Garnier Systems

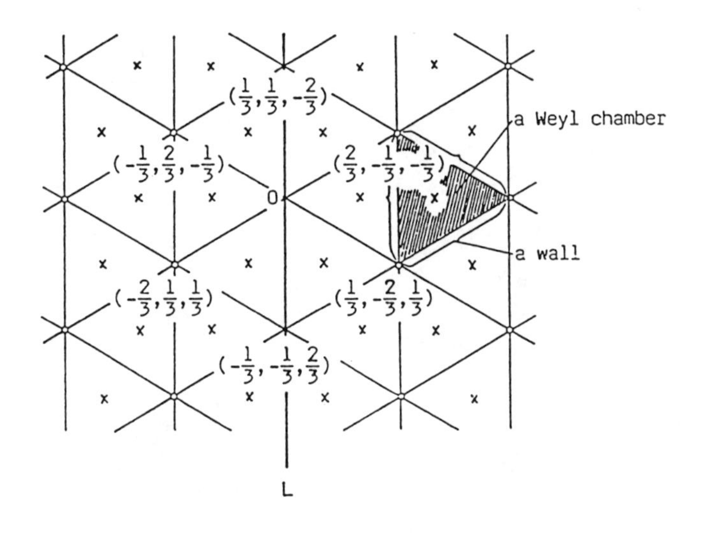

The most important non-linear ordinary differential equations are the following six Painlevé equations:

$$P_I: \frac{d^2\lambda}{dt^2} = 6\lambda^2 + t,$$

$$P_{II}: \frac{d^2\lambda}{dt^2} = 2\lambda^3 + t\lambda + \alpha,$$

$$P_{III}: \frac{d^2\lambda}{dt^2} = \frac{1}{\lambda}\Big(\frac{d\lambda}{dt}\Big)^2 - \frac{1}{t}\frac{d\lambda}{dt} + \frac{1}{t}(\alpha\lambda^2 + \beta) + \gamma\lambda^3 + \frac{\delta}{\lambda},$$

$$P_{IV}: \frac{d^2\lambda}{dt^2} = \frac{1}{2\lambda}\Big(\frac{d\lambda}{dt}\Big)^2 + \frac{3}{2}\lambda^3 + 4t\lambda^2 + 2(t^2 - \alpha)\lambda + \frac{\beta}{\lambda},$$

$$P_V: \frac{d^2\lambda}{dt^2} = \Big(\frac{1}{2\lambda} + \frac{1}{\lambda - 1}\Big)\Big(\frac{d\lambda}{dt}\Big)^2 - \frac{1}{t}\frac{d\lambda}{dt} + \frac{(\lambda-1)^2}{t}\Big(\alpha\lambda + \frac{\beta}{\lambda}\Big)$$
$$+ \gamma\frac{\lambda}{t} + \delta\frac{\lambda(\lambda+1)}{\lambda - 1},$$

$$P_{VI}: \frac{d^2\lambda}{dt^2} = \frac{1}{2}\Big(\frac{1}{\lambda} + \frac{1}{\lambda-1} + \frac{1}{\lambda - t}\Big)\Big(\frac{d\lambda}{dt}\Big)^2 - \Big(\frac{1}{t} + \frac{1}{t-1} + \frac{1}{\lambda - t}\Big)\frac{d\lambda}{dt}$$
$$+ \frac{\lambda(\lambda-1)(\lambda-t)}{t^2(t-1)^2}\Big[\alpha - \beta\frac{t}{\lambda^2} + \gamma\frac{t-1}{(\lambda-1)^2} + (\frac{1}{2} - \delta)\frac{t(t-1)}{(\lambda-t)^2}\Big],$$

where $\alpha, \beta, \gamma, \delta$ are complex constants. (Warning: The parameters of P_{VI} are slitely different from those customarily used; $-\beta$ and $\frac{1}{2} - \delta$ have been denoted by β and δ. The reason of our choice will turn out to be clear in the text.) We study, in this chapter, these differential equations first classically (§1) and secondly in the framework of the monodromy preserving deformation. After introducing the concept of monodromy preserving deformation (§2, §3), we derive the Garnier system written in the form of Hamiltonian system, which governs such deformation of a second order Fuchsian equation with $n+3$ singularities (§4). When $n = 1$, the Garnier system turns out to be equivalent to the sixth Painlevé equation. In Section 5, we derive the Schlesinger system, which governs monodromy preserving deformations of matrix equations of Schlesinger type. By using the relation between the Garnier system and the Schlesinger system, discussed in Section 6, we transform the Garnier system into the system with polynomial Hamiltonians (§7). The system thus obtained is free of movable branch points. Symmetry of the Garnier system is studied in Section 8. For some particular values of parameters, the Garnier system happens to admit solutions expressed by hypergeometric functions of Lauricella (§9).

1 Painlevé equations

1.1 Historical remarks

We shall review briefly how the Painlevé equations $\mathrm{P_J}$ are discovered. One of the important problems of analysis in the 19th century was to find "good transcendental functions" defined by non-linear algebraic differential equations. A differential equation

$$F(t, y, \frac{dy}{dt}, \cdots, \frac{d^n y}{dt^n}) = 0 \tag{1.1.1}$$

defined in the domain D is said to be *algebraic* if $F = F(t, y_0, y_1, \cdots, y_n)$ is a polynomial in $\mathbf{y} = (y_0, y_1, \cdots, y_n)$ with coefficients meromorphic in $t \in D$, *rational* if it is algebraic and is of degree one with respect to y_n, and *linear* if it is algebraic and F is a linear form in $\mathbf{y}$. Take $c := (c_0, \cdots, c_n) \in \mathbb{C}^{n+1}$ and $t_0 \in D$ so that $F(t_0, c_0, c_1, \cdots, c_n) = 0$, and denote by $\varphi(t) = \varphi(t; t_0, c)$ the holomorphic solution such that

$$\frac{d^i \varphi}{dt^i}(t_0) = c_i \quad (i = 0, \cdots, n).$$

The function obtained by an analytic continuation of $\varphi(t)$ is also denoted by $\varphi(t)$.

If an equation is non-linear, we can in general predict neither *where the singularities of solutions appear* nor *of what kind the singularities are.* In such a case, we can hardly say that the function which is a solution of the equation is controlled by the differential equation. The following examples show that solutions may have branch points or essential singular points which change their position depending on integration constants.

EXAMPLE 1 $$my'y^{m-1} = 1, \quad m \in \mathbb{N}.$$

Solution: $y(t) = (t-c)^{1/m}$, $c \in \mathbb{C}$ being an integration constant. $y(t)$ has an algebraic branch point over $t = c$.

EXAMPLE 2 $$y'' + (y')^2 = 0.$$

Solution: $y(t) = \log(t - c_1) + c_2, \quad c_1, c_2 \in \mathbb{C}.$

EXAMPLE 3 $$yy'' + (y')^2\Big(\frac{2y}{y'} - 1\Big) = 0.$$

Solutions: $y(t) = c_1 \exp(-1/(t - c_2)), \quad c_1, c_2 \in \mathbb{C}.$

Thus we face to seeking a non-linear differential equation such that the singularities (except poles) of the solutions are predictable.

An algebraic differential equation (1.1.1) is said to be *free of movable branch (resp. essential singular) points* if the solution $\varphi(t; t_0, c)$ has no branch (resp. essential singular) point which changes its position when we vary (t_0, c) under the restriction $F(t_0, c) = 0$.

PROBLEM 1.1.1. *Find all the algebraic differential equations free of movable branch points and movable essential singular points.*

We say that an algebraic differential equation enjoys the *Painlevé property* if (1.1.1) is free of movable branch points.

When $n = 1$, the problem was studied and solved by L.Fuchs and H.Poincaré; any equation of the type (1.1.1) with the Painlevé property can be transformed, by a holomorphic change of the variable t and by a linear fractional change of the unknown with coefficients in $\mathcal{O}(D)$, into the equation of the Weierstrass $\wp$ function:

$$(1.1.2) \qquad (\frac{dy}{dt})^2 = 4y^3 - g_2 y - g_3, \quad g_2, g_3 \in \mathbb{C}$$

or into the Riccati equation:

$$(1.1.3) \qquad \frac{dy}{dt} = a(t)y^2 + b(t)y + c(t), \quad a(t), b(t), c(t) \in \mathcal{O}(D),$$

where $\mathcal{O}(D)$ stands for the ring of holomorphic functions on D. The equation (1.1.3) reduces to a linear equation; in fact we have the following result.

PROPOSITION 1.1.2. *By the change of unknown*

$$y = -\frac{1}{a(t)}\frac{d}{dt}\log u,$$

(1.1.3) *is transformed into the linear equation:*

$$u'' + \left(\frac{a'(t)}{a(t)} - b(t)\right)u' + a(t)c(t)u = 0. \tag{1.1.4}$$

When the order n of (1.1.1) is one, only movable branch points appear, whereas when $n \geq 2$, movable essential singular points may appear. E.Picard pointed out this fact in his letter to Mittag-Leffler (1893), and expressed his pessimistic opinion that there might be very little hope of success to find non-linear differential equations with the Painlevé property in case $n \geq 2$. Despite of the negative prospect of Picard, P.Painlevé attacked the problem for rational differential equations of the form

$$\frac{d^2y}{dt^2} = R\left(t, y, \frac{dy}{dt}\right), \tag{1.1.5}$$

and he showed by a huge amount of computation that any equation with the Painlevé property reduces, by an appropriate transformation of the variables, to an equation which can be integrated by quadrature, or to a linear equation, or to P_J ($J = I, \cdots, VI$). Precisely speaking, Painlevé found only P_I, P_{II}, P_{III} because of errors in his computations. His student, B.O.Gambier, added the equations P_{IV}, P_V, P_{VI} to the list. The success of Painlevé for rational differential equations of the second order encouraged his students, J.Chazy and R.Garnier, to pursue further the problem for rational differential equations of order ≥ 3 and for algebraic differential equations of the second order. However, only partial results are obtained and the complete classification is not yet achieved.

After the discovery of the Painlevé equations, there was a dispute between Painlevé and J.Liouville whether they can be integrated by solutions of linear algebraic differential equations ([Pain]). The disputation has come to an end by a recent algebraic study ([Nis.2], [Ume.3]) proving that they are never integrated by such functions.

Do the Painlevé equations form an isolated island far from the continent of analysis ? (Preface of [Pain].) It turned out that they form a part of that continent. Consider a second degree Fuchsian differential equation (cf. Chapter 1, §4)

$$\frac{d^2y}{dx^2} + a_1(x)\,\frac{dy}{dx} + a_2(x)\,y = 0$$

with five singular points $x_0, \ldots, x_4 \in \mathbb{P}^1$, and suppose that the singularity at x_0 is an apparent one (this means that it is non-logarithmic in the sense of Section 4 of Chapter 1 and that the characteristic exponents at x_0 are integral, or equivalently, that the solutions of the differential equation are meromorphic at x_0). Since this description is invariant under the group $PGL(2,\mathbb{C})$ of automorphisms of $\mathbb{P}^1$, only two parameters t (=cross-ratio of $x_1, \ldots, x_4$) and λ (=cross-ratio of $x_0, \ldots, x_3$) are needed to describe the position of the x_i's, while 8 further parameters are needed to describe the possible choices of the rational functions $a_1(x)$ and $a_2(x)$. The monodromy representation of the differential equation is a homomorphism, defined up to conjugation, from the fundamental group of $\mathbb{P}^1 - \{x_1, \ldots, x_4\}$, which is a free group on three generators, to $GL(2,\mathbb{C})$, so it is described by an element of the 9-dimensional set $GL(2,\mathbb{C})^3/PGL(2,\mathbb{C})$. Since 10 is bigger than 9, there must be one-dimensional families of equations, i.e., curves in the space of functions (a_1, a_2), with constant monodromy. The beautiful discovery of R. Fuchs (1907) is that the relation between the parameters t and λ along such a curve is exactly the Painlevé equation $\mathrm{P_{VI}}$. This will be discussed in Sections 3 and 4, and its generalizations will form one of the main themes of the rest of this chapter. Notice that the above-mentioned 24-symmetry of $\mathrm{P_{VI}}$ also becomes obvious from the above description, since the points $x_1, \ldots, x_4$ can be permuted arbitrarily.

We note a connection of the Painlevé equations with a recent development of mathematical physics ; for example, Wu et al. (1973-1976) encountered, in the study of the Ising model, the third Painlevé equation [MTW].

1.2. Relations between the $\mathrm{P_J}$'s.

We associate with each $\mathrm{P_J}$ the set Ξ_J given by:

$$\begin{aligned}
&\Xi_I \ : \infty,\\
&\Xi_{II} : \infty,\\
&\Xi_{III}: 0, \infty,\\
&\Xi_{IV} : \infty,\\
&\Xi_V \ : 0, \infty,\\
&\Xi_{VI} : 0, 1, \infty.
\end{aligned}$$

It is known that $\mathrm{P_J}$ admits no singular points outside Ξ_J except poles. So, it enjoys the Painlevé property and moreover is *free of movable essential singular points.* We call Ξ_J the set of *fixed singular points* of $\mathrm{P_J}$. We deduce from this fact that the general solution is meromorphic on the universal covering space $\tilde{B}_J$ of $B_J := \mathbb{P}^1 \backslash \Xi_J$:

$$\tilde{B}_I \simeq \tilde{B}_{II} \simeq \tilde{B}_{III} \simeq \tilde{B}_{IV} \simeq \tilde{B}_V \simeq \mathbb{C} \quad \text{and} \quad \tilde{B}_{VI} \simeq D \quad \text{a disk in } \mathbb{C}.$$

We show that equations $\mathrm{P_J}$ (J =I, $\cdots$,V) are derived from $\mathrm{P_{VI}}$ by certain limit processes. Let us derive $\mathrm{P_V}$ from $\mathrm{P_{VI}}$. Substitute the independent variable t and parameters in the equation $\mathrm{P_{VI}}$ as follows:

$$t = 1 + \epsilon t_1, \quad \beta = -\beta_1, \quad \gamma = \delta_1 \epsilon^{-2} + \gamma_1 \epsilon^{-1}, \quad \delta = -\delta_1 \epsilon^{-2};$$

then $\mathrm{P_{VI}}$:

$$\frac{d^2\lambda}{dt^2} = R\big(t, \lambda, \frac{d\lambda}{dt}\big)$$

is taken into

$$\frac{d^2\lambda}{d{t_1}^2} = \epsilon^2 R\big(1 + \epsilon t_1, \lambda, \epsilon^{-1}\frac{d\lambda}{dt_1}\big). \tag{1.2.1}$$

It is easy to see that the right-hand side of the equation (1.2.1) is holomorphic in ϵ at $\epsilon = 0$. Letting $\epsilon \to 0$ and writing t, β, γ, δ in place of $t_1, \beta_1, \gamma_1, \delta_1$, we obtain the fifth Painlevé equation $\mathrm{P_V}$. For notational simplicity, let us write the above process as $P_{VI} \to P_V$:

$$t \to 1 + \epsilon t, \quad \beta \to -\beta, \quad \gamma \to \delta\epsilon^{-2} + \gamma\epsilon^{-1}, \quad \delta \to -\delta\epsilon^{-2} \qquad (\epsilon \to 0).$$

This process causes the coalescence of fixed singular points $t = 1$ and $t = \infty$. The other equations are also derived successively from $\mathrm{P_{VI}}$. In order to state the following proposition compactly, let us introduce the equation

$$P_{III}': \qquad \frac{d^2\lambda}{dt^2} = \frac{1}{\lambda}\left(\frac{d\lambda}{dt}\right)^2 - \frac{1}{t}\frac{d\lambda}{dt} + \frac{\lambda^2}{4t^2}(\gamma\lambda + \alpha) + \frac{\beta}{4t} + \frac{\delta}{4\lambda}.$$

PROPOSITION 1.2.1. *P_J $(J = I, \cdots, V)$ are obtained from P_{VI} by the successive limit process according to the following diagram:*

$$P_{VI} \to P_V \begin{array}{c} \nearrow \\ \searrow \end{array} \begin{array}{c} P_{\mathrm{IV}} \\ P_{\mathrm{III}} \end{array} \begin{array}{c} \searrow \\ \nearrow \end{array} P_{II} \to P_I.$$

These processes are given as follows:

$$P_{VI} \to P_V: \quad t \to 1 + \epsilon t,\ \beta \to -\beta,\ \gamma \to \delta\epsilon^{-2} + \gamma\epsilon^{-1},$$
$$\delta \to -\delta\epsilon^{-2} \quad (\epsilon \to 0).$$

$$P_V \to P_{IV}: \quad t \to 1 + \sqrt{2}\epsilon t,\ \lambda \to \frac{1}{\sqrt{2}}\epsilon\lambda,\ \alpha \to \frac{1}{2}\epsilon^{-4},\ \beta \to \frac{1}{4}\beta,$$
$$\gamma \to -\epsilon^{-4},\ \delta \to -\frac{1}{2}\epsilon^{-4} + \alpha\epsilon^{-2} \quad (\epsilon \to 0).$$

$$P_{IV} \to P_{II}: \quad t \to -\epsilon^{-3}(1 - 2^{-2/3}\epsilon^4 t),\ \lambda \to \epsilon^{-3}(1 + 2^{2/3}\epsilon^2\lambda),$$
$$\alpha \to -\frac{1}{2}\epsilon^{-6} - 2\alpha,\ \beta \to -\frac{1}{2}\epsilon^{-12} \quad (\epsilon \to 0).$$

$$P_V \to P_{III}': \quad t \to t,\ \lambda \to 1 + \epsilon\lambda,\ \alpha \to \frac{1}{8}\epsilon^{-2}\gamma + \frac{1}{4}\epsilon^{-1}\alpha,$$
$$\beta \to -\frac{1}{8}\epsilon^{-2}\gamma,\ \gamma \to \frac{1}{4}\epsilon\beta,\ \delta \to \frac{1}{8}\epsilon^2\delta \quad (\epsilon \to 0).$$

$$P_{III}' \to P_{III}: \quad t \to t^2,\ \lambda \to t\lambda.$$

$$P_{III} \to P_{II}: \quad t \to 1 + \epsilon^2 t,\ \lambda \to 1 + 2\epsilon\lambda,\ \alpha \to -\frac{1}{2}\epsilon^{-6},$$
$$\beta \to \frac{1}{2}\epsilon^{-6}(1 + 4\alpha\epsilon^3),\ \gamma \to \frac{1}{4}\epsilon^{-6},$$

$$\delta \to -\frac{1}{4}\epsilon^{-6} \quad (\epsilon \to 0).$$

$$P_{II} \to P_I: \qquad t \to -6\epsilon^{-10}(1 - \frac{1}{6}\epsilon^{12}t), \ \lambda \to \epsilon^{-5}(1 + \epsilon^6\lambda),$$

$$\alpha \to 4\epsilon^{-15} \quad (\epsilon \to 0).$$

REMARK 1.2.2. The reason why such changes of the variables and of the parameters in Proposition 1.2.1 arise can be explained from the viewpoint of monodromy-preserving deformation; see [Gar.1], [Okm.6].

REMARK 1.2.3. By rescaling λ and t, we see that the number of parameters in $\mathrm{P_{III}}$ and $\mathrm{P_V}$ are practically 2 and 3, respectively.

1.3. Symmetry of the Painlevé equation $\mathrm{P_{VI}}$

We study the symmetry of the sixth Painlevé equation $\mathrm{P_{VI}}$, the master equation. The results presented here are due to Painlevé, and K.Okamoto ([Pain], [Okm.1]). Let $V = \{(\alpha, \beta, \gamma, \delta); \alpha, \beta, \gamma, \delta \in \mathbb{C}\} \simeq \mathbb{C}^4$ be the space of parameters of $\mathrm{P_{VI}}$. The sixth Painlevé equation with parameters $v \in V$ is denoted by $\mathrm{P_{VI}}(v)$.

Let us consider the change of the variables $(t, \lambda) \to (t_1, \lambda_1)$ in $\mathrm{P_{VI}}$ defined by :

$$T : t = 1 - t_1, \quad \lambda = 1 - \lambda_1;$$

and write t, λ in place of t_1, λ_1. We obtain a differential equation

$$\frac{d^2\lambda}{dt^2} = \frac{1}{2}\Big(\frac{1}{\lambda} + \frac{1}{\lambda - 1} + \frac{1}{\lambda - t}\Big)\Big(\frac{d\lambda}{dt}\Big)^2 - \Big(\frac{1}{t} + \frac{1}{t-1} + \frac{1}{\lambda - t}\Big)\frac{d\lambda}{dt}$$
$$+ \frac{\lambda(\lambda-1)(\lambda-t)}{t^2(t-1)^2}\Big[\alpha - \gamma\frac{t}{\lambda^2} + \beta\frac{t-1}{(\lambda-1)^2} + (\frac{1}{2} - \delta)\frac{t(t-1)}{(\lambda-t)^2}\Big],$$

which is nothing but $\mathrm{P_{VI}}(v')$ where $v' = (\alpha, \gamma, \beta, \delta)$. To indicate this change, we define an affine transformation $\ell : V \to V$ by $v' = \ell(v)$ and write as follows:

$$T : t \to 1 - t, \quad \lambda \to 1 - \lambda.$$

$$\ell : (\alpha, \beta, \gamma, \delta) \to (\alpha, \gamma, \beta, \delta).$$

This example suggests to make the following:

DEFINITION 1.3.1. A pair $\sigma = (T, \ell)$ consisting of a birational transformation $T : (t, \lambda) \to (t_1, \lambda_1)$ and an affine transformation $\ell : V \to V$ is called a *symmetry* of P_{VI} if $T \cdot P_{VI}(v) = P_{VI}(\ell(v))$.

PROPOSITION 1.3.2. *The equation* $P_{\rm VI}$ *admits a group* G *of symmetries which is isomorphic to the symmetric group* $\mathfrak{S}_4$. *The group* G *is generated by the following three transformations* $\sigma_i = (T_i, \ell_i)$ *of order 2:*

$$T_1 : \lambda \to 1 - \lambda, \quad t \to 1 - t,$$

$$T_2 : \lambda \to \frac{1}{\lambda}, \quad t \to \frac{1}{t},$$

$$T_3 : \lambda \to \frac{\lambda - t}{1 - t}, \quad t \to \frac{t}{t - 1},$$

$$\ell_1 : (\alpha, \beta, \gamma, \delta) \to (\alpha, \gamma, \beta, \delta),$$

$$\ell_2 : (\alpha, \beta, \gamma, \delta) \to (\beta, \alpha, \gamma, \delta),$$

$$\ell_3 : (\alpha, \beta, \gamma, \delta) \to (\alpha, \delta, \gamma, \beta).$$

The proposition can be proved by straightforward computations. Another proof will be given in Section 8 with the aid of of the theory of monodromy preserving deformations.

We consider the sixth Painlevé equation $P_{\rm VI}$ in $\mathbb{P}^1 \times B$, where $B := \mathbb{P}^1 \setminus \{0, 1, \infty\}$. Set

$$S_0 : = \{(\lambda, t) \in \mathbb{P}^1 \times B; \ \lambda = 0\}$$

$$S_1 : = \{(\lambda, t) \in \mathbb{P}^1 \times B; \ \lambda = 1\}$$

$$S_\infty : = \{(\lambda, t) \in \mathbb{P}^1 \times B; \ \lambda = \infty\}$$

$$S_t : = \{(t, \lambda) \in \mathbb{P}^1 \times B; \ \lambda = t\}$$

and

$$S := \bigcup_{\xi \in \{0, 1, t, \infty\}} S_\xi .$$

We see that S is a set of poles of the right-hand side of $P_{\rm VI}$. With the help of the explicit form of the generators σ_i of G, we immediately have

COROLLARY 1.3.3. *The group G of symmetries of P_{VI} acts transitively on the set $\{S_\xi;\ \xi = 0, 1, t, \infty\}$.*

This result will be used in Section 1.4.

REMARK 1.3.4. Further investigation of symmetries for each P_J is made in [Okm.13].

1.4. Solutions of P_{VI} at singular points

In this section, we study the *singular* initial value problem:

$$\lambda(t_0) = \xi, \quad t_0 \in B := B_{VI} = \mathbb{P}^1 \setminus \{0, 1, \infty\}, \quad \xi \in \{0, 1, \infty, t_0\}$$

for P_{VI} , and transform the equation P_{VI} into a system by introducing a new unknown μ_ξ. Here "singular" means that the right-hand side of P_{VI} has a pole at (t_0, ξ).

PROPOSITION 1.4.1. *Assume $\alpha, \beta, \gamma, \delta \neq 0$. For any $t_0 \in B$ and $\xi \in \{0, 1, \infty, t_0\}$, there are two 1-parameter families of solutions λ of P_{VI} which are analytic at $t = t_0$ and satisfy $\lambda(t_0) = \xi$.*

Proof. We prove the proposition when $\xi = 0$ under the assumption $\beta \neq 0$. In the other cases, ξ is sent to 0 by an element of the group of symmetries in Proposition 1.3.3, and the condition $\beta \neq 0$ is converted into $\alpha \neq 0, \gamma \neq 0, \delta \neq 0$ for $\xi = \infty, 1, t_0$, respectively. P_{VI} can be written in the form:

$$\frac{d^2\lambda}{dt^2} = \frac{1}{2\lambda}\left(\frac{d\lambda}{dt}\right)^2 + \frac{\beta}{(t-1)^2}\frac{1}{\lambda} + S\left(t, \lambda, \frac{d\lambda}{dt}\right),$$

where S is a polynomial in $d\lambda/dt$ with coefficients holomorphic in (λ, t) at $(0, t_0)$. Substitute the expression

$$\hat{\lambda}(t) = (t - t_0)^m \sum_{j=0}^{\infty} a_j (t - t_0)^j, \quad m > 0, \quad a_0 \neq 0$$

into the above equation; then we have

$$m = 1, \quad {a_0}^2 = \frac{2\beta}{(t_0 - 1)^2},$$

and we see that a_1 can be taken arbitrarily, say h, and that a_j $(j \geq 2)$ are determined successively by $a_0, \cdots, a_{j-1}$. Thus, defining κ_0 by

$$\kappa_0^2 = 2\beta \ (\neq 0),$$

we obtain two 1-parameter families (h as the parameter) of formal power series:

$$\hat{\lambda}_\pm(t) = \frac{\pm\kappa_0}{t_0 - 1}(t - t_0) + h(t - t_0)^2 + \cdots$$

satisfying the equation formally. Let us consider $\hat{\lambda}_-(t)$. Since

$$\frac{1}{t_0 - 1} = \frac{1}{t - 1 - (t - t_0)} = \frac{1}{t - 1}\sum_{j=0}^{\infty}\left(\frac{t - t_0}{t - 1}\right)^j,$$

we have

$$\hat{\lambda}_-(t) = \frac{-\kappa_0}{t - 1}(t - t_0) + \left[h + \frac{-\kappa_0}{(t - 1)^2}\right](t - t_0)^2 + \cdots \tag{1.4.1}$$

and

$$\frac{d\hat{\lambda}_-(t)}{dt} = \frac{-\kappa_0}{t - 1} + \left[2h + \frac{-\kappa_0}{(t - 1)^2}\right](t - t_0) + \cdots. \tag{1.4.2}$$

Notice that we have from (1.4.1):

$$t - t_0 = \frac{t - 1}{-\kappa_0}\hat{\lambda}_- + (\text{a power series in } \hat{\lambda}_- \text{ of order } \geq 2).$$

On the other hand, referring to the equation (1.4.2), we introduce a new variable μ_0 by

$$\frac{d\lambda}{dt} = \frac{-\kappa_0}{t - 1} + \lambda\mu_0. \tag{1.4.3}$$

Differentiating both sides of (1.4.3) and using P_{VI} , we have

$$\frac{d\mu_0}{dt} = A_0(\lambda,t){\mu_0}^2 + B_0(\lambda,t)\mu_0 + C_0(\lambda,t), \tag{1.4.4}$$

where

$$A_0(\lambda,t) = \frac{\lambda^2 - t}{2(\lambda-1)(\lambda-t)},$$

$$B_0(\lambda,t) = \frac{-\kappa_0}{t-1}\Big(\frac{1}{\lambda-1} + \frac{1}{\lambda-t}\Big) - \frac{1}{t} - \frac{1}{t-1} - \frac{1}{\lambda-t},$$

$$C_0(\lambda,t) = \frac{1}{2}\Big(\frac{\kappa_0}{t-1}\Big)^2\Big(-\frac{1}{t} + \frac{1}{\lambda-1} + \frac{1}{t(\lambda-t)}\Big) + \frac{\kappa_0}{t-1}\frac{1}{t(\lambda-t)}$$
$$+ \frac{1}{2}\frac{(\lambda-1)(\lambda-t)}{t(t-1)}\Big[\frac{{\kappa_\infty}^2}{t(t-1)} + \frac{{\kappa_1}^2}{t(\lambda-1)^2} + \frac{1-{\kappa_t}^2}{(\lambda-t)^2}\Big],$$

and κ_1, κ_∞ and κ_t are quantities such that

$$\alpha = \frac{1}{2}{\kappa_\infty}^2, \quad \gamma = \frac{1}{2}{\kappa_1}^2, \quad \delta = \frac{1}{2}\kappa_t^2.$$

Since $A_0(\lambda,t), B_0(\lambda,t)$ and $C_0(\lambda,t)$ are holomorphic at $(\lambda,t) = (0,t_0)$, Cauchy's theorem assures the existence of a solution $(\lambda(t), \mu_0(t))$ of the system (1.4.3 and 4), holomorphic at t_0, which satisfies the initial condition:

$$\lambda(t_0) = 0, \quad \mu_0(t_0) = -2\frac{t_0-1}{\kappa_0}h + \frac{1}{t_0-1}.$$

It is clear that $\lambda(t)$ thus constructed has the same power series expansion with respect to $t - t_0$ as $\hat{\lambda}_-(t)$. We can prove the convergence of $\hat{\lambda}_+(t)$ in a similar manner. ∎

In the proof, we transformed P_{VI} into the system (1.4.3 and 4) which is holomorphic at $\lambda = 0$. Notice that the system is defined for any β and that the equation P_{VI} can be recovered by eliminating μ_0. Let us find a system, for other ξ, which is holomorphic at $\lambda = \xi$; this is done by applying the symmetries of P_{VI} to the system (1.4.3 and 4), which will be called $(E)_0$.

PROPOSITION 1.4.2. *Let the parameters* $(\alpha,\beta,\gamma,\delta)$ *and* $(\kappa_0,\kappa_1,\kappa_\infty,\kappa_t)$ *be related by*

$$\alpha=\frac{1}{2}{\kappa_\infty}^2,\quad \beta=\frac{1}{2}{\kappa_0}^2,\quad \gamma=\frac{1}{2}{\kappa_1}^2,\quad \delta=\frac{1}{2}\kappa_t^2, \tag{1.4.5}$$

Then, for each $\xi\in\{0,1,t,\infty\}$, *the equation* P_{VI} *is transformed into the system of differential equations:*

$$(E)_\xi \qquad \begin{cases} \dfrac{d\lambda}{dt}=b_\xi(\lambda,t)\mu_\xi+c_\xi(\lambda,t) \\[2mm] \dfrac{d\mu_\xi}{dt}=A_\xi(\lambda,t){\mu_\xi}^2+B_\xi(\lambda,t)\mu_\xi+C_\xi(\lambda,t), \end{cases}$$

where $a_\xi,b_\xi,A_\xi,B_\xi,C_\xi$ *are rational functions in* (λ,t), *holomorphic at* $\lambda=\xi$, *given by*

$$b_0(\lambda,t)=\lambda,\quad c_0(\lambda,t)=\frac{\kappa_0}{t-1},$$

$A_0(\lambda,t),B_0(\lambda,t)$ *and* $C_0(\lambda,t)$ *are given in the proof of Proposition* 1.4.1,

$$b_1(\lambda,t)=\lambda-1,\quad c_1(\lambda,t)=\frac{\kappa_1}{t},$$

$$A_1(\lambda,t)=\frac{\lambda^2-2\lambda+t}{2\lambda(\lambda-t)},$$

$$B_1(\lambda,t)=\frac{\kappa_1}{t}\left(\frac{1}{\lambda}+\frac{1}{\lambda-t}\right)-\frac{1}{t}-\frac{1}{t-1}-\frac{1}{\lambda-t},$$

$$C_1(\lambda,t)=\frac{1}{2}\left(\frac{\kappa_1}{t}\right)^2\left(\frac{1}{t-1}-\frac{1}{\lambda}+\frac{1}{(t-1)(\lambda-t)}\right)-\frac{\kappa_1}{t}\frac{1}{(t-1)(\lambda-t)}$$
$$+\frac{1}{2}\frac{\lambda(\lambda-t)}{t(t-1)}(U_0+U_t+U_\infty),$$

$$b_t(\lambda,t) = \lambda - t, \quad c_t(\lambda,t) = -\kappa_t + \frac{\lambda-1}{t-1},$$

$$A_t(\lambda,t) = \frac{\lambda^2 - 2t\lambda + t}{2\lambda(\lambda-1)},$$

$$B_t(\lambda,t) = c_t(\lambda,t)(\frac{1}{\lambda} + \frac{1}{\lambda-1}) - \frac{1}{t} - \frac{2}{t-1},$$

$$\begin{aligned} C_t(\lambda,t) = &\frac{1}{2}c_t^2(\lambda,t)\Big(\frac{\lambda-1}{t\lambda} + \frac{\lambda}{(t-1)(\lambda-1)}\Big) \\ &- c_t(\lambda,t)\frac{1}{t-1}\Big(\frac{\lambda-2}{t} + \frac{\lambda}{t-1}\Big) \\ &+ \frac{1}{2}\frac{\lambda(\lambda-1)}{t(t-1)}\Big(U_0 + U_1 + U_\infty - \frac{2\lambda(t-1)+(1-2t)\lambda^2}{(t-1)^2\lambda^2}\Big), \end{aligned}$$

$$b_\infty(\lambda,t) = -\lambda, \quad c_\infty(\lambda,t) = \frac{\kappa_\infty}{t(t-1)}\lambda^2,$$

$$A_\infty(\lambda,t) = \frac{t-\lambda^2}{2(\lambda-t)(\lambda-1)},$$

$$B_\infty(\lambda,t) = -\frac{c_\infty(\lambda,t)}{\lambda^2}\Big(\frac{\lambda}{1-\lambda} + \frac{t\lambda}{t-\lambda}\Big) - \frac{1}{t} - \frac{1}{t-1} + \frac{1}{t-\lambda},$$

$$\begin{aligned} C_\infty(\lambda,t) = &\frac{1}{2}\frac{c_\infty^2(\lambda,t)}{\lambda^4}\Big(\frac{\lambda}{1-\lambda} + \frac{t^2\lambda}{t-\lambda} - t\Big) - \frac{c_\infty(\lambda,t)}{\lambda^2}\frac{\lambda}{t-\lambda} \\ &- \frac{1}{2}\frac{(1-\lambda)(t-\lambda)}{t(t-1)}(U_0 + U_1 + U_t), \end{aligned}$$

where

$$U_0 = -\frac{\kappa_0^2}{(t-1)\lambda^2}, \quad U_1 = \frac{\kappa_1^2}{t(\lambda-1)^2},$$

$$U_t = \frac{1-\kappa_t^2}{(\lambda-t)^2}, \quad U_\infty = \frac{\kappa_\infty^2}{t(t-1)}.$$

Proof. For example, let us derive $(E)_1$ from $(E)_0$ by applying the symmetry $\sigma_1 = (T_1, \ell_1)$ of P_{VI} given by

$$\begin{aligned} T_1 :&(\lambda, t) \to (\lambda_1, t_1) \\ &\lambda = 1 - \lambda_1, \quad t = 1 - t_1, \\ \ell_1 :&(\alpha, \beta, \gamma, \delta) \to (\alpha, \gamma, \beta, \delta). \end{aligned}$$

Note that the permutation:

$$(\kappa_\infty, \kappa_0, \kappa_1, \kappa_t) \to (\kappa_\infty, \kappa_1, \kappa_0, \kappa_t)$$

induces the transformation ℓ_1 (see 1.4.5). The equation (1.4.3) is taken into

$$\frac{d\lambda_1}{dt_1} = \frac{\kappa_1}{t_1} - (\lambda_1 - 1)\mu_0.$$

Define $\mu_1 := -\mu_0$. Then (1.4.4) is taken into

$$\begin{aligned} \frac{d\mu_1}{dt_1} =& A_0(1 - \lambda_1, 1 - t_1)\mu_1^2 \\ &- B_0(1 - \lambda_1, 1 - t_1)\mu_1 + C_0(1 - \lambda_1, 1 - t_1). \end{aligned}$$

We set

$$\begin{aligned} A_1(\lambda_1, t_1) &:= A_0(1 - \lambda_1, 1 - t_1), \\ B_1(\lambda_1, t_1) &:= -B_0(1 - \lambda_1, 1 - t_1), \\ C_1(\lambda_1, t_1) &:= C_0(1 - \lambda_1, 1 - t_1). \end{aligned}$$

Writing λ, t in place of λ_1, t_1, we see that A_1, B_1, C_1 are given explicitly as in the proposition. Since σ_1 is a symmetry of P_{VI} , the equation P_{VI} can be recovered from $(E)_1$ by eliminating μ_1. This proves the proposition for $\xi = 1$. The other systems $(E)_\xi$ can be obtained in a similar manner. ∎

1.5. Hamiltonian structure for P_{VI}

In this section we shall glue the systems $(E)_\xi$, $\xi \in \{0,1,\infty,t\}$ in order to obtain a Hamiltonian system, which will again be derived from the theory of monodromy preserving deformations in Section 4.

Notice that we are considering the dependent and independent variables (λ, t) of P_{VI} in $\mathbb{P}^1 \times B$, where $B = \mathbb{P}^1 \setminus \{0,1,\infty\}$. Since the right-hand sides of the system $(E)_\xi$ are polynomials in μ_ξ and are holomorphic at $\lambda = \xi$, the space of dependent variables and the independent variable (μ_ξ, λ, t) of the system $(E)_\xi$ can be considered to be

$$M^\xi := \mathbb{C} \times N^\xi \subset \mathbb{C} \times \mathbb{P}^1 \times B,$$

where

$$N^\xi := \{(\lambda, t) \in \mathbb{P}^1 \times B; \lambda \notin \{0,1,\infty,t\} \setminus \{\xi\}\} \subset \mathbb{P}^1 \times B.$$

Let us look at the relation between the $(E)_\xi$'s. Let $\lambda(t)$ be a solution of P_{VI} . Define $\mu_\xi(t)$ and $\mu_\eta(t)$ so that $(\mu_\xi(t), \lambda(t))$ and $(\mu_\eta(t), \lambda(t))$ are solutions of the systems $(E)_\xi$ and $(E)_\eta$, respectively, where $\xi, \eta = 0, 1, \infty, t$. Then they relate on a common domain of definition as

$$b_\xi(\lambda(t), t)\mu_\xi(t) + c_\xi(\lambda(t), t) = b_\eta(\lambda(t), t)\mu_\eta(t) + c_\eta(\lambda(t), t).$$

or equivalently

$$\mu_\xi(t) = g_{\xi\eta}(\lambda(t), t)\mu_\eta(t) + f_{\xi\eta}(\lambda(t), t),$$

where

$$g_{\xi\eta}(\lambda, t) = \frac{b_\eta(\lambda, t)}{b_\xi(\lambda, t)}, \quad f_{\xi\eta}(\lambda, t) = \frac{c_\eta(\lambda, t) - c_\xi(\lambda, t)}{b_\xi(\lambda, t)}. \tag{1.5.1}$$

and $a_\xi(\lambda, t)$, $b_\xi(\lambda, t)$ are given in Proposition 1.4.2. Notice that the transition functions $g_{\xi\eta}$ and $f_{\xi\eta}$ satisfy the compatibility condition:

$$\begin{aligned} g_{\xi\zeta} &= g_{\xi\eta} g_{\eta\zeta}, \\ f_{\xi\zeta} &= f_{\xi\eta} + g_{\xi\eta} f_{\eta\zeta}. \end{aligned} \tag{1.5.2}$$

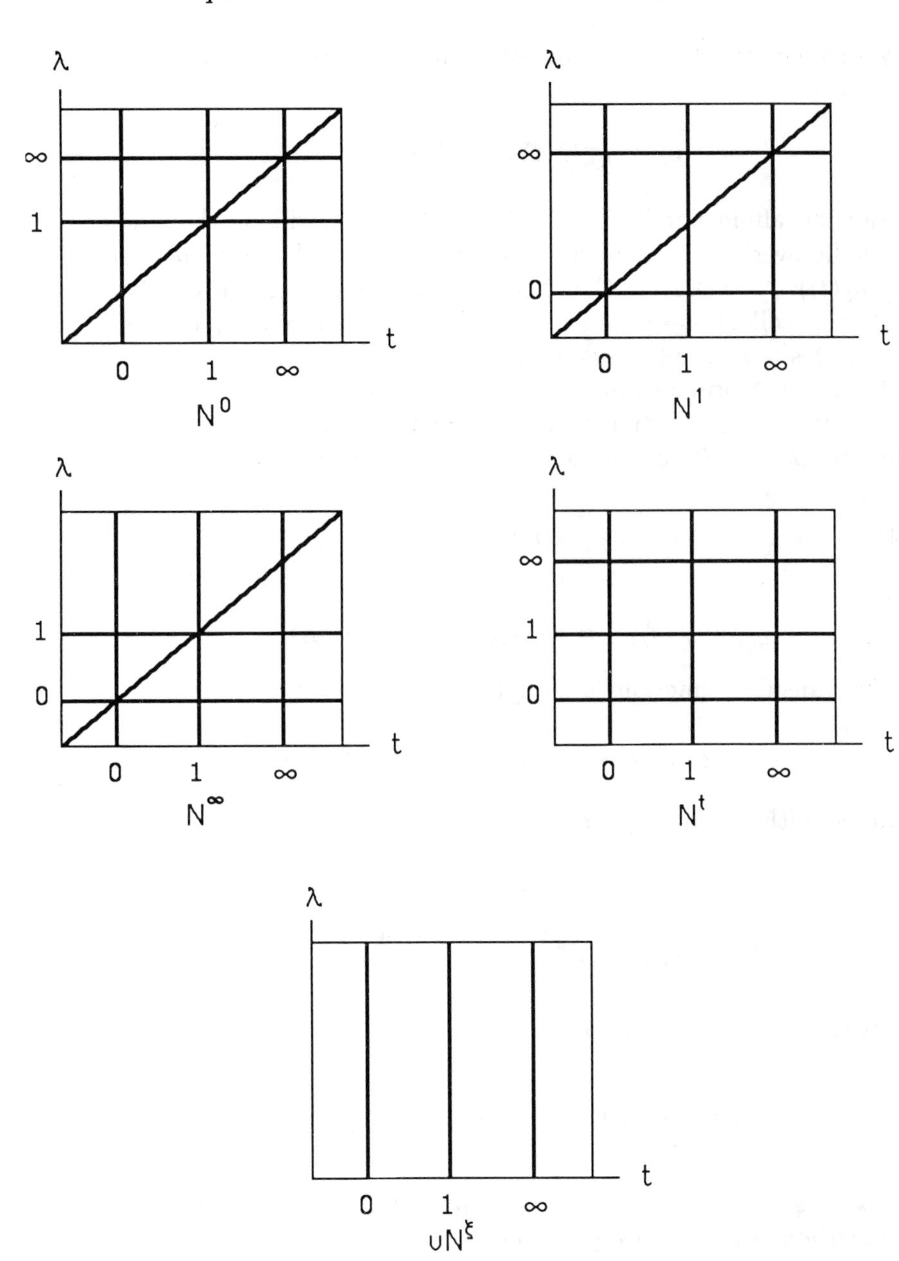

Figure

Motivated by the above observation, we patch four M^ξ ($\xi = 0, 1, \infty, t$) by

$$\mu_\xi = g_{\xi\eta}(\lambda, t)\mu_\eta + f_{\xi\eta}(\lambda, t) \tag{1.5.3}$$

to obtain an affine bundle Σ over $\mathbb{P}^1 \times B$, which will be thought of as a bundle over B in the obvious manner. Notice that the collection $\{(\lambda(t), \mu_\xi(t));\ \xi = 0, 1, \infty, t\}$ defines a local section of the bundle Σ over B. The collection of the systems $(E)_\xi$ ($\xi = 0, 1, \infty, t$), patched by (1.5.3), yields a differential system (E) on Σ.

We shall perform a change of fibre coordinates of Σ over $\mathbb{P}^1 \times B$ in order to make the transition functions independent of t. If we change coordinates μ_ξ on M^ξ into μ'_ξ by an affine transformation:

$$\mu_\xi{}' = g_\xi(\lambda, t)\mu_\xi + f_\xi(\lambda, t), \tag{1.5.4}$$

where

$$g_\xi(\lambda, t),\ f_\xi(\lambda, t) \in \mathcal{O}(N^\xi), \quad g_\xi(\lambda, t) \neq 0,$$

then the transition functions $g_{\xi\eta}{}'$ and $f_{\xi\eta}{}'$, defined by

$$\mu_\xi{}' = g_{\xi\eta}{}'(\lambda, t)\mu_\eta{}' + f_{\xi\eta}{}'(\lambda, t) \quad \text{on } M^\xi \cap M^\eta,$$

are related with $g_{\xi\eta}$ and $f_{\xi\eta}$ as follows:

$$\begin{aligned} &g_\xi{}^{-1} g_{\xi\eta}{}' g_\eta = g_{\xi\eta}, \\ &g_\xi{}^{-1} g_{\xi\eta}{}' f_\eta + g_\xi{}^{-1} f_{\xi\eta}{}' - g_\xi{}^{-1} f_\xi = f_{\xi\eta}. \end{aligned} \tag{1.5.5}$$

PROPOSITION 1.5.1. *Suppose*

$$\varepsilon := \frac{1}{2}(\kappa_0 + \kappa_1 + \kappa_t + \kappa_\infty - 1) \neq 0.$$

There is uniquely a rational change of coordinates (1.5.4) *such that the new transition functions are given by*

$$\begin{aligned} &g_{\xi\eta} = 1, && f_{\xi,\eta} = 0 && (\xi, \eta \neq \infty), \\ &g_{\xi\infty} = -1/\lambda^2, && f_{\xi\infty} = \varepsilon/\lambda && (\xi \neq \infty). \end{aligned} \tag{1.5.6}$$

Proof. Suppose that there exists a change of coordinates (1.5.4) which takes the transition functions (1.5.1) into (1.5.6). Notice that, from the expression of a_ξ and b_ξ given in Proposition 1.4.2 and the definition (1.5.1) of $g_{\xi\eta}$ and $f_{\xi\eta}$, the transition functions $g_{0\xi}$ and $f_{0\xi}$ ($\xi = 1, t, \infty$) are given by

$$
\begin{aligned}
g_{01} &= \frac{\lambda - 1}{\lambda}, & f_{01} &= \frac{\kappa_0 t + \kappa_1 (t-1)}{\lambda t (t-1)}, \\
g_{0t} &= \frac{\lambda - t}{\lambda}, & f_{0t} &= \frac{\kappa_0 - \kappa_t (t-1) + \lambda - 1}{\lambda (t-1)}, \\
g_{0\infty} &= -1, & f_{0\infty} &= \frac{\kappa_0 t + \kappa_\infty \lambda^2}{\lambda t (t-1)}.
\end{aligned}
\tag{1.5.7}
$$

First we determine the g_ξ's. From the relation (1.5.5), we have

$$
\begin{aligned}
g_1 &= g_0 g_{01} = \frac{\lambda - 1}{\lambda} g_0, \\
g_t &= g_0 g_{0t} = \frac{\lambda - t}{\lambda} g_0, \\
g_\infty &= (g'_{0\infty})^{-1} g_0 g_{0\infty} = \frac{\lambda - 1}{\lambda} g_0.
\end{aligned}
\tag{1.5.8}
$$

We show that g_ξ ($\xi = 0, 1, \infty, t$) can be written as

$$
\begin{aligned}
g_0 &= \frac{Q(t)}{(\lambda - 1)(\lambda - t)}, \\
g_1 &= \frac{Q(t)}{\lambda(\lambda - t)}, \\
g_t &= \frac{Q(t)}{\lambda(\lambda - 1)}, \\
g_\infty &= \frac{Q(t)\lambda^2}{(\lambda - 1)(\lambda - t)},
\end{aligned}
\tag{1.5.9}
$$

where $Q(t)$ is a rational function in t. Notice that g_ξ is a rational function in (λ, t) that is holomorphic in N^ξ and nowhere vanishing.

Since g_0 has poles at most at $\{(\lambda, t) \in \mathbb{P}^1 \times B; \lambda = 1, t, \infty\}$ and g_∞ is holomorphic at $\lambda = \infty$, the third relation of (1.5.8) says that g_0 has the form

$$g_0 = \frac{Q(\lambda, t)}{(\lambda - 1)^p (\lambda - t)^q},$$

where p and q are integers and $Q(\lambda, t) \in \mathbb{C}(t)[\lambda]$ such that $\deg_\lambda Q(\lambda, t) \leq p+q-2$ and $Q(1,t)\cdot Q(t,t) \neq 0$. Putting this expression into the relations (1.5.8), we have

$$g_1 = \frac{Q(\lambda, t)}{\lambda(\lambda - 1)^{p-1}(\lambda - t)^q},$$

$$g_t = \frac{Q(\lambda, t)}{\lambda(\lambda - 1)^p(\lambda - t)^{q-1}}$$

The holomorphy of g_1 at $\lambda = 1$ and of g_t at $\lambda = t$ and $g_1(1,t)\cdot g_t(t,t) \neq 0$ lead to $p = q = 1$. Since $\deg_\lambda Q(\lambda, t) \leq p + q - 2 = 0$, we have

$$Q(\lambda, t) = Q(t) \in \mathbb{C}(t),$$

which proves (1.5.9).

Let us determine f_0, f_1, f_t and f_∞. We assert that

$$Q(t) = \frac{1}{2} t(t-1),$$

$$f_0 = \frac{\kappa_0(\lambda - t - 1) + \kappa_1(\lambda - t) + (\kappa_t - 1)(\lambda - 1)}{2(\lambda - 1)(\lambda - t)},$$

$$f_1 = \frac{\kappa_0(\lambda - t) + \kappa_1(\lambda - t + 1) + (\kappa_t - 1)\lambda}{2\lambda(\lambda - t)}, \tag{1.5.10}$$

$$f_t = \frac{\kappa_0(\lambda - 1) + \kappa_1\lambda + (\kappa_t - 1)(\lambda - 1 + t) + t}{2\lambda(\lambda - 1)},$$

$$f_\infty = \frac{\kappa_1(\lambda - t) + (\kappa_t - 1)(\lambda - 1) + \kappa_\infty\{(\lambda - t)t + \lambda\}}{2(\lambda - 1)(\lambda - t)}.$$

By using the expression (1.5.7) of $f_{0\xi}$ and the expression (1.5.8) of g_ξ, the relation (1.5.4) between the new and the old transition functions gives the relation

$$
\begin{aligned}
f_1 &= f_0 + g_0 f_{01} \\
&= f_0 + \frac{Q(t)}{\lambda(\lambda-1)(\lambda-t)} \frac{\kappa_0 t + \kappa_1(t-1)}{t(t-1)}, \\
f_t &= f_0 + g_0 f_{0t} \\
&= f_0 + \frac{Q(t)}{\lambda(\lambda-1)(\lambda-t)} \frac{\kappa_0 - \kappa_t(t-1) + \lambda - 1}{t(t-1)}, \\
f_\infty &= (g'_{0\infty})^{-1}(f_0 + g_0 f_{0\infty} - f'_{0\infty}) \\
&= -\lambda^2 \left(f_0 + \frac{Q(t)}{\lambda(\lambda-1)(\lambda-t)} \frac{\kappa_0 t + \kappa_\infty \lambda^2}{t(t-1)} - \frac{\varepsilon}{\lambda} \right).
\end{aligned}
\tag{1.5.11}
$$

Since f_1 and f_t are holomorphic at $\lambda = 1$ and at $\lambda = t$, respectively, the first and the second relations of (1.5.11) say that f_0 has simple poles at $\lambda = 1$ and at $\lambda = t$, and that the residues there are given by

$$
c_1 := \operatorname*{Res}_{\lambda=1} f_0 \, d\lambda = \frac{Q(t)}{t(t-1)^2}\{\kappa_0 t + \kappa_1(t-1)\},
$$

$$
c_t := \operatorname*{Res}_{\lambda=t} f_0 \, d\lambda = \frac{Q(t)}{t(t-1)^2}\{-\kappa_0 + (\kappa_t - 1)(t-1)\}.
$$

The holomorphy of f_∞ at $\lambda = \infty$ says that f_0 has simple zero at $\lambda = \infty$, which implies

$$
\begin{aligned}
f_0 &= \frac{c_1}{\lambda - 1} + \frac{c_t}{\lambda - t} \\
&= \frac{Q(t)}{t(t-1)^2} \left\{ \frac{\kappa_0 t + \kappa_1(t-1)}{\lambda - 1} + \frac{-\kappa_0 + (\kappa_t - 1)(t-1)}{\lambda - t} \right\}
\end{aligned}
\tag{1.5.12}
$$

and that

$$
c_1 + c_t + \frac{\kappa_\infty Q(t)}{t(t-1)} - \varepsilon = 0.
$$

The last relation can be written as

$$
\varepsilon \left\{ \frac{Q(t)}{t(t-1)} - 1 \right\} = 0,
$$

and it determines $Q(t)$ as in (1.5.10) because of $\varepsilon \neq 0$. Putting the expressions of $Q(t)$ and f_0 into (1.5.11), we obtain the expressions (1.5.10) of f_ξ ($\xi = 0, 1, \infty, t$). Thus we have shown that, if there is a transformation (1.5.4) changing the transition function (1.5.1) into (1.5.6), then g_ξ and f_ξ are given by (1.5.9) and (1.5.10). Conversely if g_ξ and f_ξ are given by (1.5.9) and (1.5.10), it is immediate to show that the new transition functions defined by (1.5.4) are given by (1.5.6). ∎

Let us define an affine bundle Σ_ε over $\mathbb{P}^1$ with fibre $\mathbb{C}$ as follows: Let $z = (z_0, z_1)$ be homogeneous coordinates of $\mathbb{P}^1$ and let $U_i = \{z \in \mathbb{P}^1; z_i \neq 0\}$ $(i = 0, 1)$ be affine charts of $\mathbb{P}^1$ with coordinates $\lambda = z_1/z_0$ in U_0 and $u = z_0/z_1$ in U_1. We patch $U_0 \times \mathbb{C}$ and $U_1 \times \mathbb{C}$ by identifying two points $(\lambda, \mu) \in U_0 \times \mathbb{C}$ and $(\lambda', \mu') \in U_1 \times \mathbb{C}$ if

$$\lambda\lambda' = 1, \quad \mu' = \varepsilon\lambda - \lambda^2\mu. \tag{1.5.13}$$

This defines an affine bundle Σ_ε over $\mathbb{P}^1$. This can also be defined as

$$\Sigma_\varepsilon = \{(z, z'; \mu, \mu') \in \mathbb{P}^1 \times \mathbb{C}^2 | z^2\mu' - \varepsilon z z' + z'^2\mu = 0\}.$$

The proposition above tells us that the bundle Σ over B can be thought of the direct product $\Sigma = \Sigma_\varepsilon \times B$. Let us express the differential system (E) on Σ, no matter whether or not $\varepsilon = 0$, in terms of the coordinates (λ, μ, t) on $U_0 \times \mathbb{C} \times B$ as

$$\begin{cases} \dfrac{d\lambda}{dt} = F(\lambda, \mu, t), \\[2ex] \dfrac{d\mu}{dt} = G(\lambda, \mu, t). \end{cases} \tag{1.5.14}$$

Since the right-hand sides of $(E)_\xi$ are rational functions in (λ, μ, t) holomorphic in M^ξ, rational functions $F(\lambda, \mu, t)$ and $G(\lambda, \mu, t)$ must be holomorphic in $U_0 \times \mathbb{C} \times B$. It follows that F and G are polynomials in (λ, μ) with coefficients in $\mathbb{C}(t)$. The actual forms of F and G are known by computations, and we have the desired result ([Okm.6]):

THEOREM 1.5.2. *Suppose that a parameter $(\alpha, \beta, \gamma, \delta)$ is related with $(\kappa_0, \kappa_1, \kappa_\infty, \kappa_t)$ by (1.4.4) . Then, the sixth Painlevé equation is equivalent to the Hamiltonian system, called the sixth Painlevé system:*

$$\mathcal{H}_{VI}: \quad \begin{cases} \dfrac{d\lambda}{dt} = \dfrac{\partial H_{VI}}{\partial \mu}, \\[2ex] \dfrac{d\mu}{dt} = -\dfrac{\partial H_{VI}}{\partial \lambda} \end{cases}$$

with the Hamiltonian

$$H_{VI} := \frac{1}{t(t-1)}\big[\lambda(\lambda-1)(\lambda-t)\mu^2 - \{\kappa_0(\lambda-1)(\lambda-t) + \kappa_1\lambda(\lambda-t) + (\kappa_t-1)\lambda(\lambda-1)\}\mu + \kappa(\lambda-t)\big],$$

where

$$\kappa = \frac{1}{4}\big[(\kappa_0+\kappa_1+\kappa_t-1)^2 - \kappa_\infty^2\big].$$

The transformation $(\lambda,\mu) \to (\lambda',\mu')$ defined by (1.5.13) extends to a symplectic transformation

$$(\lambda,\mu,t,H_{VI}) \to (\lambda',\mu',t,H_{VI}{}'),$$

where

$$H_{VI}{}' = H_{VI}(\lambda'^{-1}, \varepsilon\lambda' - \lambda'^2\mu', t).$$

We can see that $H_{VI}{}'$ is a polynomial in (λ',μ').

COROLLARY 1.5.3. *The Hamiltonian system $\mathcal{H}_{VI}$ extends to the Hamiltonian system on $\Sigma_\varepsilon \times B$ on which the Hamiltonian is holomorphic.*

REMARK 1.5.4 When $\varepsilon = 0$, the manifold Σ_ε has a natural section $\mu = \mu' = 0$ and is isomorphic to the cotangent bundle $T^*\mathbb{P}^1$. The compactification of the manifold Σ_ε plays important roles in constructing the space of initial values for the $\mathrm{P_J}$'s, see [Okm.1].

For the other Painlevé equations $\mathrm{P_J}$ (J = I, $\cdots$,V), we state, without giving proofs, results corresponding to Theorem 1.5.2.

THEOREM 1.5.5. *The Painlevé equation* $\mathrm{P_J}$ (J=I,...,V) *is equivalent to the Hamiltonian system, called the* J*-th Painlevé system,:*

$$\mathcal{H}_J : \quad \begin{cases} \dfrac{d\lambda}{dt} = \dfrac{\partial H_J}{\partial \mu}, \\[2mm] \dfrac{d\mu}{dt} = -\dfrac{\partial H_J}{\partial \lambda} \end{cases}$$

with the Hamiltonian $H_J \in \mathbb{C}(t)[\lambda, \mu]$ given as follows:

$$H_I = \frac{1}{2}\mu^2 - 2\lambda^3 - t\lambda;$$

$$H_{II} = \frac{1}{2}\mu^2 - (\lambda^2 + \frac{t}{2})\mu - (\alpha + \frac{1}{2})\lambda;$$

$$H_{III} = \frac{1}{t}\big[2\lambda^2\mu^2 - \{2\eta_\infty t\lambda^2 + (2\kappa_0 + 1)\lambda - 2\eta_0 t\}\mu + \eta_\infty(\kappa_0 + \kappa_\infty)t\lambda\big],$$

where

$$\alpha = -4\eta_\infty\kappa_\infty, \quad \beta = 4\eta_0(\kappa_0 + 1), \quad \gamma = 4{\eta_\infty}^2, \quad \delta = -4{\eta_0}^2;$$

$$H_{IV} = 2\lambda\mu^2 - \{\lambda^2 + 2t\lambda + 2\kappa_0\}\mu + \kappa_\infty\lambda,$$

where

$$\alpha = -\kappa_0 + 2\kappa_\infty + 1, \quad \beta = -2{\kappa_0}^2;$$

$$H_V = \frac{1}{t}\big[\lambda(\lambda - 1)^2\mu^2 - \{\kappa_0(\lambda - 1)^2 + \kappa_t\lambda(\lambda - 1) - \eta t\lambda\}\mu + \kappa(\lambda - 1)\big],$$

where

$$\alpha = \frac{1}{2}{\kappa_\infty}^2, \quad \beta = -\frac{1}{2}{\kappa_0}^2,$$

$$\gamma = -\eta(1 + \kappa_t), \quad \delta = -\frac{1}{2}\eta^2,$$

$$\kappa = \frac{1}{4}(\kappa_0 + \kappa_t)^2 - \frac{1}{4}{\kappa_\infty}^2.$$

The limit processes given in Section 1.2 can be translated into those for the Hamiltonian systems. For example, we shall derive $\mathcal{H}_V$ from $\mathcal{H}_{VI}$.

Replace, in H_{VI}, variables and parameters $\lambda, \mu, t, \kappa_1$ and κ_t by $\lambda^*, \mu^*, 1 + \epsilon t^*, \eta^*\epsilon^{-1} + \kappa_t^* + 1$ and $-\eta^*\epsilon^{-1}$, respectively, and put $H(\epsilon) = \epsilon H_{VI}$. Then we have

$$H(\epsilon) = \frac{1}{t^*(1+\epsilon t^*)}\big[\lambda^*(\lambda^*-1)(\lambda^*-1-\epsilon t^*)\mu^{*2}$$
$$-\{\kappa_0(\lambda^*-1)(\lambda^*-1-\epsilon t^*) + \kappa_t^*\lambda^*(\lambda^*-1)$$
$$-(\eta^* + \epsilon\kappa_t^* + \epsilon)t^*\lambda^*\}\mu^*$$
$$+\{\frac{1}{4}(\kappa_0+\kappa_t^*)^2 - \frac{1}{4}\kappa_\infty^2\}(\lambda^*-1-\epsilon t^*)\big].$$

Notice that the change of variables is a symplectic transformation, i.e.:

$$d\lambda \wedge d\mu + dH \wedge dt = d\lambda^* \wedge d\mu^* + dH(\epsilon) \wedge dt^*.$$

Letting ϵ tends to zero in $H(\epsilon)$, and writing $(\lambda, \mu, t, \kappa_t, \eta)$ in place of $(\lambda^*, \mu^*, t^*, \kappa_t^*, \eta^*)$, we get the Hamiltonian H_V. For the sake of simplicity, we write the above process as

$$H_{VI} \to H_V : \quad \begin{cases} (\lambda, \mu, H_{VI}, t) \to (\lambda, \mu, \epsilon^{-1}H_V, 1+\epsilon t), \\ \kappa_1 \to \eta\epsilon^{-1} + \kappa_t + 1, \quad \kappa_t \to -\eta\epsilon^{-1}, (\epsilon \to 0). \end{cases}$$

PROPOSITION 1.5.5. *The Hamiltonian systems $\mathcal{H}_J$ ($J = I, ..., V$) are obtained from $\mathcal{H}_{VI}$ by successive limit processes according to the diagram:*

$$\mathcal{H}_{VI} \to \mathcal{H}_V \begin{matrix} \nearrow \mathcal{H}_{III} \searrow \\ \searrow \mathcal{H}_{IV} \nearrow \end{matrix} \mathcal{H}_{II} \to \mathcal{H}_I.$$

where $\mathcal{H}_J \to \mathcal{H}_{J'}$ denotes a symplectic transformation, containing a parameter ϵ, followed by the limit $\epsilon \to 0$. The actual limit processes are given as follows:

$$H_V \to H_{IV} : \begin{cases} (\lambda, \mu, H_V + \kappa, t) \to (\frac{\epsilon}{\sqrt{2}}\lambda, \frac{\sqrt{2}}{\epsilon}\mu, \frac{1}{\sqrt{2}\epsilon}H_{IV}, 1+\sqrt{2}\epsilon t), \\ \eta \to -\epsilon^{-2}, \quad \kappa_t \to \epsilon^{-2} + 2\kappa_\infty - \kappa_0, \quad \kappa_\infty \to \epsilon^{-2}, \\ (\epsilon \to 0), \end{cases}$$

$$H_{IV} \to H_{II} : \quad \begin{cases} (\lambda, \mu, H_{IV}, t) \to (\epsilon^{-3}(1+2^{2/3}\epsilon^2\lambda), 2^{-2/3}\epsilon\mu, \\ \quad 2^{2/3}\epsilon^{-1}H_{II} - \frac{1}{2}(2\alpha+1)\epsilon^{-3}, -\epsilon^{-3} + 2^{-2/3}\epsilon t), \\ \kappa_0 \to \frac{1}{2}\epsilon^{-6}, \quad \kappa_\infty \to -\frac{1}{2}(2\alpha+1), \quad (\epsilon \to 0), \end{cases}$$

$H_{II} \to H_I$:

$$\begin{cases} (\lambda, \mu - \lambda^2 - \frac{1}{2}t, H_{II} + \frac{1}{2}\lambda + \frac{1}{8}t^2, t) \\ \to (\epsilon^{-5} + \epsilon\lambda, \epsilon^{-1}\mu, \epsilon^{-2}H_I - \frac{1}{2}t\epsilon^{-8} - \frac{3}{2}\epsilon^{-20}, -6\epsilon^{-10} + \epsilon^2 t), \\ \alpha \to 4\epsilon^{-15}, \quad (\epsilon \to 0), \end{cases}$$

$$H_V \to H_{III'} : \qquad \begin{cases} (\lambda, \mu, H_V, t) \to (1 + \epsilon\lambda, \epsilon^{-1}\mu, H_{III'}, t), \\ \kappa_0 \to \epsilon^{-1}\eta_\infty, \quad \eta \to \epsilon\eta_0, \quad \kappa_t \to \kappa_0, \\ \kappa_\infty \to \eta_\infty\epsilon^{-1} - \kappa_\infty, \quad (\epsilon \to 0). \end{cases}$$

The Hamiltonian $H_{III'}$ is given by

$$H_{III'} = \frac{1}{t}[\lambda^2\mu^2 - \{\eta_\infty\lambda^2 + \kappa_0\lambda - \eta_0 t\}\mu + \frac{1}{2}\eta_\infty(\kappa_0 + \kappa_\infty)\lambda].$$

The Hamiltonian system $\mathcal{H}_{III}$ is obtained from $\mathcal{H}_{III'}$ by the symplectic transformation

$$H_{III'} \to H_{III} : \qquad (\lambda, \mu, H_{III'}, t) \to (t\lambda, t^{-1}\mu, \frac{1}{2t}(H_{III} + \frac{\lambda\mu}{t}), t^2),$$

$H_{III} \to H_{II}$:

$$\begin{cases} (\lambda, \mu, H_{III} - \eta_\infty(\kappa_0 + \kappa_\infty), t) \to (1 + 2\epsilon\lambda, \frac{1}{2}\epsilon^{-1}\mu, \epsilon^{-2}H_{II}, 1 + \epsilon^2 t), \\ \eta_0 \to -\frac{1}{4}\epsilon^{-3}, \quad \eta_\infty \to \frac{1}{4}\epsilon^{-3}, \\ \kappa_0 \to -\frac{1}{2}\epsilon^{-3} - 2\alpha - 1, \quad \kappa_\infty \to -\frac{1}{2}\epsilon^{-3}, \quad (\epsilon \to 0). \end{cases}$$

1.6. Particular solutions of the systems P_J

We shall show that, when the parameters in $\mathcal{H}_J$ are restricted to a certain hyperplanes in V_J, each Painlevé system $\mathcal{H}_J$ (J=II,...,VI) admits particular solutions which are expressed by classical special functions.

We first treat $\mathcal{H}_{VI}$.

PROPOSITION 1.6.1. *Suppose that the parameters $(\kappa_0, \kappa_1, \kappa_\infty, \theta)$ in $\mathcal{H}_{VI}$ satisfy the conditions:*

$$\kappa = \frac{1}{4}(\kappa_0 + \kappa_1 + \theta - 1 + \kappa_\infty)(\kappa_0 + \kappa_1 + \theta - 1 - \kappa_\infty) = 0,$$

$$A := \kappa_0 + \kappa_1 + \theta - 1 \neq 0,$$

then the system $\mathcal{H}_{VI}$ has solutions $(\lambda(t), \mu(t))$ of the form

$$\begin{cases} \lambda(t) = A^{-1} t(t-1)\dfrac{d}{dt}\log((t-1)^{\kappa_0} u(t)), \\ \mu(t) = 0, \end{cases}$$

where $u(t)$ is an arbitrary solution of the Gauss hypergeometric equation $E(\alpha, \beta, \gamma)$ with

$$\alpha = 1 - \kappa_1, \quad \beta = \theta + 1, \quad \gamma = \kappa_0 + \theta - 1. \tag{1.6.1}$$

Proof. Since $\kappa = 0$, the right-hand side of the second equation of $\mathcal{H}_{VI}$ is obviously satisfied if we put $\mu = 0$. The first equation of $\mathcal{H}_{VI}$ then becomes

$$t(t-1)\frac{d\lambda}{dt} = -\kappa_0(\lambda - 1)(\lambda - t) - \kappa_1 \lambda(\lambda - t) - (\theta - 1)\lambda(\lambda - 1). \tag{1.6.2}$$

Notice that this is the Riccati equation. Apply Proposition 1.1.2 to (1.6.2) by introducing a new dependent variable u:

$$\lambda = A^{-1} t(t-1)\frac{d}{dt}\log((t-1)^{\kappa_0} u(t));$$

and we obtain the Gauss hypergeometric equation $E(\alpha,\beta,\gamma)$:

$$t(1-t)\frac{d^2u}{dt^2}+\{\gamma-(\alpha+\beta+1)t\}\frac{du}{dt}-\alpha\beta u=0,$$

with parameters α, β and γ given in (1.6.1). ∎

For other Painlevé systems $\mathcal{H}_J$ $(J=II,\cdots,V)$, we obtain a similar result to the one above.

PROPOSITION 1.6.2. *If the parameters in H_J satisfy a condition so that H_J is divisible by μ, then $\mathcal{H}_J$ $(J=II,\cdots,V)$ has particular solutions $(\lambda,\mu)=(\lambda_J(t),0)$ given by*

$$\lambda_{II}(t)=\frac{d}{dt}\log u_{II}(t), \qquad \text{if } \alpha+\tfrac{1}{2}=0,$$

$$\lambda_{III}(t)=\frac{d}{dt}\log(t^{\kappa_0}u_{III}(t)), \qquad \text{if } \eta_0=\eta_\infty=\tfrac{1}{2},\ \kappa_0+\kappa_\infty=0,$$

$$\lambda_{IV}(t)=\frac{d}{dt}\log u_{IV}(t), \qquad \text{if } \kappa_\infty=0,$$

$$\lambda_{V}(t)=(\kappa_0+\theta)^{-1}\frac{d}{dt}\log u_{V}(t)+1, \text{ if } \eta=-1,\ \kappa=0,\ \kappa_0+\theta\neq 0,$$

where $u_J(t)$ satisfies the second order linear differential equation LP_J:
LP_{II} (Airy's equation)

$$\frac{d^2u}{dt^2}+\frac{1}{2}tu=0,$$

LP_{III} (Bessel's equation)

$$t^2\frac{d^2u}{dt^2}+t\frac{du}{dt}+(t^2-{\kappa_0}^2)u=0,$$

LP_{IV} (Hermite's equation)

$$\frac{d^2u}{dt^2}-2t\frac{du}{dt}+2\kappa_0u=0,$$

LP_V (Laguerre's equation)

$$t\frac{d^2u}{dt^2} + (\theta + 1 + t)\frac{du}{dt} + (\kappa_0 + \theta)u = 0.$$

REMARK 1.6.3. In Section 9, we shall show that under certain conditions on parameters, the Hamiltonian system $\mathcal{H}_n$ (a generalization of $\mathcal{H}_{VI}$ in n variables) admits solutions expressible in terms of solutions of the Lauricella hypergeometric equation $E_D(\alpha, \beta_1, \cdots, \beta_n, \gamma)$, a generalization of the Gauss hypergeometric equation in n variables.

2 The Riemann-Hilbert Problem for second order linear differential equations

We are principally interested in nonlinear differential equations which govern the monodromy preserving deformation of linear ordinary differential equations. This section poses a problem which will be discussed in the following sections.

2.1. Spaces of Fuchsian differential equations and those of representations of π_1

We consider a second order linear differential equation defined on the Riemann sphere $\mathbb{P}^1 := \mathbb{C} \cup \{\infty\}$

$$D^2 y + p_1(x) D y + p_2(x) y = 0, \quad D = d/dx, \tag{2.1.1}$$

where $p_1(x)$ and $p_2(x)$ are rational functions in $x \in \mathbb{C}$. Let $S = \{a_1, \cdots, a_m, a_{m+1} = \infty\}$ be a subset of $\mathbb{P}^1$, and let x_0 be a point in $X := \mathbb{P}^1 - S$.

DEFINITION 2.1.1. A differential equation (2.1.1) is said to be *d-reducible* if the differential operator $L := D^2 + p_1(x)D + p_2(x)$ decomposes into the product of two operators of the first order:

$$L = (D + q(x)) \cdot (D + r(x)), \quad q(x),\ r(x) \in \mathbb{C}(x); \tag{2.1.2}$$

otherwise it is said to be *d-irreducible.*

If the operator L is decomposed as in (2.1.2), we have

$$\begin{aligned} p_1(x) &= q(x) + r(x), \\ p_2(x) &= q(x) r(x) + \frac{dr(x)}{dx}. \end{aligned} \tag{2.1.3}$$

PROPOSITION 2.1.2. *A Fuchsian differential equation (2.1.1) with singular points in S has irreducible monodromy if and only if it is d-irreducible.*

Proof. Assume that the monodromy is reducible. There is a fundamental system $\mathcal{Y}(x) = (y_1(x), y_2(x))$ of solutions such that the monodromy representation $\rho : \pi_1(X, x_0) \to GL(2, \mathbb{C})$, with respect to $\mathcal{Y}(x)$, has the form

$$\rho(\gamma) = \begin{pmatrix} c_1(\gamma) & 0 \\ c_2(\gamma) & c_3(\gamma) \end{pmatrix}, \quad \gamma \in \pi_1(X - S, x_0).$$

It follows that $r(x) := -Dy_2 \cdot y_2^{-1}$ is single valued in X. Furthermore, since (2.1.1) is Fuchsian, we see that $Dy_2 \cdot y_2^{-1}$ has at most poles in S and hence it is a rational function. We divide the differential operator L by $D + r(x)$:

$$L = (D + q(x)) \cdot (D + r(x)) + R(x),$$

where $q(x)$ and $R(x)$ are rational functions in x. Since $L \cdot y_2 = 0$ and $(D + r(x))y_2 = 0$, we see that $R(x) = 0$. Conversely, if the equation is d-reducible and is decomposed as in (2.1.2), the space of solutions of the equation $(D + r(x))y = 0$ forms a proper ρ-invariant space. ∎

From now on we say simply "*(ir)reducible*" in place of "d-(ir)reducible". Since we treat only Fuchsian equations of second order, by virtue of the proposition, this convention is compatible with Definition 4.3.1 in Chapter 2.

We introduce the following notation.

$\tilde{\mathcal{E}}(S) :=$ {second order Fuchsian differential equations having singular points at most in S},
$\mathcal{E}(S) :=$ { second order irreducible Fuchsian differential equations having singular points at most in S},
$\mathcal{M}(S) :=$ {conjugacy classes of irreducible linear representations of $\pi_1(X, x_0)$ of rank 2 }.

Proposition 4.2 in Chapter 1 tells us that $\tilde{\mathcal{E}}(S)$ admits a natural structure of affine space, and it is easy to see that $\mathcal{E}(S)$ is an open dense subset of $\tilde{\mathcal{E}}(S)$. It is known [Gunn, Theorem 27] that $\mathcal{M}(S)$ is a complex manifold. So that we can speak of their complex dimensions.

$e(S) :=$ complex dimension of the manifold $\mathcal{E}(S)$ $(= \dim \tilde{\mathcal{E}}(S))$;
$m(S) :=$ complex dimension of the manifold $\mathcal{M}(S)$.

We shall compute the numbers $e(S)$ and $m(S)$.

PROPOSITION 2.1.3. $e(S) = 3m - 1$.

Proof. By Chapter 1, Proposition 4.2, we have

$$\begin{aligned} e(S) &= \sum_{j=1,2} \{j(m-1) + 1\} \\ &= 3m - 1. \end{aligned}$$ ∎

PROPOSITION 2.1.4. $m(S) = 4m - 3$.

Proof. The fundamental group $\pi_1(X, x_0)$ of $X = \mathbb{P}^1 \setminus S$ with base point x_0 is a free group generated by m elements, say, $\gamma_1, \cdots, \gamma_m$. Thus any representation $\rho : \pi_1(X, x_0) \to GL(2, \mathbb{C})$ is determined by the images of the γ_j's, so by m arbitrary 2×2 nonsingular matrices. The group $GL(2, \mathbb{C})$ acts on the space of representations by the adjoint action :

$$\rho \to g\rho g^{-1}, \quad g \in GL(2, \mathbb{C}).$$

If ρ is an irreducible representation, then Schur's lemma (see Remark 2.1.5) asserts that the adjoint action is trivial if and only if $g = cI_2$, $c \in \mathbb{C}^\times$ (I_2 is the 2×2 identity matrix). Hence the quotient group $GL(2, \mathbb{C})$ $/\{cI_2 | c \in \mathbb{C}^\times\}$ acts freely on the open set of $GL(2, \mathbb{C})^m$ consisting of m-tuples $(g_1, \cdots, g_m)$ such that $\gamma_i \mapsto g_i$ $(i = 1, \cdots, m)$ gives an irreducible representation. Thus the dimension of the orbit space $\mathcal{M}(S)$ is given as follows:

$$\begin{aligned} m(S) &= 2^2 m - (\dim GL(2, \mathbb{C}) - 1) \\ &= 4m - 3. \quad \blacksquare \end{aligned}$$

REMARK 2.1.5. Let V be a finite-dimensional $\mathbb{C}$-vector space and let $\rho : G \to GL(V)$ be an irreducible representation of a group G in V. If an endomorphism L of V satisfies

$$L \cdot \rho(g) = \rho(g) L \quad (g \in G),$$

then L is a scalar endomorphism (Schur's Lemma). In fact, let $\lambda \in \mathbb{C}$ be an eigenvalue of L and v be an eigenvector of L corresponding to the eigenvalue λ. By assumption, we have

$$(L - \lambda)\rho(g) = \rho(g)(L - \lambda) \quad (g \in G),$$

from which we have $(L - \lambda)\rho(g)v = 0$ $(g \in G)$. Since ρ is irreducible (see Chapter 2, §4.2), $\{\rho(g)v; \ g \in G\}$ spans the linear space V. Hence $L = \lambda I$, where I is the identity endomorphism of V.

REMARK 2.1.6. For second-order differential equations defined on a compact Riemann surface, analogous results to Propositions 2.1.3 and 2.1.4 are known, [Sait.1], [Iws.3], [Oht.2].

2.2. The Riemann-Hilbert problem

For a Fuchsian differential equation (2.1.1) we associate its monodromy, and we have the mapping $\Phi : \mathcal{E}(S) \to \mathcal{M}(S)$, which leads to the problems: "Is the map Φ surjective ?" and "Is it one-to-one ?" To state the problem more precisely, let us make some definitions.

DEFINITION 2.2.1. A pair (S, ρ) consisting of a subset $S = \{a_1, \cdots, a_m, a_{m+1} = \infty\}$ of $\mathbb{P}^1$ and of the conjugacy class ρ of an irreducible representation of $\pi_1(\mathbb{P}^1 \setminus S, x_0)$ of rank two is called a *Riemann datum.*

The Riemann-Hilbert problem: *Given a Riemann datum (S, ρ), find a Fuchsian differential equation with singular points in S which has ρ as its monodromy.*

PROPOSITION 2.2.2. *The mapping Φ is not surjective if $m \geq 3$.*

Proof. By Propositions 2.1.3 and 2.1.4, we have

$$m(S) - e(S) = m - 2,$$

which implies that, if $m \geq 3$, then $m(S) > e(S)$. ∎

When $m = 2$, by Corollary 4.3.4 in Chapter 2, we know that Φ is surjective.

REMARK 2.2.3. When $m \leq 1$, the problem can be easily solved. In fact, if $m = 0$, $\mathcal{E}(S)$ is empty and $\mathcal{M}(S)$ contains trivial representation only. When $m = 1$, any Fuchsian differential equation can be integrated by quadrature.

Since we cannot in general find a solution of the Riemann-Hilbert problem in the class of differential equation $\mathcal{E}(S)$, we ask whether we can solve the problem in a larger class of differential equations which includes $\mathcal{E}(S)$. We consider Fuchsian differential equations with singular points in S and additional singular points in $X = \mathbb{P}^1 \setminus S$ at which the circuit matrices are trivial.

DEFINITION 2.2.4. A regular singular point $x = a$ of (2.1.1) is said to be *apparent* if it is non-logarithmic in the sense of Section 3 of Chapter 1 and if the exponents α_1 and α_2 at $x = a$ are integers.

A singular point $x = a$ of (2.1.1) is apparent if and only if it admits two linearly independent solutions which are meromorphic at $x = a$.

Let us define, for a non-negative integer q, a class $\mathcal{E}(S, q)$ of linear differential equations of the form (2.1.1) by

$\mathcal{E}(S, q)$:= {second order irreducible Fuchsian linear differential equations having at most q apparent singular points besides S}.

By counting parameters we can expect that $\dim \mathcal{E}(S, q) = e(S) + q$, and this can in fact be shown to be true. Notice that we have

$$\mathcal{E}(S) = \mathcal{E}(S, 0) \subset \mathcal{E}(S, 1) \subset \cdots \subset \mathcal{E}(S, q) \subset \cdots.$$

Set $\mathcal{E}(S, \infty) := \bigcup_{q=0}^{\infty} \mathcal{E}(S, q)$. By associating to a differential equation its monodromy, we define the mappings:

$$\Phi_q : \mathcal{E}(S, q) \to \mathcal{M}(S), \quad \Phi_0 = \Phi,$$
$$\Phi_\infty : \mathcal{E}(S, \infty) \to \mathcal{M}(S).$$

The following two theorems give a solution to the Riemann-Hilbert problem in an abstract manner. Since the proofs of the theorems require a knowledge of complex analytic geometry, we do not give them in this book.

THEOREM 2.2.5. ([Plem],[Röh]) *The mapping $\Phi_\infty : \mathcal{E}(S, \infty) \to \mathcal{M}(S)$ is surjective.*

This theorem says nothing about the number of apparent singular points necessary to solve the Riemann-Hilbert problem. Since

$$\dim \mathcal{E}(S, m(S) - e(S)) = m(S),$$

one may expect that $\Phi_{m(S)-e(S)}$ is already surjective; this is not quite true. We have the more precise result:

THEOREM 2.2.6. ([Oht.2])

(i) $\Phi_{m(S)-e(S)+1} : \mathcal{E}(S, m(S)-e(S)+1) \to \mathcal{M}(S)$ *is surjective.*

(ii) *Let* $\mathcal{M}'(S) = \{\rho \in \mathcal{M}(S)$; *the image by* ρ *of a loop encircling a point of* S *is diagonalizable*$\}$. *Then* $\mathcal{M}'(S)$ *is an open dense subset of* $\mathcal{M}(S)$ *and is contained in the image of* $\Phi_{m(S)-e(S)}$.

Let $S(t) = \{a_1(t), \cdots, a_m(t), a_{m+1} = \infty\}$ $(t \in U \subset \mathbb{C}^N)$ be a small deformation of $S = S(0)$. Since the space $\mathcal{M}(S)$ is stable under a small deformation of S, we can define the map

$$\Phi_q^U : \bigcup_{t \in U} \mathcal{E}(S(t), q) \longrightarrow \mathcal{M}(S)$$

by collecting maps $\Phi_q : \mathcal{E}(S(t), q) \to \mathcal{M}(S)$ for $t \in U$. Notice that $\bigcup_{t \in U} \mathcal{E}(S(t), q)$ carries a natural complex structure. We are led to the following problem.

PROBLEM 2.2.7. *For a given* $\rho \in \mathcal{M}(S)$, *describe the variety* $(\Phi_q^U)^{-1}(\rho)$. *That is, describe the family of differential equations with* ρ *as its monodromy.*

We call this the *problem of monodromy-preserving deformation*, which we shall study in the rest of this chapter.

3 Monodromy-preserving deformations

3.1. M-invariant fundamental solutions

Let U be a simply connected domain in $\mathbb{C}^n$ with coordinates $t = (t_1, \cdots, t_n)$, and put $X := \mathbb{P}^1 \times U$. A function f on X is said to be *uniform* if there exist finitely many subsets S_i of X given by the graph of holomorphic maps $\xi_i : U \to \mathbb{P}^1$, i.e.,

$$S_i = \{(\xi_i(t), t) | t \in U\}.$$

such that f is single valued and holomorphic outside $S = \bigcup_i S_i$. Let us fix an inhomogeneous coordinate x on $\mathbb{P}^1$. We say that a matrix or a vector is uniform if its components are uniform functions on X. A 1-form $\omega = a_0 dx + a_1 dt_1 + \cdots + a_n dt_n$ is said to be uniform if the a_i are uniform on X. Put $X_S := X \setminus S$ and denote by π the natural projection of X on U: $\pi(x, t) = t$. For $t \in U$, set $S(t) := \pi^{-1}(t) \cap S$ and $X_S(t) := \pi^{-1}(t) \setminus S(t)$.

Given a 2×2 uniform matrix $P(x, t)$, consider a family of ordinary differential equations of the form

$$D\vec{y} = P(x, t)\vec{y}, \quad D = \frac{d}{dx} \tag{3.1.1}$$

with a parameter $t \in U$, where $\vec{y}$ is an unknown 2-vector. Let us choose $t_0 \in U$ so that there exists a simply connected neighbourhood $U' \subset U$ of t_0, such that $X_S(t)$ $(t \in U')$ are homeomorphic. For simplicity, we write U in place of U'. Let us choose $x_0 \in X_S(t_0)$ and fix a homeomorphism

$$\Psi : X_S(t_0) \times U \to X_S$$

such that

$$\Psi(X_S(t_0) \times \{t\}) = X_S(t), \quad \Psi((x_0, t)) = (x_0, t).$$

The restriction of Ψ on $X_S(t_0) \times \{t\}$ defines a homeomorphism

$$\Psi_t : X_S(t_0) \to X_S(t),$$

and it induces an isomorphism

$$(\Psi_t)_* : \pi_1\big(X_S(t_0), (x_0, t_0)\big) \xrightarrow{\sim} \pi_1\big(X_S(t), (x_0, t)\big).$$

For $\gamma_0 \in \pi_1(X_S(t_0),(x_0,t_0))$, define $\gamma(t;t_0,\gamma_0) \in \pi_1(X_S(t),(x_0,t))$ by

$$\gamma(t;t_0,\gamma_0) := (\Psi_t)_*\gamma_0.$$

In what follows, we write $\gamma(t)$ in place of $\gamma(t;t_0,\gamma_0)$. Note that $\gamma(t_0) = \gamma_0$.

Let $Y(x,t_0) := Y(x,(x_0,t_0))$ be a *fundamental solution* of (3.1.1), which is by definition a holomorphic non-singular matrix solution at (x_0,t_0). Let $Y(x,t) := Y(x,(x_0,t))$ be a fundamental solution of (3.1.1) which is holomorphic in a neighbourhood of $\{x_0\} \times U \subset X_S$ and coincides $Y(x,t_0)$ on $X_S(t_0)$. For $\gamma \in \pi_1(X_S(t),(x_0,t))$, let us denote by $Y^\gamma(x,t)$ a solution of (3.1.1) obtained by the analytic continuation of $Y(x,t)$ along γ. Since $Y^\gamma(x,t)$ is a fundamental solution holomorphic at (x_0,t), there exists a matrix $\rho(t,\gamma) \in GL(2,\mathbb{C})$ such that

$$Y^\gamma(x,t) = Y(x,t)\rho(t,\gamma).$$

Thus we have an analytic family of monodromy representations:

$$\rho(t,\cdot) : \pi_1(X_S(t),(x_0,t)) \to GL(2,\mathbb{C}), \quad t \in U.$$

DEFINITION 3.1.1. $(Y(x,t),(x_0,t))$, or simply $Y(x,t)$, is said to be *M-invariant* if, for any $\gamma(t_0) \in \pi_1(X_S(t_0),(x_0,t_0))$, the monodromy matrices $\rho(t,\gamma(t))$ are independent of t.

DEFINITION 3.1.2. A family (3.1.1) of differential equations is said to be *monodromy preserving* if there exists an M-invariant fundamental solution of (3.1.1). We also say that (3.1.1) gives a *monodromy preserving deformation.*

PROPOSITION 3.1.3. *A fundamental solution $Y(x,t)$ is M-invariant if and only if the matrix valued* 1*-form*

$$\Omega(x,t) = dY(x,t)Y(x,t)^{-1} \tag{3.1.2}$$

is uniform on X, where d denotes exterior differentiation with respect to t.

Proof. Suppose that $Y(x,t)$ is M-invariant. Define $\Omega_i(x,t)$ by $\Omega(x,t) = \sum_i \Omega_i(x,t)dt_i := dY \cdot Y^{-1}$. We show that $\Omega_i(x,t)$, $i =$

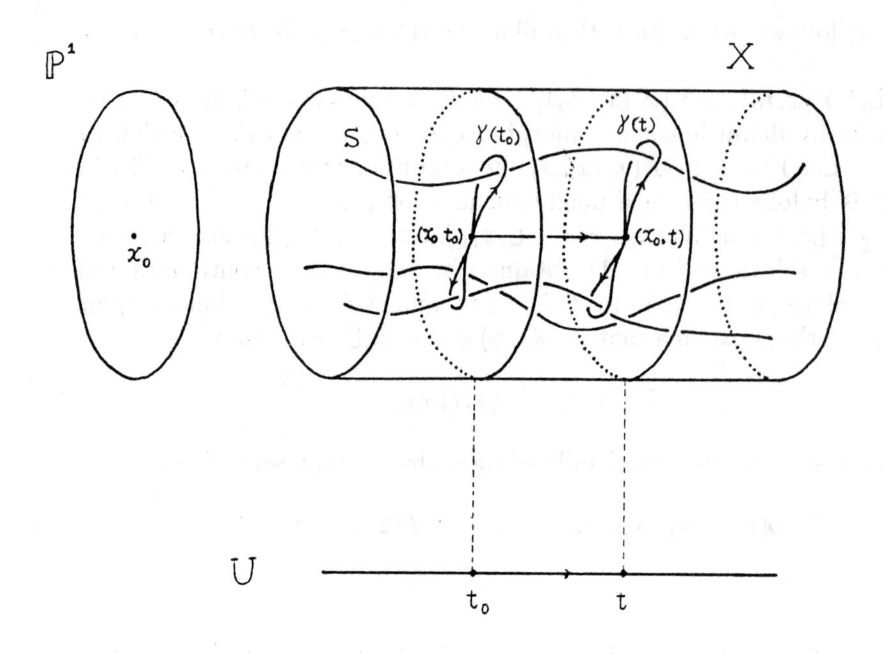

Figure

$1, \cdots, n$, are uniform on X. Let $\gamma(t) \in \pi_1(X_S(t), (x_0, t))$ and $Y(x, t)$ be as above. We have

$$\begin{aligned}
\Omega^{\gamma(t)}(x,t) :&= dY^{\gamma(t)}(x,t) \cdot Y^{\gamma(t)}(x,t)^{-1} \\
&= d(Y\rho(t,\gamma(t))) \cdot \rho(t,\gamma(t))^{-1}Y^{-1} \\
&= dY \cdot \rho(t,\gamma(t))\rho(t,\gamma(t))^{-1}Y^{-1} \\
&= dY \cdot Y^{-1} \\
&= \Omega(x,t),
\end{aligned}$$

since $\rho(t, \gamma(t))$ is independent of t. This identity implies that the $\Omega_i(x, t)$ are single valued on $X_S(t)$.

Conversely, suppose that $\Omega(x,t)$ is uniform on X. Let $\rho(t,\cdot)$ be the monodromy representation with respect to $Y(x,t)$. Since the analytic continuation along $\gamma(t)$ and the exterior differentiation d with respect to t are commutative, for any $\gamma(t) \in \pi_1(X_S(t),(x_0,t))$, we have

$$\begin{aligned} dY^{\gamma(t)}(x,t) &= d\{Y(x,t)\rho(t,\gamma(t))\} \\ &= Y d\rho(t,\gamma(t)) + dY\cdot\rho(t,\gamma(t)) \\ &= Y d\rho(t,\gamma(t)) + \Omega Y\rho(t,\gamma(t)). \end{aligned}$$

and

$$dY^{\gamma(t)}(x,t) = \Omega^{\gamma(t)}(x,t)Y^{\gamma(t)}(x,t) = \Omega(x,t)Y(x,t)\rho(t,\gamma(t)),$$

because $\Omega(x,t)$ is single valued on $X_S(t)$. It follows that

$$Y(x,t)d\rho(t,\gamma(t)) = 0.$$

Hence $d\rho(t,\gamma(t)) = 0$ because $\det Y(x,t) \neq 0$. This shows that the monodromy representation $\rho(t,\cdot)$ is independent of t. ∎

LEMMA 3.1.4. *The Pfaffian system*

$$D\vec{y} = P(x,t)\vec{y}, \tag{3.1.1}$$

$$d\vec{y} = \Omega(x,t)\vec{y}, \tag{3.1.3}$$

is completely integrable (cf. Chapter 1, §5) if and only if

$$\begin{aligned} dP(x,t) &= D\Omega(x,t) + [\Omega(x,t),P(x,t)], \\ d\Omega(x,t) &= \Omega(x,t)\wedge\Omega(x,t) \end{aligned} \tag{3.1.4}$$

holds.

Proof. By virtue of the Frobenius theorem (Chapter 1 Corollary 5.3) we have only to check that the integrability condition of (3.1.1 and 3) is equivalent to (3.1.4). The Pfaffian system (3.1.1 and 3) can be written as

$$d_{(x,t)}\vec{y} = \omega\vec{y},$$

where $\omega = Pdx + \Omega$ and $d_{(x,t)}$ is the exterior differentiation with respect to (x,t), i.e., $d_{(x,t)} = dxD + d$. We compute the identity $d_{(x,t)}\omega - \omega \wedge \omega = 0$. The following computation leads to the conclusion.

$$\begin{aligned} d_{(x,t)}\omega &= dP \wedge dx + dx \wedge D\Omega + d\Omega \\ &= (dP - D\Omega) \wedge dx + d\Omega, \\ \omega \wedge \omega &= (Pdx + \Omega) \wedge (Pdx + \Omega) \\ &= (-P\Omega + \Omega P) \wedge dx + \Omega \wedge \Omega \\ &= [\Omega, P] \wedge dx + \Omega \wedge \Omega. \quad \blacksquare \end{aligned}$$

Therefore we have

PROPOSITION 3.1.5. *If $Y(x,t)$ is an M-invariant fundamental solution of* (3.1.1), *then the matrix 1-form $\Omega(x,t) = dY(x,t) \cdot Y(x,t)^{-1}$ is uniform on X and satisfies* (3.1.4). *Conversely, if the system* (3.1.4) *admits a solution $\Omega(x,t)$ which is uniform on X, then any matrix $Y(x,t)$ satisfying*

$$DY = PY, \quad dY = \Omega Y$$

is an M-invariant fundamental solution of (3.1.1).

The system (3.1.4) is called the *deformation equation* of (3.1.1).

Since $\Omega(x,t) = dY(x,t) \cdot Y(x,t)^{-1}$ depends on the choice of M-invariant fundamental solution $Y(x,t)$, we discuss in the next section the ambiguity of $Y(x,t)$. As a result, we know the ambiguity of the matrix $\Omega(x,t)$.

3.2. Totality of M-invariant fundamental solutions

Let $Y^{(1)}(x,t) = Y^{(1)}(x,(x_0,t))$ and $Y^{(2)}(x,t) = Y^{(2)}(x,(x_0,t))$ be two fundamental solutions of the equation (3.1.1). There is a matrix $C(t) \in GL(2, \mathcal{O}(U))$ such that

$$Y^{(1)}(x,t) = Y^{(2)}(x,t)C(t), \tag{3.2.1}$$

where $\mathcal{O}(U)$ stands for the ring of holomorphic functions on U. If $Y^{(2)}(x,t)$ is an M-invariant fundamental solution of (3.1.1), $Y^{(1)}(x,t)$ defined by (3.2.1) is not necessarily M-invariant. We find a condition for the matrix $C(t)$ so that $Y^{(1)}(x,t)$ is also M-invariant.

PROPOSITION 3.2.1. *Assume the monodromy of* (3.1.1) *is irreducible. Let $Y^{(1)}(x,t)$ and $Y^{(2)}(x,t)$ be as above and suppose that $Y^{(2)}(x,t)$ is M-invariant. Then*

(i) *$Y^{(1)}(x,t)$ is M-invariant if and only if $C(t)$ has the form:*

$$C(t) = \mu(t)C, \quad \mu(t) \in \mathcal{O}(U), \ \mu(t) \neq 0, \ C \in GL(2,\mathbb{C}), \tag{3.2.2}$$

(ii) *if $Y^{(1)}(x,t)$ is M-invariant, then we have*

$$\Omega^{(1)}(x,t) - \Omega^{(2)}(x,t) = d\log\mu(t) \cdot I_2 \tag{3.2.3}$$

for some $\mu(t) \in \mathcal{O}(U)$, $\mu(t) \neq 0$, where I_2 is the 2×2 identity matrix and $\Omega^{(i)}(x,t) := dY^{(i)}(x,t) \cdot Y^{(i)}(x,t)^{-1}$, $i = 1,2$.

REMARK 3.2.2. The (1,2)-component of $\Omega(x,t) := dY(x,t) \cdot Y(x,t)^{-1}$ is independent of the choice of M-invariant fundamental solution Y.

To prove the above proposition we make use of the following lemma. Consider a differential equation of the form

$$DL = [Q(x), L], \tag{3.2.4}$$

where L is a 2×2 unknown matrix and $Q(x)$ is a 2×2 matrix valued function. We associate with (3.2.4) the following differential equation

$$D\vec{y} = Q(x)\vec{y}, \tag{3.2.5}$$

where $\vec{y}$ is an unknown 2-vector. Let $Y(x)$ be a fundamental solution of (3.2.5).

LEMMA 3.2.3. *A general solution $L(x)$ of (3.2.4) is given by*

$$L(x) = Y(x)L_0Y(x)^{-1}, \tag{3.2.6}$$

where $Y(x)$ is a fundamental solution of (3.2.5) and L_0 is an arbitrary constant matrix.

Proof. We show that $L(x)$ satisfies the equation (3.2.4). In fact, noting that $Y(x)$ is a fundamental solution of (3.2.5) and that

$$D(Y(x)^{-1}) = -Y(x)^{-1}DY(x)\cdot Y(x)^{-1},$$

we get

$$\begin{aligned} DL(x) &= D(Y(x)L_0Y(x)^{-1}) \\ &= DY(x)\cdot L_0Y(x)^{-1} - Y(x)L_0Y(x)^{-1}DY(x)\cdot Y(x)^{-1} \\ &= Q(x)Y(x)L_0Y(x)^{-1} - Y(x)L_0Y(x)^{-1}Q(x) \\ &= [Q(x), L(x)]. \end{aligned}$$

Since $L(x)$ is a solution of (3.2.4) which contains $2\times 2 = 4$ parameters, it gives a general solution. ∎

Proof of Proposition 3.2.1. The proof will be carried out in four steps.

1°) Let $Y^{(i)}(x,t)$ $(i = 1,2)$ be M-invariant fundamental solutions of the equation (3.1.1), and set

$$\Omega^{(i)}(x,t) := dY^{(i)}(x,t)\cdot Y^{(i)}(x,t)^{-1}. \tag{3.2.7}$$

Define $L_i(x,t)$ $(i = 1,\cdots,n)$ by

$$\Omega^{(1)}(x,t) - \Omega^{(2)}(x,t) = \sum_i L_i(x,t)dt_i.$$

We show that each $L_i(x,t)$ satisfies the equation (3.2.4) with $Q(x) = P(x,t)$. Since $Y^{(i)}(x,t)$ is M-invariant, it follows from Proposition 3.1.5 that

$$dP(x,t) = D\Omega^{(i)}(x,t) + [\Omega^{(i)}(x,t), P(x,t)], \tag{3.2.8}$$

$$d\Omega^{(i)}(x,t) = \Omega^{(i)}(x,t)\wedge\Omega^{(i)}(x,t). \tag{3.2.9}$$

Subtracting (3.2.8) for $i = 2$ from that for $i = 1$, we have

$$DL_i(x,t) = [P(x,t), L_i(x,t)]. \tag{3.2.10}$$

2°) Let us show that $L_i(x,t) = \lambda_i(t)I_2$ $(i = 1, \cdots, n)$ for some $\lambda_i(t) \in \mathcal{O}(U)$. Applying Lemma 3.2.3 to the equation (3.2.10), we see that $L_i(x,t)$ can be written as

$$L_i(x,t) = Y^{(2)}(x,t)L_{i0}(t)Y^{(2)}(x,t)^{-1}, \tag{3.2.11}$$

by choosing $L_{i0}(t) \in M(2, \mathcal{O}(U))$ appropriately.
Let $\rho(t,\cdot) : \pi_1(X_S(t),(x_0,t)) \to GL(2,\mathbb{C})$ be the monodromy representation with respect to $Y^{(2)}(x,t)$. Analytic continuation along $\gamma(t) \in \pi_1(X_S(t),(x_0,t))$ yields

$$L_i(x,t) = Y^{(2)}(x,t)\rho(t,\gamma(t))L_{i0}(t)\rho(t,\gamma(t))^{-1}Y^{(2)}(x,t)^{-1}, \tag{3.2.12}$$

because $L_i(x,t)$ is uniform. It follows from (3.2.11) and (3.2.12) that

$$\rho(t,\gamma(t))L_{i0}(t) = L_{i0}(t)\rho(t,\gamma(t))$$

for any $\gamma(t)$. Since the representation $\rho(t,\cdot)$ is irreducible by assumption, using Schur's lemma (see Remark 2.1.5), we see that $L_{i0}(t)$ are scalar matrices: $L_{i0}(t) = \lambda_i(t)I_2$, $\lambda_i(t) \in \mathcal{O}(U)$. Thus we have

$$L_i(x,t) = L_{i0}(t) = \lambda_i(t)I_2,$$

and hence

$$\Omega^{(1)}(x,t) - \Omega^{(2)}(x,t) = \theta I_2, \tag{3.2.13}$$

where $\theta = \sum_i \lambda_i(t)dt_i$.

3°) We show that the 1-form θ is closed: $d\theta = 0$. Differentiating both sides of (3.2.13) and using (3.2.9), we see

$$\begin{aligned}
d\theta I_2 &= d\Omega^{(1)} - d\Omega^{(2)} \\
&= \Omega^{(1)} \wedge \Omega^{(1)} - \Omega^{(2)} \wedge \Omega^{(2)} \\
&= \Omega^{(1)} \wedge (\Omega^{(1)} - \Omega^{(2)}) + (\Omega^{(1)} - \Omega^{(2)}) \wedge \Omega^{(2)} \\
&= \Omega^{(1)} \wedge \theta I_2 + \theta I_2 \wedge \Omega^{(2)} \\
&= \theta I_2 \wedge (\Omega^{(2)} - \Omega^{(1)}) \\
&= (\theta \wedge \theta) I_2 \\
&= 0.
\end{aligned}$$

Since U is simply connected, there is a holomorphic function $\nu(t)$ in U such that $d\nu = \theta$. Define $\mu(t)$ by $\mu(t) := \exp(\nu(t))$, then (3.2.13) implies (3.2.3) proving the assertion (ii).

4°) Substituting $\Omega^{(i)} = dY^{(i)} \cdot Y^{(i)^{-1}}$ into (3.2.13), we see

$$\begin{aligned}\theta I_2 &= dY^{(1)} \cdot Y^{(1)^{-1}} - dY^{(2)} \cdot Y^{(2)^{-1}} \\ &= d(Y^{(2)}C(t))C(t)^{-1}Y^{(2)^{-1}} - dY^{(2)} \cdot Y^{(2)^{-1}} \\ &= Y^{(2)}dC(t) \cdot C(t)^{-1}Y^{(2)^{-1}},\end{aligned}$$

and hence

$$dC(t) = \theta C(t).$$

Thus we have $C(t) = \mu(t)C$ for some $C \in GL(2, \mathbb{C})$. Conversely, if the matrix $C(t)$ in (3.2.1) has the form (3.2.2), it is obvious that $Y^{(1)}(x,t)$ is M-invariant. Thus the assertion (i) is proved. ∎

3.3. Monodromy preserving deformation of second order differential equations

Let us consider a monodromy preserving family of second order differential equations:

$$D^2y + p_1(x,t)Dy + p_2(x,t)y = 0, \tag{3.3.1}$$

where $p_i(x,t)$ $(i = 1,2)$ are uniform functions on $X = \mathbb{P}^1 \times U$, U being a simply connected open set in $\mathbb{C}^n$. Let us explain how the monodromy preserving deformation of the equation (3.3.1) is related to that of a system (3.1.1).

Set

$$P(x,t) = \begin{pmatrix} 0 & 1 \\ -p_2 & -p_1 \end{pmatrix}. \tag{3.3.2}$$

The equation (3.3.1) is equivalent to the system (3.1.1) with (3.3.2). In fact, if $y(x,t)$ is a solution of (3.3.1), $\vec{y}(x,t) := {}^t(y(x,t), Dy(x,t))$ gives a solution of (3.1.1) with (3.3.2); conversely if $\vec{y}(x,t) = {}^t(y(x,t), z(x,t))$

is a solution of (3.1.1) with (3.3.2), then $z(x,t)$ equals to $Dy(x,t)$ and $y(x,t)$ is a solution of (3.3.1).

Let

$$\mathcal{Y}(x,t) = (y_1(x,t), y_2(x,t))$$

be a *fundamental system of solutions* of (3.3.1), i.e. $y_1(x,t), y_2(x,t)$ are linearly independent solutions of (3.3.1). Let $Y(x,t)$ be the Wronskian matrix of $\mathcal{Y}(x,t)$:

$$Y(x,t) = \begin{pmatrix} \mathcal{Y}(x,t) \\ D\mathcal{Y}(x,t) \end{pmatrix} = \begin{pmatrix} y_1(x,t) & y_2(x,t) \\ Dy_1(x,t) & Dy_2(x,t) \end{pmatrix}.$$

Then $Y(x,t)$ gives a fundamental solution of the system (3.1.1) with (3.3.2). A fundamental system $\mathcal{Y}(x,t)$ of solutions of (3.3.1) is said to be M-*invariant* if the associated Wronskian matrix $Y(x,t)$ is M-invariant.

Suppose that (3.3.1) has an M-invariant fundamental system $\mathcal{Y}(x,t)$ of solutions. Let us show that the 2×2 matrix 1-form

$$\Omega(x,t) := dY(x,t) \cdot Y^{-1}(x,t) \tag{3.3.3}$$

is determined by its first row. Define a linear operator ∇ acting on a row 2-vector $\vec{f}(x,t)$ by

$$\nabla \vec{f}(x,t) = D\vec{f}(x,t) + \vec{f}(x,t)P(x,t).$$

PROPOSITION 3.3.1. *The matrix $\Omega(x,t)$ can be written in the form*

$$\Omega(x,t) = \begin{pmatrix} \vec{A}(x,t) \\ \nabla \vec{A}(x,t) \end{pmatrix}. \tag{3.3.4}$$

Proof. Let $\vec{A}(x,t)$ be the first row of $\Omega(x,t)$. The first row of (3.3.3) reads

$$d\mathcal{Y}(x,t) = \vec{A}(x,t)Y(x,t). \tag{3.3.5}$$

Differentiate both sides of (3.3.5) with respect to x:

$$\begin{aligned} dD\mathcal{Y}(x,t) &= D\{\vec{A}(x,t)Y(x,t)\} \\ &= D\vec{A} \cdot Y + \vec{A}DY \\ &= \nabla \vec{A} \cdot Y, \end{aligned}$$

and combine it with (3.3.5). Then we have

$$dY = \begin{pmatrix} \vec{A} \\ \nabla \vec{A} \end{pmatrix} Y. \blacksquare$$

PROPOSITION 3.3.2. *If $\mathcal{Y}(x,t)$ is an M-invariant fundamental system of solutions of* (3.3.1), *then the vector 1-form $\vec{A}$, defined by* (3.3.4) *using the matrix 1-form*

$$\Omega := dY \cdot dY^{-1}, \; Y := \begin{pmatrix} \mathcal{Y} \\ D\mathcal{Y} \end{pmatrix},$$

is uniform and satisfies the system:

$$\nabla^2 \vec{A} + p_1(x,t)\nabla \vec{A} + p_2(x,t)\vec{A} + d\vec{P}(x,t) = 0, \tag{3.3.6}$$

$$d\begin{pmatrix} \vec{A} \\ \nabla \vec{A} \end{pmatrix} - \begin{pmatrix} \vec{A} \\ \nabla \vec{A} \end{pmatrix} \wedge \begin{pmatrix} \vec{A} \\ \nabla \vec{A} \end{pmatrix} = 0, \tag{3.3.7}$$

where

$$\vec{P}(x,t) = (p_2(x,t), p_1(x,t)).$$

Conversely, if (3.3.6 and 7) *admits a uniform solution $\vec{A}$, any solution Y of the system*

$$DY = PY, \quad dY = \Omega Y,$$

where

$$\Omega := \begin{pmatrix} \vec{A} \\ \nabla \vec{A} \end{pmatrix},$$

can be written in the form

$$Y = \begin{pmatrix} \mathcal{Y} \\ D\mathcal{Y} \end{pmatrix},$$

and $\mathcal{Y}$ is an M-invariant fundamental system of solutions of (3.3.1).

Proof. It is sufficient to show that (3.3.6) is equivalent to the first equation of (3.1.4). The first equation of (3.1.4) can be written as

$$\nabla\Omega(x,t) - P(x,t)\Omega(x,t) - dP(x,t) = 0. \tag{3.3.8}$$

Noting that $P(x,t)$ and $\Omega(x,t)$ have the forms (3.3.2) and (3.3.4), we see that the first row of (3.3.8) yields nothing; from the second row, we obtain (3.3.6). ∎

Name the entries of the 2-vector $\vec{A}$ as

$$\vec{A} =: (B, A).$$

The following result is the consequence of Remark 3.2.2.

COROLLARY 3.3.3. *Suppose that the family (3.3.1) of differential equations is monodromy preserving and that its monodromy is irreducible. Then the uniform 1-form $A(x,t)$ is independent of the choice of M-invariant fundamental system of solutions.*

3.4. SL-equations

As a special case let us consider a linear differential equation of the form

$$D^2 w = p(x,t)w, \tag{3.4.1}$$

where $p(x,t)$ is a uniform function on X. Such differential equations are said to be *of SL-type* and are called *SL-equations* (this name comes from the fact that a monodromy representation has its image in $SL(2,\mathbb{C})$). They are characterized by the property: the Wronskian of a fundamental system of solutions is independent of x. We show that the monodromy preserving deformation of an equation (3.3.1) reduces to that of an SL-equation (3.4.1).

Let $\phi(x,t)$ be a non-zero solution of the equation

$$2D\phi = -p_1(x,t)\phi. \tag{3.4.2}$$

Then by the change of unknown $y \to w$:

$$y = \phi(x,t)w, \tag{3.4.3}$$

the equation (3.3.1) is transformed into an equation of the form (3.4.1) with $p(x,t)$ given by

$$p(x,t) = -p_2(x,t) + \frac{1}{4}p_1(x,t)^2 + \frac{1}{2}Dp_1(x,t). \tag{3.4.4}$$

PROPOSITION 3.4.1. *If the equation* (3.3.1) *has an M-invariant fundamental system of solutions, then so does the SL-equation* (3.4.1) *with $p(x,t)$ given by* (3.4.4).

Proof. Suppose that the equation (3.3.1) has an M-invariant solution $\mathcal{Y}(x,t)$. Let $Y(x,t)$ be the Wronskian matrix of $\mathcal{Y}(x,t)$ and let $r(x,t)$ be its Wronskian:

$$r(x,t) := \det\begin{pmatrix} \mathcal{Y}(x,t) \\ D\mathcal{Y}(x,t) \end{pmatrix}.$$

Then $r(x,t)$ satisfies the differential equation

$$Dr = -p_1(x,t)r.$$

We take $\phi(x,t)$ in (3.4.3) as

$$\phi(x,t) = \sqrt{r(x,t)}.$$

Since $\mathcal{Y}(x,t)$ is M-invariant, we see that the monodromy representation with respect to $r(x,t)$ is t-invariant and hence so is the monodromy representation with respect to $\phi(x,t)$. This shows, through the transformation (3.4.3), that the SL-equation (3.4.1) with (3.4.4) also has an M-invariant fundamental system of solutions. ∎

3.5. Deformation equations for second order SL-equations

Define $A_i(x,t)$ and $B_i(x,t)$ by

$$\vec{A}(x,t) = \sum_{i=1}^{n} (B_i(x,t), A_i(x,t))dt_i. \tag{3.5.1}$$

For an SL-equation, we can simplify the conditions (3.3.6) and (3.3.7) by eliminating the $B_i(x,t)$'s.

PROPOSITION 3.5.1. *If $\mathcal{Y}(x,t)$ is an M-invariant fundamental system of solutions of* (3.4.1), *the functions $A_i(x,t)$ $(i = 1, \cdots, n)$, defined by* (3.5.1) *and* (3.3.4) *using*

$$\Omega := dY \cdot Y^{-1},\ Y := \begin{pmatrix} \mathcal{Y} \\ D\mathcal{Y} \end{pmatrix},$$

are uniform and satisfy the system:

$$D^3 A_i - 4pDA_i - 2Dp \cdot A_i + 2D_i p = 0, \tag{3.5.2}$$

$$D_i A_j - D_j A_i = A_i DA_j - A_j DA_i \tag{3.5.3}$$

for $i, j = 1, \cdots, n$, where $D_i = \partial/\partial t_i$. Conversely, if the system (3.5.2 and 3) *admits uniform solutions $A_i(x,t)$ $(i = 1, \ldots, n)$, then setting*

$$\Omega := \begin{pmatrix} \vec{A} \\ \nabla \vec{A} \end{pmatrix},\ \vec{A} := \sum_{i=1}^{n} (-\frac{1}{2} DA_i(x,t), A_i(x,t))dt_i,$$

any solution of

$$DY = PY, \quad dY = \Omega Y$$

can be written in the form

$$Y = \begin{pmatrix} \mathcal{Y} \\ D\mathcal{Y} \end{pmatrix},$$

and $\mathcal{Y}$ *is an M-invariant fundamental system of solutions of* (3.4.1).

Proof. We apply Proposition 3.3.2 for $\vec{A}_i(x,t) = (B_i(x,t), A_i(x,t))$, $(i = 1, \cdots, n)$ and

$$P(x,t) = \begin{pmatrix} 0 & 1 \\ p(x,t) & 0 \end{pmatrix}.$$

Writing down the equations (3.3.6) we have

$$D(2B_i + DA_i) = 0, \tag{3.5.4}$$

$$D^2B_i + 2pDA_i + DpA_i - D_ip = 0. \tag{3.5.5}$$

The equation (3.3.7) can be written as

$$D_iA_j + A_jDA_i = D_jA_i + A_iDA_j, \tag{3.5.6}$$

$$D_iB_j + A_jDB_i = D_jB_i + A_iDB_j \tag{3.5.7}$$

for $i, j = 1, \cdots, n$. Suppose the existence of an M-invariant fundamental system of solutions of the SL-equation (3.4.1). Then from (3.5.4), we have

$$DB_i(x,t) = -\frac{1}{2}D^2A_i(x,t), \quad i = 1, \cdots, n. \tag{3.5.8}$$

Putting this into (3.5.5), we have (3.5.2).

Conversely suppose the system (3.5.2 and 3) admits a uniform solution $(A_1(x,t), \cdots, A_n(x,t))$. We define uniform functions $B_i(x,t)$ $(i = 1, \cdots, n)$ by

$$B_i(x,t) = -\frac{1}{2}DA_i(x,t), \quad i = 1, \cdots, n. \tag{3.5.9}$$

Then functions A_i and B_i solve $(3.5.4, \cdots, 7)$. In fact, (3.5.4 and 5) follow from (3.5.9 and 3), respectively; and we have (3.5.7) by differentiating (3.5.6) with respect to x. This proves the proposition. ∎

4 The Garnier system $\mathcal{G}_n$

4.1 Main theorem

We consider a Fuchsian differential equation

$$D^2 y + p_1(x) D y + p_2(x) y = 0 \tag{4.1.1}$$

with singularities at $t_1, \cdots, t_n, t_{n+1} = 0, t_{n+2} = 1, t_{n+3} = \infty, \lambda_1, \cdots, \lambda_n$. We assume:

(i) The singular points are all distinct and the Riemann scheme is given as follows:

$$\begin{pmatrix} x = t_i & x = t_{n+3} & x = \lambda_k \\ 0 & \alpha & 0 \\ \theta_i & \alpha + \theta_{n+3} & 2 \end{pmatrix} \tag{4.1.2}$$
$$i = 1, \cdots, n+2; k = 1, \cdots, n.$$

(ii) $\theta_i \notin \mathbb{Z}, \quad i = 1, \cdots, n+3.$

(iii) The singular points λ_k are apparent in the sense of Defintion 2.2.4.

The Fuchs relation implies

$$\alpha = -\frac{1}{2} \left\{ \sum_{(i)} \theta_i + \theta_{n+3} - 1 \right\}, \tag{4.1.3}$$

where from now on we use the symbols $\sum_i$ and $\sum_{(i)}$ to stand for the summation over $i = 1, \ldots, n$ and over $i = 1, \ldots, n+2$, respectively. For notational simplicity we sometimes write

$$\theta_\infty \quad \text{in place of} \quad \theta_{n+3}.$$

By assumption (i) and Proposition 4.2 of Chapter 1, the coefficients of (4.1.1) can be written as follows:

$$
(4.1.4) \qquad \begin{cases} p_1(x) = \displaystyle\sum_{(i)} \frac{1-\theta_i}{x-t_i} - \sum_k \frac{1}{x-\lambda_k}, \\ p_2(x) = \displaystyle\frac{\kappa}{x(x-1)} - \sum_i \frac{t_i(t_i-1)K_i}{x(x-1)(x-t_i)} \\ \qquad + \displaystyle\sum_k \frac{\lambda_k(\lambda_k-1)\mu_k}{x(x-1)(x-\lambda_k)}. \end{cases}
$$

where μ_k, K_i $(k, i = 1, \cdots, n)$ are constants and

$$
\kappa = \frac{1}{4}\left\{(\sum_{(i)} \theta_i - 1)^2 - \theta_\infty^2\right\}.
$$

In fact, since the singular points of (4.1.1) are $t_1, \cdots, t_{n+3}, \lambda_1, \cdots, \lambda_n$ and all are regular singular, Theorem 2.3 says that $p_1(x)$ and $p_2(x)$ are of the form:

$$
\begin{aligned}
p_1(x) &= \sum_{(i)} \frac{a_i}{x-t_i} + \sum_k \frac{c_k}{x-\lambda_k}, \\
p_2(x) &= \sum_{(i)} \frac{b_i}{(x-t_i)^2} + \sum_k \frac{d_k}{(x-\lambda_k)^2} \\
&\quad - \sum_{(i)} \frac{K_i}{x-t_i} + \sum_k \frac{\mu_k}{x-\lambda_k},
\end{aligned}
$$

where $a_i, b_i, K_i, c_k, d_k, \mu_k$ are independent of x. The characteristic equation at $x = t_i$ is

$$
s(s-1) + a_i s + b_i = 0
$$

and its roots are characteristic exponents, 0 and θ_i. It follows that

$$
a_i = 1 - \theta_i, \quad b_i = 0.
$$

The same argument for the singular point $x = \lambda_k$ leads to:

$$
c_k = -1, \quad d_k = 0.
$$

Moreover the expansion of p_1, p_2 at $x = \infty$ is

$$p_1(x) = (2 - \sum_{(i)} \theta_i) x^{-1} + \mathcal{O}(x^{-2})$$

$$p_2(x) = (-\sum_{(i)} K_i + \sum_k \mu_k) x^{-1} + (-\sum_{(i)} t_i K_i + \sum_k \lambda_k \mu_k) x^{-2} + \mathcal{O}(x^{-3}).$$

Since $x = \infty$ is regular singular, we have

$$-\sum_{(i)} K_i + \sum_k \mu_k = 0. \tag{4.1.5}$$

The characteristic equation at $x = \infty$ reads:

$$s(s+1) - (2 - \sum_{(i)} \theta_i) s + (-\sum_{(i)} t_i K_i + \sum_k \lambda_k \mu_k) = 0,$$

whose roots are α and $\alpha + \theta_\infty$. It follows that

$$2\alpha + \theta_\infty = 1 - \sum_{(i)} \theta_i, \tag{4.1.6}$$

$$\alpha(\alpha + \theta_\infty) = -\sum_{(i)} t_i K_i + \sum_k \lambda_k \mu_k. \tag{4.1.7}$$

Set $\kappa := \alpha(\alpha + \theta_\infty)$ and eliminate K_{n+1}, K_{n+2} in $p_2(x)$ by using (4.1.5 and 7). Finally we arrive at (4.1.4).

Notice that

$$\begin{aligned} \mu_i &= \operatorname*{Res}_{x=\lambda_i} p_2(x,t), \\ K_i &= -\operatorname*{Res}_{x=t_i} p_2(x,t). \end{aligned} \tag{4.1.8}$$

By assumption (iii), as we shall see in Proposition 4.3.2, there are n independent relations among $\{\lambda_i, \mu_i, K_i\}$, which determine K_i $(i = 1, \ldots, n)$ as rational functions in

$$(\theta, \lambda, \mu, t) = (\theta_1, \cdots, \theta_{n+3}, \lambda_1, \ldots, \lambda_n, \mu_1, \ldots, \mu_n, t_1, \ldots, t_n).$$

Explicit forms of K_i will be given in (4.3.7).

Thus our differential equation depends on parameters $(\theta, \lambda, \mu, t)$; let us denote by $E_\theta(\lambda, \mu, t)$ the equation (4.1.1) with parameters $(\theta, \lambda, \mu, t)$. To solve Problem 2.2.7, we shall find submanifolds M in $(\theta, \lambda, \mu, t)$-space such that the family $E_\theta(\lambda, \mu, t)$ $((\theta, \lambda, \mu, t) \in M)$ on M is monodromy preserving. On each such manifold, the parameters $\theta = (\theta_i)_{i=1,\dots,n+3}$ are constants; thus for each θ satisfying assumption (ii) we find submanifolds M in (λ, μ, t)-space such that the family $E_\theta(\lambda, \mu, t)$ $((\lambda, \mu, t) \in M)$ is monodromy preserving. The case $n = 1$ will correspond to the Painlevé equation P_{VI}.

Before stating the main theorem we introduce a Hamiltonian system which will play a central role in the following.

DEFINITION 4.1.1. The n-dimensional Garnier system $\mathcal{G}_n$ is the Hamiltonian system

$$\tag{4.1.9} \begin{cases} d\lambda_i = \sum_j \{K_j, \lambda_i\} dt_j, \\ d\mu_i = \sum_j \{K_j, \mu_i\} dt_j, \end{cases}$$

$i = 1, \cdots, n$, where $\{\,\cdot, \cdot\,\}$ stands for the Poisson bracket

$$\{f, g\} = \sum_i \left(\frac{\partial f}{\partial \mu_i}\frac{\partial g}{\partial \lambda_i} - \frac{\partial g}{\partial \mu_i}\frac{\partial f}{\partial \lambda_i}\right).$$

The Hamiltonians

$$K_i = K_i(\lambda, \mu, t) = K_i(\theta, \lambda, \mu, t)$$

are given in (4.3.7).

THEOREM 4.1.2 *Assume* (i), (ii) *and* (iii) *above, and let* $\theta = (\theta_1, \cdots, \theta_{n+3})$ *be fixed.*

(i) *The Hamiltonian system* $\mathcal{G}_n$ *is completely integrable.*

(ii) *Let* M *be an* (n*-dimensional*) *integral manifold in* (λ, μ, t)*-space of* $\mathcal{G}_n$. *Then the family* $E_\theta(\lambda, \mu, t)$ $((\lambda, \mu, t) \in M)$ *on* M *is monodromy preserving.*

(iii) *Let M be a manifold in (λ, μ, t)-space such that the family $E_\theta(\lambda, \mu, t)$ on M is monodromy preserving. Then M is a submanifold of an integral manifold of $\mathcal{G}_n$.*

Consider the Garnier system in case $n = 1$. Write (λ, μ, t, K) in place of $(\lambda_1, \mu_1, t_1, K_1)$. Then

$$K(\lambda, \mu, t) = \frac{1}{t(t-1)}[\lambda(\lambda-1)(\lambda-t)\mu^2 - \{\theta_2(\lambda-1)(\lambda-t) + \theta_3\lambda(\lambda-t) + (\theta_1-1)\lambda(\lambda-1)\}\mu + \kappa\lambda],$$

which is $H_{VI}(\lambda, \mu, t)$ with parameters $(\kappa_t, \kappa_0, \kappa_1, \kappa_\infty) = (\theta_1, \theta_2, \theta_3, \theta_4)$. So the 1-dimensional Garnier system $\mathcal{G}_1$ is just the sixth Painlevé system $\mathcal{H}_{VI}$. In this sense we can regard $\mathcal{G}_n$ as a generalization of the sixth Painlevé system to a completely integrable system of partial differential equations.

4.2 Reduction to SL-equations

Since the monodromy preserving deformation of (4.1.1) and that of the corresponding SL-equation are equivalent to each other, we shall reduce Theorem 4.1.2 to a similar theorem for SL-equation.

Let

$$D^2 w = p(x) w \tag{4.2.1}$$

be the SL-equation derived from (4.1.1) with (4.1.4). To obtain this SL-equation, it is sufficient to make the transformation:

$$y = \phi(x) w,$$
$$\phi(x) = \prod_{(i)} (x - t_i)^{-(1-\theta_i)/2} \prod_k (x - \lambda_k)^{1/2};$$

we have

$$p(x) = \sum_{(i)} \frac{a_i}{(x-t_i)^2} + \frac{a_{n+3}}{x(x-1)} + \sum_i \frac{t_i(t_i-1)L_i}{x(x-1)(x-t_i)} + \sum_k \Big[\frac{3}{4(x-\lambda_k)^2} - \frac{\lambda_k(\lambda_k-1)\nu_k}{x(x-1)(x-\lambda_k)}\Big], \tag{4.2.2}$$

where the constants a_i $(i = 1, \cdots, n+3)$ are related to the exponents θ_i $(i = 1, \cdots, n+3)$ of the equation (4.1.1) by

$$
(4.2.3) \qquad \begin{cases} a_i = \dfrac{1}{4}(\theta_i^2 - 1), \quad i = 1, \cdots, n+2, \\ a_{n+3} = -\dfrac{1}{4}\Big(\sum_{(i)} \theta_i^2 - \theta_{n+3}^2 - 1\Big) - \dfrac{1}{2}n. \end{cases}
$$

Notice that

$$
(4.2.4) \qquad \begin{cases} L_i = \operatorname*{Res}_{x=t_i} p(x,t), \\ \nu_i = -\operatorname*{Res}_{x=\lambda_i} p(x,t) \end{cases} \qquad (i = 1, \cdots, n).
$$

The SL-equation thus obtained from (4.1.1) has the Riemann scheme

$$
\begin{pmatrix} x = t_i & x = \infty & x = \lambda_k \\ \dfrac{1-\theta_i}{2} & \dfrac{-1-\theta_\infty}{2} & -\dfrac{1}{2} \\ \dfrac{1+\theta_i}{2} & \dfrac{-1-\theta_\infty}{2} & \dfrac{3}{2} \end{pmatrix},
$$

and, by the assumptions (ii) and (iii) in Section 4.1, the singular points $x = t_i$, λ_k are non-logarithmic. The quantities L_i, ν_k $(i, k = 1, \cdots, n)$ are related to K_i, μ_k by the formula

$$
(4.2.5) \qquad \begin{cases} \nu_k = \mu_k + \dfrac{1}{2}\omega_k, \\ L_i = K_i + \dfrac{1}{2}(1 - \theta_i) W_i, \end{cases}
$$

where

$$
(4.2.6) \qquad \begin{cases} W_i = \sum_{(j)}^{i} \dfrac{1-\theta_j}{t_i - t_j} + \sum_{k} \dfrac{1}{\lambda_k - t_i}, \\ \omega_k = \sum_{(i)} \dfrac{1-\theta_i}{\lambda_k - t_i} - \sum_{\ell}^{k} \dfrac{1}{\lambda_k - \lambda_\ell}. \end{cases}
$$

Here $\sum_{(m)}^{i}$ stands for the summation over $m = 1, \cdots, i-1, i+1, \ldots, n+2$.

PROPOSITION 4.2.1. *The transformation*

$$(\lambda, \mu, t, K) \to (\lambda, \nu, t, L),$$

defined by (4.2.5) *and* (4.2.6), *is symplectic, i.e.,*

$$\sum_k d\mu_k \wedge d\lambda_k - \sum_i dK_i \wedge dt_i = \sum_k d\nu_k \wedge d\lambda_k - \sum_i dL_i \wedge dt_i. \tag{4.2.7}$$

Proof. By (4.2.6), we have

$$d\nu_k = d\mu_k + \frac{1}{2}\sum_j \frac{\partial \omega_k}{\partial t_j} dt_j + \frac{1}{2}\sum_\ell \frac{\partial \omega_k}{\partial \lambda_\ell} d\lambda_\ell$$

$$dL_i = dK_i + \frac{1}{2}(1-\theta_i)\left[\sum_j \frac{\partial W_i}{\partial t_j} dt_j + \sum_k \frac{\partial W_i}{\partial \lambda_k} d\lambda_k\right].$$

Since

$$(1-\theta_i)\frac{\partial W_i}{\partial \lambda_k} = -\frac{\partial \omega_k}{\partial t_i}, \quad \frac{\partial \omega_i}{\partial \lambda_k} = \frac{\partial \omega_k}{\partial \lambda_i}, \quad \frac{\partial W_k}{\partial t_i} = \frac{\partial W_i}{\partial t_k},$$

for $i, k = 1, \ldots, n$, we get

$$\begin{aligned} &\sum_k d\lambda_k \wedge d\nu_k \\ &= \sum_k d\lambda_k \wedge d\mu_k - \frac{1}{2}\sum_k \sum_i (1-\theta_i)\frac{\partial W_i}{\partial \lambda_k} d\lambda_k \wedge dt_i, \end{aligned} \tag{4.2.8}$$

$$\begin{aligned} &\sum_i dL_i \wedge dt_i \\ &= \sum_k dK_i \wedge dt_i + \frac{1}{2}\sum_k \sum_i (1-\theta_i)\frac{\partial W_i}{\partial \lambda_k} d\lambda_k \wedge dt_i. \end{aligned} \tag{4.2.9}$$

Hence (4.2.8) and (4.2.9) give the desired identity (4.2.7). ∎

Since λ_k $(k = 1, \cdots, n)$ are non-logarithmic singular points of our SL-equation, as we shall show in Proposition 4.3.4, L_i $(i = 1, \cdots, n)$ are rational functions in

$$(a, \lambda, \nu, t) = (a_1, \cdots, a_{n+3}; \lambda_1, \cdots, \lambda_n; \nu_1, \cdots, \nu_n; t_1, \cdots, t_n).$$

Thus our SL-equation depends on parameters (a, λ, ν, t). Let us denote by $E_a(\lambda, \nu, t)$ the equation (4.2.1) with parameters (a, λ, ν, t). By the help of Proposition 4.2.1, we see that Theorem 4.1.2 is equivalent to the following theorem for the SL-equation (4.2.1) with (4.2.2).

THEOREM 4.2.2. *Under the same assumption as in Theorem 4.1.2, let* $a = (a_i)_{i=1,\dots,n+3}$ *be fixed.*

(i) *The Hamiltonian system*

$$(4.2.10) \qquad \begin{cases} d\lambda_k = \sum_i \{L_i, \lambda_k\} dt_i, \\ d\nu_k = \sum_i \{L_i, \nu_k\} dt_i \end{cases} \qquad (k = 1, \cdots, n),$$

with the Hamiltonians

$$L_i = L_i(\lambda, \nu, t) = L_i(a, \lambda, \nu, t)$$

given in (4.3.15), *is completely integrable.*

(ii) *Let* M *be an* (n*-dimesional*) *integral manifold in* (λ, ν, t)*-space of the Hamiltonian system* (4.2.10). *Then the family* $E_a(\lambda, \nu, t)$ $((\lambda, \nu, t) \in M)$ *on* M *is monodromy preserving.*

(iii) *Let* M *be a manifold in* (λ, ν, t)*-space such that the family* $E_a(\lambda, \nu, t)$ *on* M *is monodromy preserving. Then* M *is a submanifold of an integral manifold of the Hamiltonian system* (4.2.10).

In Subsection 4.3, in order to complete the statement of Theorems 4.1.1 and 4.2.2, we express K_i and L_i as rational functions in (λ, μ, t) and (λ, ν, t), respectively. In Subsections 4.4 and 4.5, we prove Theorem 4.2.2.

The proof will be carried out in the following way. We regard (λ, ν) as holomorphic functions of t and consider the family of differential equations (4.2.1) with parameters $t = (t_i)_{i=1,\dots,n}$. We assume that the family $E_a(\lambda(t), \nu(t), t)$ $(t \in U \subset B)$ is monodromy preserving. Let

$\mathcal{W} = \mathcal{W}(x,t)$ be an M-invariant fundamental system of solutions of (4.2.1) with (4.2.2). Putting $W = \begin{pmatrix} \mathcal{W} \\ D\mathcal{W} \end{pmatrix}$ and $\Omega = dW \cdot W^{-1}$, we define $A_i = A_i(x,t)$ by

$$\Omega = \sum_i \begin{pmatrix} B_i & A_i \\ \nabla(B_i & A_i) \end{pmatrix} dt_i.$$

By using local expressions of $\mathcal{W}$, we will know the poles and zeros of the rational functions A_i in x. Since the A_i satisfy (3.5.2 and 3) in Proposition 3.5.1, we will know, in Subsection 4.4, the explicit expression of A_i as rational functions in (x, λ, t). In Subsection 4.5, we substitute these expressions of A_i into the conditions (3.5.2 and 3) and show that the conditions (3.5.2 and 3) are equivalent to a system of differential equations with unknowns λ, ν and variables t, which will turn out to be the completely integrable Hamiltonian system with the Hamiltonians L_i $(i = 1, \cdots, n)$. The integrability of the Hamiltonian system implies that for each point (λ_0, ν_0, t_0), there is a unique n-dimensional submanifold $M \ni (\lambda_0, \nu_0, t_0)$ in the (λ, ν, t)-space such that the family $E_a(\lambda, \nu, t)$ $((\lambda, \nu, t) \in M)$ is monodromy preserving.

4.3 Explicit forms of K_i and L_i

Let us express L_i and K_i $(i = 1, \cdots, n)$ as rational functions in (λ, μ, t) and (λ, ν, t), which will turn out to be Hamiltonians for the systems $\mathcal{G}_n$ and (4.2.10), respectively.

We start by showing the following lemma. Let $E = (E_{ki})_{k,i=1,\cdots,n}$ be the $n \times n$ matrix defined by

$$E_{ki} = \frac{t_i(t_i - 1)}{\lambda_k(\lambda_k - 1)(\lambda_k - t_i)}. \tag{4.3.1}$$

LEMMA 4.3.1. *The inverse matrix $F = (F_{ik})_{i,k=1,\cdots,n}$ of E is given by*

$$F_{ik} = M_i M^{k,i}, \quad i, k = 1, \cdots, n,$$

where M_i and $M^{k,i}$ are defined by

$$
(4.3.2)\qquad \begin{cases} M_i = -\dfrac{\Lambda(t_i)}{T'(t_i)}, \\ M^{k,i} = \dfrac{T(\lambda_k)}{(\lambda_k - t_i)\Lambda'(\lambda_k)}, \end{cases} \qquad (i, k = 1, \cdots, n, n+1, n+2),
$$

and

$$
(4.3.3)\qquad T(x) = \prod_{(i)} (x - t_i), \quad \Lambda(x) = \prod_k (x - \lambda_k).
$$

Proof. Define the rational functions $Z_i(x)$ $(i = 1, \cdots, n)$ by

$$
(4.3.4)\qquad Z_i(x) = \frac{T(x)}{x(x-1)(x-t_i)\Lambda(x)}.
$$

Notice that the poles of $Z_i(x)$ are $\lambda_1, \cdots, \lambda_n$, which are all simple, and that ∞ is its simple zero. Expand $Z_i(x)$ into partial fractions:

$$
\begin{aligned}
(4.3.5)\quad Z_i(x) &= \sum_k \frac{T(\lambda_k)}{\lambda_k(\lambda_k - 1)(\lambda_k - t_i)\Lambda'(\lambda_k)} \frac{1}{x - \lambda_k} \\
&= \frac{1}{t_i(t_i - 1)} \sum_k M^{k,j} E_{ki} \frac{\lambda_k - t_j}{x - \lambda_k} \quad (j = 1, \cdots, n).
\end{aligned}
$$

On the other hand, since $Z_i(x)$ has simple zeros at $x = t_j$ $(j = 1, \cdots, n; j \neq i)$, we have

$$
(4.3.6)\qquad Z_i(t_j) = -\frac{\delta_{ij}}{t_i(t_i - 1)M_j}.
$$

It follows from (4.3.5) and (4.3.6) that

$$
\sum_k M_j M^{k,j} E_{ki} = \delta_{ij}.
$$

These identities show that the inverse matrix of E is F. ∎

PROPOSITION 4.3.2. *The points $\lambda_1, \cdots, \lambda_n$ are non-logarithmic singular points of* (4.1.1) *with* (4.1.6) *if and only if*

$$K_i = M_i \sum_k M^{k,i} \Big\{ {\mu_k}^2 - \sum_{(m)} \frac{\theta_m - \delta_{im}}{\lambda_k - t_m} \mu_k + \frac{\kappa}{\lambda_k(\lambda_k - 1)} \Big\}, \tag{4.3.7}$$

where

$$\kappa := \frac{1}{4} \Big\{ \Big(\sum_{(i)} \theta_i - 1 \Big)^2 - {\theta_\infty}^2 \Big\}.$$

Proof. Take a local coordinate $z := x - \lambda_k$ at λ_k, and rewrite (4.1.1) into

$$z^2 D^2 y + z q_1(z) D y + z q_2(z) y = 0, \tag{4.3.8}$$

where $q_1(z)$ and $q_2(z)$ are functions holomorphic at $z = 0$. Let

$$q_1(z) = -1 + \sum_{m=1}^{\infty} a_{k,m} z^m,$$

$$q_2(z) = \mu_k + \sum_{m=1}^{\infty} b_{k,m} z^m$$

be expansions at $z = 0$. Since $z = 0$ is a non-logarithmic singular point having 0 and 2 as its exponents, the equation (4.3.8) must have a power series solution:

$$y(z) = 1 + \sum_{m=1}^{\infty} y_m z^m. \tag{4.3.9}$$

Putting this into (4.3.8) and comparing the coefficients of like powers, one has a recursive formula which determines the y_m's:

$$\begin{aligned} y_1 &= \mu_k, \\ m(m-2) y_m &= R_{k,m}(y_1, \cdots, y_{m-1}), \quad m = 2, 3, \cdots. \end{aligned} \tag{4.3.10}$$

When $m = 2$, the left-hand side of (4.3.10) vanishes, and therefore the existence of the solution (4.3.9) implies $R_{k,2}(\mu_k) = 0$. Conversely, if $R_{k,2}(\mu_k) = 0$, there is a solution of the form (4.3.9) of (4.1.1) with (4.1.6) and λ_k is a non-logarithmic singular point of (4.1.1) with (4.1.6). Using (4.3.8), we see that the conditions $R_{k,2}(\mu_k) = 0$ $(k = 1, \cdots, n)$ are written as

$$\mu_k{}^2 + a_{k,1}\mu_k + b_{k,1} = 0, \quad k = 1, \dots, n, \tag{4.3.11}$$

where $a_{k,1}$ and $b_{k,1}$ are determined by $p_1(x)$ and $p_2(x)$ as follows:

$$\begin{aligned} a_{k,1} &= \Big(p_1(x) + \frac{1}{x-\lambda_k}\Big)\Big|_{x=\lambda_k} \\ &= \sum_{(i)} \frac{1-\theta_i}{\lambda_k - t_i} - \sum_m^k \frac{1}{\lambda_k - \lambda_m}, \\ b_{k,1} &= \Big(p_2(x) - \frac{\mu_k}{x-\lambda_k}\Big)\Big|_{x=\lambda_k} \\ &= \frac{\kappa}{\lambda_k(\lambda_k - 1)} - \sum_i \frac{t_i(t_i-1)K_i}{\lambda_k(\lambda_k-1)(\lambda_k-t_i)} \\ &\quad + \sum_m^k \frac{\lambda_m(\lambda_m-1)\mu_m}{\lambda_k(\lambda_k-1)(\lambda_k-\lambda_m)} - \Big(\frac{1}{\lambda_k} + \frac{1}{\lambda_k - 1}\Big)\mu_k, \end{aligned}$$

where the symbol $\sum_m^k$ stands for the summation over $m = 1, \dots, k-1, k+1, \dots, n$.

Using the matrix E defined by (4.3.1), the conditions (4.3.11) can be written in the form

$$E\ {}^t(K_1, \cdots, K_n) = {}^t(V_1, \cdots, V_n), \tag{4.3.12}$$

where

$$\begin{aligned} V_k &= \mu_k{}^2 + \Big(a_{k,1} - \frac{1}{\lambda_k} - \frac{1}{\lambda_k - 1}\Big)\mu_k \\ &\quad - \sum_m^k \frac{\lambda_m(\lambda_m-1)\mu_m}{\lambda_k(\lambda_k-1)(\lambda_m-\lambda_k)} + \frac{\kappa}{\lambda_k(\lambda_k-1)}. \end{aligned}$$

By the help of Lemma 4.3.1, the equation (4.3.12) can be solved:

$$\begin{aligned}K_i &= M_i \sum_k M^{k,i} V_k \\ &= M_i \sum_k \Big[M^{k,i} {\mu_k}^2 + (M^{k,i} a_{k,1} - M^{k,i,0}) \mu_k + \kappa \frac{M^{k,i}}{\lambda_k(\lambda_k - 1)} \Big],\end{aligned}$$

where

$$M^{k,i,0} = \Big(\frac{1}{\lambda_k} + \frac{1}{\lambda_k - 1} \Big) M^{k,i} + \sum_m^k \frac{\lambda_k(\lambda_k - 1) M^{m,i}}{\lambda_m(\lambda_m - 1)(\lambda_k - \lambda_m)}.$$

To complete the proof of the proposition, it suffices to show that

$$M^{k,i,0} = M^{k,i} \Big[\sum_{(m)}^i \frac{1}{\lambda_k - t_m} - \sum_m^k \frac{1}{\lambda_k - \lambda_m} \Big]. \tag{4.3.13}$$

We make use of a similar technique as in the proof of Lemma 4.3.1. Let us consider the rational functions $Z_i(x)/(x - \lambda_k)$, where $Z_i(x)$ are defined by (4.3.4), and let us apply the residue theorem to them. By definition, it is seen that

$$\operatorname*{Res}_{x=\lambda_m} \frac{Z_i(x)}{x - \lambda_k} = \begin{cases} \dfrac{M^{m,i}}{\lambda_m(\lambda_m - 1)(\lambda_m - \lambda_k)} & \text{if } m \neq k \\ \dfrac{M^{k,i}}{\lambda_k(\lambda_k - 1)} \Big[\displaystyle\sum_\ell^i \frac{1}{\lambda_k - t_\ell} - \sum_\ell^k \frac{1}{\lambda_m - \lambda_\ell} \Big] & \text{if } m = k \end{cases}$$

and

$$\operatorname*{Res}_{x=\infty} \frac{Z_i(x)}{x - \lambda_k} dx = 0.$$

Since the residue theorem tells us

$$\sum_m \operatorname*{Res}_{x=\lambda_m} \frac{Z_i(x)}{x - \lambda_k} dx + \operatorname*{Res}_{x=\infty} \frac{Z_i(x)}{x - \lambda_k} dx = 0,$$

it follows that

$$M^{k,i,0} = \Big(\frac{1}{\lambda_k} + \frac{1}{\lambda_k - 1}\Big) M^{k,i} - \lambda_k(\lambda_k - 1) \sum_m^k \operatorname*{Res}_{x=\lambda_m} \frac{Z_i(x)}{x - \lambda_k}$$

$$= \Big(\frac{1}{\lambda_k} + \frac{1}{\lambda_k - 1}\Big) M^{k,i} + \lambda_k(\lambda_k - 1) \operatorname*{Res}_{x=\lambda_k} \frac{Z_i(x)}{x - \lambda_k}$$

$$= M^{k,i} \Big(\sum_{(m)}^{i} \frac{1}{\lambda_k - t_m} - \sum_m^k \frac{1}{\lambda_k - \lambda_m} \Big). \quad \blacksquare$$

REMARK 4.3.3. By applying the residue theorem to $Z_i(x)$ given by (4.3.4), we have

$$\sum_k \frac{M^{k,i}}{\lambda_k(\lambda_k - 1)} = 1. \tag{4.3.14}$$

Let us determine $L_1, \cdots, L_n$.

PROPOSITION 4.3.4. *The singular points $\lambda_1, \cdots, \lambda_n$ of the SL-equation* (4.2.1) *with* (4.2.2) *are non-logarithmic if and only if*

$$L_i = M_i \sum_k \big[M^{k,i} {\nu_k}^2 - M^{k,i,0} \nu_k - M^{k,i} U_k \big], \tag{4.3.15}$$

where $M^{k,i,0}$ is given by (4.3.13) *and*

$$U_k = \frac{a_{n+3}}{\lambda_k(\lambda_k - 1)} + \sum_{(m)} \frac{a_m}{(\lambda_k - t_m)^2} + \sum_m^k \frac{3}{4(\lambda_k - \lambda_m)^2}. \tag{4.3.16}$$

The proof of this proposition can be carried out in exactly the same manner as in the proof of Proposition 4.3.2, so we leave it to the reader as an exercise.

4.4 Explicit expression of $A_i(x,t)$

The proof of Theorem 4.2.2 proceeds along the scheme stated at the end of Section 4.2. We find an explicit form of $A_i(x;t)$ $(i = 1, \cdots, n)$, which is the second component of $\vec{A}_i$, defined in (3.5.1) for SL-equations (4.2.1) with (4.2.2).

PROPOSITION 4.4.1. *The functions $A_i(x,t)$ $(i = 1, \cdots, n)$ are given by*

$$A_i(x,t) = M_i \frac{T(x)}{(x - t_i)\Lambda(x)}, \tag{4.4.1}$$

where $T(x)$, $\Lambda(x)$ *and* M_i *are given in* (4.3.3) *and* (4.3.2), *respectively.*

In order to prove this, we shall first study the singularity of $A_i(x,t)$.

LEMMA 4.4.2. *For any fixed* t, $A_i(x,t)$ *satisfies the following properties:*

(i) *It is holomorphic outside the set* $\{\lambda_1, \cdots, \lambda_n, \infty\}$,
(ii) $x = \lambda_k$ $(k = 1, \cdots, n)$ *and* $x = \infty$ *are poles of order at most* 1,
(iii) $x = t_j$ $(j = 1, \cdots, n+2, \neq i)$ *are zeros of order at least* 1.

Proof. Let $\mathcal{W}(x,t) = (w_1(x,t), w_2(x,t))$ be an M-invariant fundamental system of solutions of (4.2.1) with (4.2.2), and let $W(x,t)$ be its Wronskian matrix :

$$W(x,t) = \begin{pmatrix} \mathcal{W}(x,t) \\ D\mathcal{W}(x,t) \end{pmatrix}.$$

Recall that $\vec{A}_i(x,t) = (B_i(x,t), A_i(x,t))$ $(i = 1, \cdots, n)$ is determined by

$$D_i \mathcal{W}(x,t) = \vec{A}_i(x,t) W(x,t). \tag{4.4.2}$$

Solving this with respect to $A_i(x,t)$ we have

$$\det W(x,t) A_i(x,t) = \det \begin{pmatrix} \mathcal{W}(x,t) \\ D_i \mathcal{W}(x,t) \end{pmatrix}, \tag{4.4.3}$$

where $\det W(x,t)$ is the Wronskian of $\mathcal{W}(x,t)$. Since the equation (4.2.1) is of SL-type, $\det W(x,t)$ is independent of x. Set $w(t) :=$ $\det W(x,t)$. Notice that, for any fundamental system of solutions $\mathcal{V}(x,t)$ $= (v_1(x,t), v_2(x,t))$ of (4.2.1), there is a matrix $C(t) = (c_{k\ell}(t))_{k,\ell=1,2}$ $\in GL(2, \mathcal{O}(U))$ such that

$$\mathcal{W}(x,t) = \mathcal{V}(x,t) C(t). \tag{4.4.4}$$

Substituting (4.4.4) into (4.4.3), we have

$$(4.4.5)\quad w(t)A_i(x,t) = \det\begin{pmatrix}\mathcal{V}\\ D_i\mathcal{V}\end{pmatrix}\det C(t) + \sum_{k,\ell}\det\begin{pmatrix}c_{k1} & c_{\ell 2}\\ D_ic_{k1} & D_ic_{\ell 2}\end{pmatrix} v_k(x,t)v_\ell(x,t).$$

First we consider (4.4.5) at $x=\lambda_k(t)$. Since the exponents of (4.2.1) at $x=\lambda_k$ are $-1/2$ and $3/2$, and since $x=\lambda_k$ is a non-logarithmic singular point, we can take a fundamental system of solutions $\mathcal{V}(x,t)$ of the form

$$v_1(x,t) = (x-\lambda_k)^{-1/2}f_1,\quad v_2(x,t) = (x-\lambda_k)^{3/2}f_2,$$

where f_i $(i=1,2,)$ are convergent power series in $x-\lambda_k$ with coefficients holomorphic in t such that $f_i(\lambda_k)\neq 0$. Put these into (4.4.5); we see the first term

$$\det\begin{pmatrix}\mathcal{V}\\ D_i\mathcal{V}\end{pmatrix} = v_1D_iv_2 - v_2D_iv_1$$

is holomorphic at $x=\lambda_k$, and a simple pole appears from the second term of the right-hand side of (4.4.5). Thus we have

$$A_i(x,t) = (x-\lambda_k)^{-1}f_3,$$

where f_3 is a convergent power series in $x-\lambda_k$ with coefficients holomorphic in t; this shows the assertion (ii). By virtue of the assumption (ii) in Section 4.1, the assertions (i) and (iii) are proved in a similar way. ∎

Proof of Proposition 4.4.1. The above lemma says that $A_i(x,t)$ can be written in the form

$$(4.4.6)\qquad A_i(x,t) = X_i\frac{T(x)}{(x-t_i)\Lambda(x)},$$

where X_i is a function independent of x. Let us determine X_i as a rational function of t and λ. Multiply (3.5.2) by A_i and rewrite it as

$$(4.4.7)\qquad A_iD^3A_i - 2D(pA_i^2) + 2A_iD_ip = 0.$$

Denote by $\Xi_i(x)$ the left-hand side of this equation. Expanding $p(x,t)$ and $A_i(x,t)$ in Laurent series in $x - t_i$ and putting them into $\Xi_i(x)$, we get

$$\Xi_i(x) = \frac{A_i(t_i,t)+1}{(x-t_i)^3} + \frac{*}{(x-t_i)^2} + \cdots .$$

It follows from the equation (4.4.7) that $A_i(t_i,t)+1=0$; which yields $X_i = M_i$, where M_i is given in (4.3.2). ∎

4.5 Proof of Theorem 4.2.2

In this subsection, we show that the conditions (3.5.2 and 3) imply that $\lambda(t)$ and $\nu(t)$ satisfy a system of differential equations with variables t. Furthermore, it will be shown that the system thus obtained is the completely integrable Hamiltonian system with the Hamiltonians $L_1, \cdots, L_n$ which are given in (4.3.15).

LEMMA 4.5.1. *The equation* (3.5.3) *is equivalent to the system of differential equations:*

$$\frac{\lambda_k - t_i}{M_i} D_i\lambda_k - \frac{\lambda_k - t_j}{M_j} D_j\lambda_k + \frac{(t_j - t_i)T(\lambda_k)}{(\lambda_k - t_i)(\lambda_k - t_j)\Lambda'(\lambda_k)} = 0 \tag{4.5.1}$$

$$(i,j,k = 1,\cdots,n).$$

Proof. Rewrite (3.5.3) into the form

$$\frac{1}{A_i} D_i \log A_j - \frac{1}{A_j} D_j \log A_i + D(\log A_i - \log A_j) = 0, \tag{4.5.2}$$

and denote by $\Psi_{ij}(x)$ the left-hand side of (4.5.2). With the help of the explicit form (4.4.1) of A_i in Proposition 4.4.1, it is seen that

$$D\log A_i = \sum_{(m)}^{i} \frac{1}{x-t_m} - \sum_k \frac{1}{x-\lambda_k},$$

$$D_j \log A_i = D_j \log M_i - \frac{1}{x-t_j} - \sum_k \frac{D_j\lambda_k}{x-\lambda_k},$$

$$A_i(t_i,t) = -1,$$

for $i, j = 1, \cdots, n$; $i \neq j$. It follows from these identities and the expressions (4.4.1) of A_i that $\Psi_{ij}(x)$ has simple poles at 0, 1 and t_m ($m = 1, \cdots, n$; $m \neq i, j$) and a simple zero at ∞. Since a non-trivial rational function has the same number of poles and zeros counting their multiplicity, it is seen that $\Psi_{ij}(x) = 0$ ($i, j = 1, \cdots, n$) are equivalent to the conditions

$$\Psi_{ij}(\lambda_k) = 0, \quad i, j, k = 1, \cdots, n;$$

it is immediate that these conditions are equal to (4.5.1). ∎

Consider next the equations (4.4.7), which are equivalent to (3.5.2). As in Section 4.4, let us denote $\Xi_i(x)$ ($i = 1, \cdots, n$) the left-hand side of the equations (4.4.7):

$$\Xi_i(x) = A_i D^3 A_i - 2D(pA_i^2) + 2A_i D_i p.$$

LEMMA 4.5.2. (i) *The rational functions* $\Xi_i(x)$ ($i = 1, \cdots, n$) *are expanded in partial fractions as follows:*

$$\Xi_i(x) = \frac{w_i}{x - t_i} + \sum_k \sum_{m=1}^{4} \frac{w_{i,k,m}}{(x - \lambda_k)^m},$$

where $w_i, w_{i,k,m}$ *are independent of* x,

(ii) *The equations* $\Xi_i(x) = 0$ ($i = 1, \cdots, n$) *are equivalent to*

$$w_{i,k,m} = 0, \quad i, k = 1, \cdots, n; m = 1, \cdots, 4 \tag{4.5.3}$$

Proof. The assertion (i) is proved as follows. Note that the poles of $\Xi_i(x)$ are contained in $\{t_1, \cdots, t_{n+2}, \lambda_1, \cdots, \lambda_n\}$.

First we consider $\Xi_i(x)$ at t_j ($j \neq i$). By the explicit form of $A_i(x)$ given by (4.4.1), we see that $A_i(x)$ has a simple zero at t_j and

$$p(x, t) = \frac{a_j}{(x - t_j)^2} + \frac{L_j}{x - t_j} + (\text{a term holomorphic at } x = t_j),$$

where a_j is a constant independent of t. Then

$$D_i p(x, t) = \frac{D_i L_j}{x - t_j} + (\text{a term holomorphic at } x = t_j).$$

It follows that $\Xi_i(x)$ is holomorphic at $x = t_j$.

Next we consider $\Xi_i(x)$ at $x = t_i$. The Laurent series expansion of $A_i(x,t)$ and $p(x,t)$ at $x = t_i$ has the form

$$A_i(x,t) = -1 + b_1(x - t_i) + b_2(x - t_i)^2 + \cdots$$

$$p(x,t) = \frac{a_i}{(x - t_i)^2} + \frac{L_i}{x - t_i} + \cdots.$$

Then

$$pA_i^2 = \frac{a_i}{(x - t_i)^2} + \frac{L_i - 2a_i}{x - t_i} + \cdots$$

and

$$D(pA_i^2) = -\frac{2a_i}{(x - t_i)^3} - \frac{L_i - 2a_i b_1}{(x - t_i)^2} + \cdots.$$

Furthermore we have

$$D_i p = \frac{2a_i}{(x - t_i)^3} + \frac{L_i}{(x - t_i)^2} + \cdots$$

and

$$A_i D_i p = -\frac{2a_i}{(x - t_i)^3} - \frac{L_i - 2a_i b_1}{(x - t_i)^2} + \cdots.$$

So t_i is a simple pole of $\Xi_i(x)$.

Finally let us consider $\Xi_i(x)$ at $x = \lambda_k$. Let $z := x - \lambda_k$ be a local coordinate at $x = \lambda_k$, and let us expand $A_i(x,t)$ and $p(x,t)$ at $z = 0$ in Laurent series:

$$\begin{aligned} A_i(x,t) &= M_i\Big[\frac{M^{k,i}}{z} + \sum_{m=0}^{\infty} M^{k,i,m} z^m\Big], \\ p(x,t) &= \frac{3}{4z^2} - \frac{\nu_k}{z} + \sum_{m=0}^{\infty} u_{k,m} z^m. \end{aligned} \tag{4.5.4}$$

Noting that

$$A_i D^3 A_i = 6(M_i M^{k,i})^2 z^{-5} + \mathcal{O}(z^{-4})$$

$$D(pA_i^2) = 3(M_i M^{k,i})^2 z^{-5} + \mathcal{O}(z^{-4}),$$

we see that $\Xi_i(x)$ has a pole λ_k of order at most 4. Now the assertion is obvious.

Let us show (ii). It is clear that $\Xi_i(x) = 0$ $(i = 1, \cdots, n)$ imply (4.5.3). So we prove the converse. If (4.5.3) holds, $\Xi_i(x)$ has the form $\Xi_i(x) = w_i/(x - t_i)$. On the other hand, by the help of the expressions (4.4.1) for the A_i's and (4.2.2) for $p(x,t)$, it is easily seen that ∞ is a zero of order more than two. It follows that $w_i = 0$ $(i = 1, \cdots, n)$. Hence we have $\Xi_i(x) = 0$ $(i = 1, \cdots, n)$. ∎

We shall write down the conditions (4.5.3) explicitly in terms of (λ, ν, t). Putting the Laurent expansions of A_i and p at $x = \lambda_k$ into the equation $\Xi_i(x) = 0$ and equating the coefficients of like powers in z, we see that $w_{i,k,4} = 0$ implies

$$D_i \lambda_k = M_i \big[2M^{k,i} \nu_k - M^{k,i,0} \big], \tag{4.5.5}$$

that $w_{i,k,2} = 0$ implies

$$D_i \nu_k = M_i \big[M^{k,i} u_{k,1} + M^{k,i,1} \nu_k - \frac{3}{2} M^{k,i,2} \big], \tag{4.5.6}$$

and that $w_{i,k,1} = 0$ implies

$$\begin{aligned} &M^{k,i} D_i u_{k,0} - M^{k,i,0} D_i \nu_k \\ &= \big[M^{k,i} u_{k,1} + M^{k,i,1} \nu_k - \frac{3}{2} M^{k,i,2} \big] D_i \lambda_k, \end{aligned} \tag{4.5.7}$$

where $u_{k,m}$ are the coefficients of the Laurent expansion (4.5.4) of $p(x,t)$. We see that $w_{i,k,3} = 0$ is derived from $w_{i,k,4} = 0$ and

$$u_{k,0} = {\nu_k}^2, \tag{4.5.8}$$

which is obtained by analogous computations as in the proof of Proposition 4.3.2 under the condition that the singular points $\lambda_1, \cdots, \lambda_n$ are non-logarithmic.

LEMMA 4.5.3. (4.5.5), (4.5.6) *and* (4.5.8) *imply* (4.5.7).

Proof. Differentiating (4.5.8) with respect to t_i, we have

$$D_i u_{k,0} = 2 \nu_k D_i \nu_k.$$

Then (4.5.7) is written as

$$\begin{aligned}&(2M^{k,i}\nu_k - M^{k,i,0})D_i\nu_k\\ =&(M^{k,i}u_{k,1} + M^{k,i,1}\nu_k - \frac{3}{2}M^{k,i,2})D_i\lambda_k.\end{aligned}$$

This is a consequence of (4.5.5) and (4.5.6). ∎

LEMMA 4.5.4. (4.5.5) *implies* (4.5.1).

Proof. If $\lambda_k(t)$ satisfies (4.5.5), the equations (4.5.1) is written as

$$(4.5.9)\quad (\lambda_k - t_i)M^{k,i,0} - (\lambda_k - t_j)M^{k,j,0} = \frac{(t_j - t_i)T(\lambda_k)}{(\lambda_k - t_i)(\lambda_k - t_j)\Lambda'(\lambda_k)}.$$

To show this, we consider rational functions

$$W_i(x) := \frac{\lambda_k - t_i}{x - t_i}\frac{T(x)}{\Lambda(x)}.$$

Set $z := x - \lambda_k$. Since

$$W_i = M_i^{-1}(\lambda_k - t_i)A_i(x,t),$$

it is expanded in Laurent series at $x = \lambda_k$ as

$$W_i(x) = (\lambda_k - t_i)M^{k,i}z^{-1} + (\lambda_k - t_i)M^{k,i,0} + \mathcal{O}(z).$$

Then

$$\left\{W_i(x) - W_j(x)\right\}\Big|_{x=\lambda_k} = (\lambda_k - t_i)M^{k,i,0} - (\lambda_k - t_j)M^{k,j,0}.$$

On the other hand, since

$$\begin{aligned}W_i(x) - W_j(x) &= \Big(\frac{\lambda_k - t_i}{x - t_i} - \frac{\lambda_k - t_j}{x - t_j}\Big)\frac{T(x)}{\Lambda(x)}\\ &= \frac{(t_j - t_i)(x - \lambda_k)}{(x - t_i)(x - t_j)}\frac{T(x)}{\Lambda(x)},\end{aligned}$$

we have

$$\left\{W_i(x) - W_j(x)\right\}\Big|_{x=\lambda_k} = \frac{(t_j - t_i)(x - \lambda_k)}{(x - t_i)(x - t_j)}\frac{T(x)}{\Lambda(x)}.$$

This proves (4.5.9), and hence the lemma. ∎

Let us summarize the lemmas obtained in this subsection:

PROPOSITION 4.5.5. *The deformation equations* (3.5.2 and 3) *for SL-equation* (4.2.1) *with* (4.2.2) *are equivalent to the system of differential equations:*

$$D_i\lambda_k = M_i\big[2M^{k,i}\nu_k - M^{k,i,0}\big], \tag{4.5.5}$$

$$D_i\nu_k = M_i\big[M^{k,i}u_{k,1} + M^{k,i,1}\nu_k - \frac{3}{2}M^{k,i,2}\big]. \tag{4.5.6}$$

To complete the proof of Theorem 4.2.2, it is sufficient to prove the following two propositions.

PROPOSITION 4.5.6. *The system of differential equations* (4.5.5 and 6) *is the Hamiltonian system* (4.2.10) *with the Hamiltonians* $L_1, \cdots, L_n$ *given by* (4.3.15).

PROPOSITION 4.5.7. *The Hamiltonian system* (4.2.10) *with* (4.3.15) *is completely integrable.*

Proof of Proposition 4.5.6. It is easily checked that (4.5.5) is just the equations

$$D_i\lambda_k = \{L_i, \lambda_k\}, \quad i, k = 1, \cdots, n.$$

We shall show that (4.5.6) is equivalent to

$$D_i\nu_k = \{L_i, \nu_k\} = -\frac{\partial L_i}{\partial \lambda_k}, \quad i, k = 1, \cdots, n. \tag{4.5.10}$$

Recall that the singular points $\lambda_1, \cdots, \lambda_n$ of the equation (4.2.1) are non-logarithmic if and only if (4.5.8) is satisfied; the condition can be written as

$$\sum_i E_{mi}L_i = W_m, \tag*{(4.5.11)$_m$}$$

$m = 1, \cdots, n$, where

$$W_m = {\nu_m}^2 - \Big(\frac{1}{\lambda_m} + \frac{1}{\lambda_m - 1}\Big)\nu_m - \sum_{\ell}^{m} \frac{\lambda_\ell(\lambda_\ell - 1)\nu_\ell}{\lambda_m(\lambda_m - 1)(\lambda_\ell - \lambda_m)} - U_m,$$

and U_m is given in (4.3.16). Differentiating $(4.5.11)_m$, $m \neq k$, with respect to λ_k, we get

$$(4.5.12)_m \quad \sum_i E_{mi}\frac{\partial L_i}{\partial \lambda_k}$$
$$= -\frac{\partial}{\partial \lambda_k}\Big\{\frac{\lambda_k(\lambda_k-1)}{\lambda_m(\lambda_m-1)(\lambda_k-\lambda_m)}\Big\}\nu_k + \frac{3}{2(\lambda_k-\lambda_m)^3},$$

$m \neq k$, and differentiating $(4.5.11)_k$ with respect to λ_k, we get

$$\sum_i E_{ki}\frac{\partial L_i}{\partial \lambda_k} = \frac{\partial W_k}{\partial \lambda_k} - \sum_i L_i \frac{\partial E_{ki}}{\partial \lambda_k}.$$

The above equation can be written as

$$(4.5.13)_k \qquad \sum_i E_{ki}\frac{\partial L_i}{\partial \lambda_k} = -u_{k,1} - \frac{\nu_k}{\lambda_k(\lambda_k-1)};$$

in fact, since $u_{k,1}$ is the coefficient of the power $(x-\lambda_k)$ in the Laurent expansion of (4.5.4) of $p(x,t)$ at $x=\lambda_k$, we have

$$u_{k,1} = \frac{\partial}{\partial x}\left[p(x,t) - \frac{3}{4(x-\lambda_k)^2} + \frac{\nu_k}{x-\lambda_k}\right]\Bigg|_{x=\lambda_k}$$
$$= \frac{\partial}{\partial x}\Big[\sum_{(i)} \frac{a_i}{(x-t_i)^2} + \frac{a_\infty}{x(x-1)} + \sum_\ell^k \frac{3}{4(x-\lambda_\ell)^2}$$
$$+ \frac{x+\lambda_k-1}{x(x-1)}\nu_k - \sum_\ell^k \frac{l_\ell(l_\ell-1)\nu_\ell}{x(x-1)(x-\lambda_\ell)}$$
$$+ \sum_i \frac{t_i(t_i-1)L_i}{x(x-1)(x-t_i)}\Big]\Bigg|_{x=\lambda_k}$$
$$= -\frac{\partial W_k}{\partial \lambda_k} + \sum_i L_i\frac{\partial E_{ki}}{\partial \lambda_k} - \frac{\nu_k}{\lambda_k(\lambda_k-1)};$$

and hence we obtain $(4.5.13)_k$.

By using $(4.5.13)_m$ $(m = 1, \cdots, n)$ and Lemma 4.3.1, $-\partial L_i/\partial\lambda_k$ $(i, k = 1, \cdots, n)$ can be written as

$$(4.5.14)\qquad -\frac{\partial L_i}{\partial\lambda_k}$$

$$=M_i\Big[M^{k,i}u_{k,1} + \Big\{\frac{M^{k,i}}{\lambda_k(\lambda_k-1)}$$

$$+\sum_m^k M^{m,i}\frac{\partial}{\partial\lambda_k}\Big(\frac{\lambda_k(\lambda_k-1)}{\lambda_m(\lambda_m-1)(\lambda_k-\lambda_m)}\Big)\Big\}\nu_k$$

$$-\sum_m^k \frac{3M^{m,i}}{2(\lambda_k-\lambda_m)^3}\Big].$$

On the other hand, we have

$$(4.5.15)\qquad \begin{cases} M^{k,i,1} = \dfrac{M^{k,i}}{\lambda_k(\lambda_k-1)} \\ \qquad\qquad + \displaystyle\sum_\ell^k M^{\ell,i}\frac{\partial}{\partial\lambda_k}\Big\{\frac{\lambda_k(\lambda_k-1)}{\lambda_\ell(\lambda_\ell-1)(\lambda_k-\lambda_\ell)}\Big\}, \\ M^{k,i,2} = \displaystyle\sum_\ell^k \frac{M^{\ell,i}}{(\lambda_k-\lambda_\ell)^3}. \end{cases}$$

In fact, if we put $Z_i(x) = T(x)/\{x(x-1)(x-t_i)\Lambda(x)\}$, we have

$$Z_i(x) = \sum_\ell \frac{M^{\ell,i}}{\lambda_\ell(\lambda_\ell-1)(x-\lambda_\ell)},$$

and, by the definition of $M^{k,i,m}$, we have

$$M^{k,i,1}$$

$$=\frac{\partial}{\partial x}\Big[x(x-1)Z_i(x) - \frac{M^{k,i}}{x-\lambda_k}\Big]\Big|_{x=\lambda_k}$$

$$=\sum_\ell^k M^{\ell,i}\frac{\partial}{\partial x}\Big[\frac{x(x-1)}{\lambda_\ell(\lambda_\ell-1)(x-\lambda_\ell)}\Big]\Big|_{x=\lambda_k} + \frac{M^{k,i}}{\lambda_k(\lambda_k-1)},$$

which shows the first identity of (4.5.15). The second identity of (4.5.15) is the consequence of the partial fraction expansion:

$$\begin{aligned}
&x(x-1)Z_i(x)\\
=&\frac{T(x)}{(x-t_i)\Lambda(x)}\\
=&\sum_\ell M^{\ell,i}\Big\{\frac{1}{x-\lambda_\ell}+\frac{x+\lambda_\ell-1}{\lambda_\ell(\lambda_\ell-1)}\Big\}.
\end{aligned}$$

Put the identity (4.5.15) into the equation (4.5.6) and compare it with (4.5.14). Then we obtain the latter half (4.5.10) of the Hamiltonian system (4.2.10), which completes the proof. ∎

Proof of Proposition 4.5.7. Let

$$\Gamma=\sum_k d\lambda_k\wedge d\nu_k+\sum_i dL_i\wedge dt_i$$

be the fundamental 2-form associated with the Hamiltonian system (4.2.9) and let $\mathcal{I}$ be the ideal (of the exterior differential algebra) generated by 1-forms

$$d\lambda_i-\sum_j\{L_j,\lambda_j\}dt_j,\quad d\nu_i-\sum_j\{L_j,\nu_j\}dt_j,\quad i=1,\ldots,n.$$

Then, by Lemma 6.3 in Chapter 1, we have

$$2\Gamma\equiv\sum_{i,j}\Gamma_{ij}dt_i\wedge dt_j\quad\text{modulo }\mathcal{I},$$

where

$$\Gamma_{ij}:=\partial_iL_j-\partial_jL_i+\{L_i,L_j\},$$

and ∂_i denotes the partial differentiation with respect to t_i.
Let us define operators $\hat{D}_j$ $(j=1,\cdots,n)$ acting on functions f in (λ,ν,t) by

$$\hat{D}_j:=\partial_j+\{L_j,f\}.$$

(Note that if $(\lambda(t), \nu(t))$ is a solution of the Hamiltonian system, then

$$D_j f(\lambda(t), \nu(t), t) = (\hat{D}_j f)(\lambda(t), \nu(t), t)$$

holds.) Then we have

$$2\Gamma_{ij} = \hat{D}_j L_i - \hat{D}_i L_j - \partial_i L_j + \partial_j L_i. \tag{4.5.16}$$

By virtue of Proposition 6.4 in Chapter 1, it is sufficient to show that Γ_{ij} vanishes.

Let

$$A_j(x,t) = M_j(x - t_i)\big(m_{j,i,0} + m_{j,i,1}(x - t_i) + \cdots\big),$$
$$p(x,t) = \frac{a_i}{(x - t_i)^2} + \frac{L_i}{x - t_i} + \cdots$$

be local expansions at $x = t_i$ $(i \neq j)$, where

$$m_{j,i,0} = \frac{1}{M_i(t_j - t_i)},$$
$$m_{j,i,1} = -\sum_k \frac{M^{k,j}}{(\lambda_k - t_i)^3}.$$

Since the left-hand side $\Xi_j(x)$ of (4.4.7) vanishes identically, in particular, we have $\Xi_j(t_i) = 0$. This gives the differential equations

$$\hat{D}_j L_i = M_j\big[m_{j,i,0} L_i + 2m_{j,i,0} a_i\big], \quad i, j = 1, \cdots, n.$$

By using these equations, we can see that the right-hand side of (4.5.16) vanishes. ∎

REMARK 4.5.8 The fundamental 2-form Γ of the Hamiltonian system (4.2.10) vanishes along any solution i.e., for a solution $(\lambda(t), \nu(t))$, defining $\varphi : t \mapsto (\lambda(t), \nu(t), t)$, we have

$$\varphi^* \Gamma = 0.$$

5. Schlesinger Systems

So far, we have studied the monodromy preserving deformation for the second order Fuchsian differential equations on $\mathbb{P}^1$. As we shall see in Section 7, the Garnier system $\mathcal{G}_n$ does not enjoy the Painlevé property. So we intend to transform $\mathcal{G}_n$ into a Hamiltonian system with the Painlevé property by an appropriate symplectic transformation. In finding such a transformation, we make use of the Schlesinger system, which will be discussed in this section.

Let B be the open subset of $\mathbb{C}^n$ defined by

$$B := \{t = (t_1, \ldots, t_n) \in \mathbb{C}^n;\ t_i \neq 0, 1, t_j\ (i \neq j)\},$$

and let U be the simply connected open subset of B. Consider a system of differential equations of the first order:

$$D\vec{y} = Q(x,t)\vec{y}, \qquad Q(x,t) = \sum_{(i)} \frac{Q_i(t)}{x - t_i}, \quad t \in U \tag{5.1}$$

with singular points $x = t_1, \ldots, t_n, t_{n+1} = 0, t_{n+2} = 1$ and $t_{n+3} = \infty$, where $Q_i(t)$ $(i = 1, \ldots, n+2)$ are 2×2 matrices whose components are holomorphic in U. Here the symbol $\sum_{(i)}$ stands for the summation over $i = 1, \ldots, n+2$. In order to consider the system (5.1) at $x = \infty$, we put $z = 1/x$; the system (5.1) can be written as

$$\frac{d\vec{y}}{dz} = \left\{\frac{L_\infty}{z} + \text{(a term holomorphic at } z = 0)\right\}\vec{y},$$

where

$$L_\infty := -\sum_{(i)} Q_i(t).$$

A system of differential equations of the form (5.1) is said to be of *Schlesinger type.*

Let us denote by $S(t)$ the set of singular points of (5.1) for $t = (t_1, \ldots, t_n) \in U$:

$$S(t) := \{t_1, \ldots, t_{n+3}\} \subset \mathbb{P}^1.$$

The system (5.1) is considered to be defined on

$$X = \bigcup_{t \in U} X_t, \qquad X_t = \mathbb{P}^1 \setminus S(t).$$

For the system (5.1), we assume:

(i) *For each* $i = 1, \dots, n+2$, *the eigenvalues* θ_i^+ *and* θ_i^- *of* $Q_i(t)$ *do not differ by an integer,*

(ii) *the system is normalized at* $x = \infty$ *as*

$$L_\infty = -\sum_{(i)} Q_i(t) = \begin{pmatrix} \theta_\infty^+ & 0 \\ 0 & \theta_\infty^- \end{pmatrix},$$

and $\theta_\infty^+ - \theta_\infty^- \notin \mathbf{Z}$.

The assumption (ii) implies that at $x = \infty$ (5.1) has a unique local fundamental solution $Z_\infty(x,t)$ of the form

$$Z_\infty(x,t) = R_\infty(z,t) z^{L_\infty}, \qquad R_\infty(0,t) = I_2,$$

where $z = 1/x$ and $R_\infty(z,t)$ is a power series in z. Regarding $t = (t_1, \dots, t_n) \in U$ as a parameter, we study the monodromy preserving deformation of (5.1). It follows that $\theta_i^\pm$, $\theta_\infty^\pm$ should remain invariant under the deformation. Now we state the theorem:

THEOREM 5.1. *Suppose that the assumptions* (i) *and* (ii) *hold. There is an M-invariant fundamental solution* $Y(x,t)$ *of* (5.1) *such that* $Y(x,t) = Z_\infty(x,t) C_\infty$ *with* $C_\infty \in GL(n, \mathbb{C})$, *if and only if* $Q_j(t)$ $(j = 1, \dots, n+2)$ *satisfy the system of differential equations:*

$$dQ_i = \sum_{(j)}^{i} [Q_j, Q_i] d \log(t_i - t_j), \qquad i = 1, \dots, n+2, \tag{5.2}$$

where d *denotes the exterior differentiation with respect to* t.

The above system of nonlinear differential equations (5.2) is called the *Schlesinger system.*

To prove the theorem, we recall the results in Section 3: The linear equation (5.1) has an M-invariant fundamental solution if and only if there is a 2×2 matrix of 1-forms

$$\Omega(x,t) = \sum_i \Omega_i(x,t)\, dt_i,$$

uniform on $\mathbb{P}^1 \times U$, such that the system of differential equations

$$DY = Q(x,t)Y, \qquad dY = \Omega(x,t)Y \tag{5.3}$$

is completely integrable; the integrability condition is given by

$$dQ = D\Omega - [Q, \Omega], \qquad d\Omega = \Omega \wedge \Omega. \tag{5.4}$$

To deduce the conditions for $Q_i(t)$ $(i = 1, \ldots, n+2)$ from (5.4), we give an explicit form of the 1-form $\Omega(x,t)$.

LEMMA 5.2. *The 1-form $\Omega(x,t)$ is given by*

$$\Omega(x,t) = -\sum_i \frac{Q_i(t)}{x - t_i}\, dt_i. \tag{5.5}$$

Proof Let $Y(x,t)$ be an M-invariant fundamental solution of (5.1) such that

$$Y(x,t) = Z_\infty(x,t)C_\infty, \quad \det C_\infty \neq 0 \tag{5.6}$$

with a matrix C_∞ independent of t. We define $\Omega(x,t)$ by $\Omega(x,t) = dY(x,t) \cdot Y(x,t)^{-1}$. By virtue of the assumption (i), the system (5.1) has a local fundamental solution $Z_i(x,t)$ of the form

$$Z_i(x,t) = R_i(x,t)(x - t_i)^{L_i}, \qquad i = 1, \ldots, n+2.$$

where $R_i(x,t)$ is a 2×2 matrix which is holomorphic at $x = t_i$ such that

$$\det R_i(t_i, t) \neq 0, \qquad Q_i(t) = R_i L_i R_i^{-1}|_{x=t_i}$$

and

$$L_i := \begin{pmatrix} \theta_i^+ & 0 \\ 0 & \theta_i^- \end{pmatrix}$$

for $i = 1, \dots, n+2$. There exist nonsingular matrices $C_i(t)$ ($i = 1, \dots, n+2$) holomorphic in U such that

$$Y(x,t) = Z_i(x,t)C_i(t).$$

Note here that the matrices L_i ($i = 1, \dots, n+2, \infty$) are independent of t, and the circuit matrices $C_i^{-1}e^{2\sqrt{-1}\pi L_i}C_i$ around $x = t_i$ ($i = 1, \dots, n+2$) with respect to the fundamental solution $Y(x,t)$ are also independent of t, since $Y(x,t)$ is an M-invariant fundamental solution. Since

$$D_j(C_i^{-1}e^{2\sqrt{-1}\pi L_i}C_i) = 0, \qquad j = 1, \dots, n+2,$$

we have

$$[D_jC_i \cdot C_i^{-1},\, e^{2\sqrt{-1}\pi L_i}] = 0.$$

So, taking account of the assumption, we see from Lemma 4.2.2 in Chapter 2, that $D_jC_i \cdot C_i^{-1}$ ($i = 1, \dots, n+2$) are diagonal matrices. Thus we have, at $x = t_i$,

$$\begin{aligned} D_jY \cdot Y^{-1} &= D_jR_i \cdot R_i^{-1} - \delta_{ij}\frac{R_iL_iR_i^{-1}}{x - t_i} \\ &\quad + R_i(x-t_i)^{L_i}D_jC_i \cdot C_i^{-1} \cdot (x-t_i)^{-L_i}R_i^{-1} \\ &= -\delta_{ij}\frac{Q_i}{x - t_i} \\ &\quad + (\text{a term holomorphic at } x = t_i). \end{aligned}$$

In a similar way, we can show that $D_jY \cdot Y^{-1} = \mathcal{O}(1/x)$ by taking into consideration $dC_\infty = 0$, and we have

$$\Omega_j(x,t) = D_jY \cdot Y^{-1} = -\frac{Q_j}{x - t_j}, \qquad j = 1, \dots, n.$$

Hence we obtain the lemma. ∎

Proof of Theorem 5.1 Suppose that $Y(x,t) = Z_\infty(x,t)C_\infty$ ($\det C_\infty \neq 0$) is an M-invariant fundamental solution of (5.1) such that $dC_\infty = 0$. Then substitute the expression of $Q(x,t)$ and (5.5) into the first equation of (5.4):

$$\sum_i \frac{Q_i\,dt_i}{(x-t_i)^2} + \sum_{(i)} \frac{dQ_i}{x-t_i}$$

$$= \sum_i \frac{Q_i\,dt_i}{(x-t_i)^2} + \left[\sum_{(i)} \frac{Q_i}{x-t_i}, \sum_j \frac{Q_j\,dt_j}{x-t_j}\right].$$

Bilinearity of the bracket $[\cdot,\cdot]$ yields

$$\sum_{(i)} \frac{dQ_i}{x-t_i} = -\sum_{(i)}\sum_j \frac{[Q_j, Q_i]}{(x-t_i)(x-t_j)}\,dt_j;$$

taking the residue at $x = t_i$, we obtain (5.2):

$$dQ_i = \sum_{(j)}^{i} [Q_j, Q_i]\, d\log(t_i - t_j), \qquad i = 1, \cdots, n+2.$$

Conversely, we have (5.4) with (5.5), provided that Q_i's satisfy the system (5.2). In fact, we readily obtain the first equation of (5.4) by tracing back the above computations. On the other hand, the second one reads:

$$dQ_i \wedge dt_i = -\left(\sum_j^i \frac{[Q_j, Q_i]}{t_i - t_j} dt_j\right) \wedge dt_i, \quad i = 1, \cdots, n,$$

which is a immediate consequence of (5.2). Thus the system (5.3) is completely integrable; so there exists an M-invariant fundamental solution $Y(x,t)$. Let us express it, near $x = \infty$, as

$$Y(x,t) = Z_\infty(x,t)C_\infty(t).$$

We have

$$\Omega_j(x,t) = D_j Y \cdot Y^{-1} = Y dC_\infty \cdot C_\infty^{-1} Y^{-1} + \mathcal{O}(x^{-1}).$$

It follows from (5.5) that $dC_\infty = 0$.

PROPOSITION 5.3. *The Schlesinger system* (5.2) *is completely integrable.*

Proof Put

$$\omega_i := dQ_i + \sum_{(j)}^{i} [Q_i, Q_j] d\log(t_i - t_j), \qquad i = 1, \cdots, n+2.$$

The complete integrability of the Schlesinger system (5.2) is equivalent to $d\omega_i \equiv 0$ modulo $\mathcal{I}$ ($i = 1, \cdots, n+2$), where $\mathcal{I}$ is the ideal of the exterior algebra generated by $\omega_1, \ldots, \omega_{n+2}$ (Chapter 1, Corollary 5.2). Using the Jacobi identity for the bracket $[\cdot, \cdot]$, we have

$$\begin{aligned}
-d\omega_i &= \sum_{(j)}^{i} \{[dQ_j, Q_i] + [Q_j, dQ_i]\} \wedge d\log(t_i - t_j) \\
&\equiv \sum_{(j)}^{i} \sum_{(k)}^{j} [[Q_k, Q_j], Q_i] d\log(t_k - t_j) \wedge d\log(t_j - t_i) \\
&\quad + \sum_{(j)}^{i} \sum_{(k)}^{i} [Q_j, [Q_k, Q_i]] d\log(t_k - t_i) \wedge d\log(t_i - t_j) \\
&= \sum_{1 \le j < k \le n+2, j,k \ne i} [[Q_k, Q_j], Q_i] \{ d\log(t_k - t_j) \wedge d\log(t_j - t_i) \\
&\quad - d\log(t_j - t_k) \wedge d\log(t_k - t_i) \\
&\quad - d\log(t_k - t_i) \wedge d\log(t_i - t_j) \} \\
&= 0.
\end{aligned}$$

The last equality is true because the expression in curly brackets vanishes identically, as can be checked by direct computations. ∎

Now we state a celebrated theorem (see Section 1.1).

THEOREM 5.4 ([Malg.2],[Miw]). *The Schlesinger system enjoys the Painlevé property. Moreover, any solution of the system is meromorphic on the universal covering space of B.*

This theorem is not proved in this book, since it requires a long preparation.

6. The Schlesinger system and the Garnier system $\mathcal{G}_n$

To transform the Garnier system $\mathcal{G}_n$ into a system with the Painlevé property in Section 7, we first transform, in this section, linear equations of Schlesinger type into second order linear equations, and then we get a transformation formula taking the Schlesinger system (the deformation equation for the former) into the Garnier system (the deformation equation for the latter). We also study a transformation of Garnier systems into Schlesinger systems; this transformation is not used later.

6.1. Transformation of systems of equations into second order equations

Consider a system of differential equations for an unknown vector $\vec{y} = {}^t(y, z)$:

$$D\vec{y} = Q(x)\vec{y}, \qquad Q(x) = \begin{pmatrix} q_{11}(x) & q_{12}(x) \\ q_{21}(x) & q_{22}(x) \end{pmatrix}, \tag{6.1.1}$$

where $q_{ij}(x) \in \mathbb{C}(x)$, $i, j = 1, 2$.

LEMMA 6.1.1. *Suppose that $q_{12}(x)$ does not vanish identically. Then the first component $y(x)$ of a solution $\vec{y}(x)$ of (6.1.1) satisfies the differential equation*

$$D^2y + p_1(x)Dy + p_2(x)y = 0, \tag{6.1.2}$$

where

$$\begin{cases} p_1(x) = -D\log q_{12} - \operatorname{Tr} Q(x), \\ p_2(x) = \det Q(x) - Dq_{11} + q_{11}D\log q_{12}. \end{cases} \tag{6.1.3}$$

Proof We write the system (6.1.1) componentwise:

$$\begin{cases} Dy = q_{11}y + q_{12}z, \\ Dz = q_{21}y + q_{22}z. \end{cases} \tag{6.1.4}$$

Let us eliminate z in (6.1.4): Differentiating the first equation of (6.1.4), and using the second equation, we have

$$\begin{aligned} D^2y &= q_{11}Dy + q_{12}Dz + Dq_{11}\cdot y + Dq_{12}\cdot z \\ &= q_{11}Dy + (Dq_{11} + q_{12}q_{21})y + (Dq_{12} + q_{12}q_{22})z. \end{aligned}$$

Solving the first equation of (6.1.4) with respect to z as

$$z = \frac{1}{q_{12}}(Dy - q_{11}y), \tag{6.1.5}$$

and substituting it into the equation obtained above, we see that $y(x)$ satisfies the equation (6.1.2) with $p_1(x)$, $p_2(x)$ given by (6.1.3). ∎

LEMMA 6.1.2. *Suppose that λ is not a singular point of the system (6.1.1). If $x = \lambda$ is a zero of $q_{12}(x)$ of order m, then $x = \lambda$ is an apparent singular point of (6.1.2) with the exponents 0 and $m+1$.*

Proof Let $x = \lambda$ be a zero of $q_{12}(x)$ of order m; the function $q_{12}(x)$ admits the following expression:

$$q_{12}(x) = (x-\lambda)^m b(x), \qquad b(\lambda) \neq 0, \qquad b(x) \in \mathbb{C}\{x-\lambda\}$$

in the neighbourhood of $x = \lambda$; and hence

$$D\log q_{12}(x) = \frac{m}{x-\lambda} + \frac{b'(x)}{b(x)}.$$

Substituting it into (6.1.3), we get

$$\begin{aligned} p_1(x) &= -\frac{m}{x-\lambda} + (\text{a term holomorphic at } x = \lambda), \\ p_2(x) &= q_{11}(\lambda)\frac{m}{x-\lambda} + (\text{a term holomorphic at } x = \lambda). \end{aligned}$$

It follows that $x = \lambda$ is a regular singular point of (6.1.2) at which solutions are holomorphic, and the characteristic exponents are obtained by solving

$$\rho(\rho-1) - m\rho = 0.$$

Hence, $x = \lambda$ is an apparent singular point of (6.1.2), and the exponents are 0 and $m+1$. ∎

The following two propositions are easy to see.

PROPOSITION 6.1.3. *Let*

$$Y(x) = (\vec{y}_1(x), \vec{y}_2(x)), \quad \vec{y}_i(x) = {}^t(y_i(x), z_i(x)), \qquad i = 1, 2$$

be a fundamental solution of (6.1.1). *Then* $y_1(x)$ *and* $y_2(x)$ *form a fundamental system of solutions of the equation* (6.1.2) *with* $p_1(x)$, $p_2(x)$ *given by* (6.1.3).

Now let us suppose that the matrix $Q(x)$ in (6.1.1) depends holomorphically on the parameters $t = (t_1, \ldots, t_n) \in U$, where U is a simply connected domain in $\mathbb{C}^n$.

PROPOSITION 6.1.4. *If the system* (6.1.1) *admits a monodromy preserving deformation, then so does the second order equation* (6.1.2) *with* (6.1.3).

6.2. Transformation of the Schlesinger system into the Garnier system

We start by considering the system (6.1.1) of Schlesinger type with coefficients

$$Q(x,t) = \begin{pmatrix} q_{11}(x,t) & q_{12}(x,t) \\ q_{21}(x,t) & q_{22}(x,t) \end{pmatrix} = \sum_{(i)} \frac{Q_i(t)}{x - t_i}, \tag{6.2.1}$$

where

$$Q_i(t) = \begin{pmatrix} q_{11}^i(t) & q_{12}^i(t) \\ q_{21}^i(t) & q_{22}^i(t) \end{pmatrix},$$

and $t_{n+1} = 0, t_{n+2} = 1$.
We assume the following:

(i) *The eigenvalues of* $Q_i(t)$ *are* 0 *and* $\theta_i \notin \mathbb{Z}$, θ_i *being independent of* t,

(ii) *The system* (6.1.1) *is normalized at* $x = t_{n+3} = \infty$, *i.e.*,

$$-\sum_{(i)} Q_i(t) = \begin{pmatrix} \alpha & 0 \\ 0 & \alpha + \theta_\infty - 1 \end{pmatrix}, \qquad \theta_\infty \notin \mathbb{Z},$$

where θ_∞ *is independent of* t *and*

$$\alpha = -\frac{1}{2}\left(\sum_{(i)} \theta_i + \theta_\infty - 1\right),$$

(iii) *all zeros of* $q_{12}(x,t)$ *are simple.* (Logically, θ_∞ should be denoted by θ_{n+3}, but θ_∞ is easier to read.)

By the assumptions (ii) and (iii), we see that $q_{12}(x,t)$ is a rational function in x having $n+2$ simple poles and n simple zeros: $\lambda_1, \ldots, \lambda_n$. We set

$$q_{12}(x,t) = \sum_{(i)} \frac{q_{12}^i}{x - t_i} = X\frac{\Lambda(x)}{T(x)}, \tag{6.2.2}$$

where

$$X = \sum_{(i)} t_i q_{12}^i, \quad \Lambda(x) = \prod_k (x - \lambda_k), \quad T(x) = \prod_{(i)} (x - t_i)$$

and the symbol $\prod_{(i)}$ stands for the product over $i = 1, \cdots, n+2$. We transform the equation of Schlesinger type into a second order equation,

$$D^2 y + p_1(x,t) Dy + p_2(x,t) y = 0, \tag{6.2.3}$$

as in Lemma 6.1.1. We see, by Lemmas 6.1.2 and by the assumptions, that the Riemann scheme of (6.2.3) is

$$\begin{pmatrix} x = t_i & x = \infty & x = \lambda_k \\ 0 & \alpha & 0 \\ \theta_i & \alpha + \theta_\infty & 2 \end{pmatrix} \quad i = 1, \cdots, n+2; k = 1, \cdots, n.$$

Thus the coefficients $p_1(x,t)$ and $p_2(x,t)$ can be written in the form

$$\begin{cases} p_1(x,t) = \displaystyle\sum_{(i)} \frac{1 - \theta_i}{x - t_i} - \sum_k \frac{1}{x - \lambda_k}, \\ p_2(x,t) = \displaystyle\frac{\kappa}{x(x-1)} - \sum_i \frac{t_i(t_i - 1)K_i}{x(x-1)(x-t_i)} \\ \qquad\qquad + \displaystyle\sum_k \frac{\lambda_k(\lambda_k - 1)\mu_k}{x(x-1)(x-\lambda_k)}, \end{cases} \tag{6.2.4}$$

where

$$\kappa = \frac{1}{4}\left[\left(\sum_{(i)} \theta_i - 1\right)^2 - \theta_\infty{}^2\right].$$

PROPOSITION 6.2.1. *A system of equations of Schlesinger type with* (6.2.1) *corresponds to a second order equation* (6.2.3) *with* (6.2.4) *as follows:*

$$(6.2.5)\quad \begin{cases} \lambda_k \quad (k = 1, \dots, n): \text{ zeros of } q_{12}(x) \\ \mu_k := \sum_{(i)} \frac{q_{11}^i}{\lambda_k - t_i}, \qquad k = 1, \dots, n, \\ K_i := \sum_k \frac{q_{11}^i}{\lambda_k - t_i} + \sum_{(j)}^{i} \frac{1}{t_i - t_j}(q_{11}^i + q_{11}^j - \theta_i \theta_j) \\ \qquad + \operatorname{Tr}\left[Q_i \sum_{(j)}^{i} \frac{Q_j}{t_i - t_j}\right], \qquad i = 1, \dots, n. \end{cases}$$

Proof Notice that $p_1(x,t)$ and $p_2(x,t)$ are obtained from $Q(x,t)$ by (6.1.3), and that K_i, μ_k are written as

$$K_i = -\operatorname*{Res}_{x=t_i} \; p_2(x,t)dx, \quad \mu_k = \operatorname*{Res}_{x=\lambda_k} \; p_2(x,t)dx.$$

We have

$$\begin{aligned} \mu_k &= \operatorname*{Res}_{x=\lambda_k} \; \{\det Q(x,t) - Dq_{11}(x,t) + q_{11}(x,t) D \log q_{12}(x,t)\}\, dx \\ &= \operatorname*{Res}_{x=\lambda_k} \frac{q_{11}(x,t)\Lambda'(x)}{\Lambda(x)}\, dx \\ &= \sum_{(i)} \frac{q_{11}^i}{\lambda_k - t_i}, \end{aligned}$$

and

$$K_i = -\operatorname*{Res}_{x=t_i} \{\det Q(x,t) - Dq_{11}(x,t) + q_{11}(x,t) D \log q_{12}(x,t)\}\, dx$$

$$= -q_{11}^i \left(\frac{\Lambda'(x)}{\Lambda(x)} - \frac{T'(x)}{T(x)} + \frac{1}{x-t_i} \right) |_{x=t_i} + \left(q_{11}(x,t) - \frac{q_{11}^i}{x-t_i} \right) |_{x=t_i}$$

$$- q_{22}^i \left(q_{11}(x,t) - \frac{q_{11}^i}{x-t_i} \right) |_{x=t_i} - q_{11}^i \left(q_{22}(x,t) - \frac{q_{22}^i}{x-t_i} \right) |_{x=t_i}$$

$$+ q_{12}^i \left(q_{21}(x,t) - \frac{q_{21}^i}{x-t_i} \right) |_{x=t_i} + q_{21}^i \left(q_{12}(x,t) - \frac{q_{12}^i}{x-t_i} \right) |_{x=t_i}$$

$$= \sum_k \frac{q_{11}^i}{\lambda_k - t_i} + \sum_{(j)}^{i} \frac{q_{11}^i + q_{11}^j}{t_i - t_j}$$

$$+ (q_{11}^i - \theta_i) \sum_{(j)}^{i} \frac{q_{11}^j}{t_i - t_j} + (q_{22}^i - \theta_i) \sum_{(j)}^{i} \frac{q_{22}^j}{t_i - t_j}$$

$$+ q_{12}^i \sum_{(j)}^{i} \frac{q_{21}^j}{t_i - t_j} + q_{21}^i \sum_{(j)}^{i} \frac{q_{12}^j}{t_i - t_j}$$

$$= \sum_k \frac{q_{11}^i}{\lambda_k - t_i} + \sum_{(j)}^{i} \frac{1}{t_i - t_j} (q_{11}^i + q_{11}^j - \theta_i \theta_j)$$

$$+ \operatorname{Tr}\left(Q_j \sum_{(j)}^{i} \frac{Q_j}{t_i - t_j} \right),$$

where we used $\operatorname{Tr} Q_i(t) = \theta_i$ to show the second and third equality. ∎

The following is the consequence of Theorem 4.1.2, Propositions 6.1.4 and 6.2.1.

COROLLARY 6.2.2. *If $(Q_1(t), \ldots, Q_{n+2}(t))$ satisfies the Schlesinger system, then $(\lambda_1(t), \ldots, \lambda_n(t), \mu_1(t), \ldots, \mu_n(t))$ defined by* (6.2.5) *satisfies the Garnier system $\mathcal{G}_n$.*

REMARK 6.2.3. The Schlesinger system enjoys the Painlevé property, whereas the Garnier system does not. This follows from Theorem 5.4

and the fact that $\lambda_1(t),\dots,\lambda_n(t)$ are n zeros of a rational function $q_{12}(x,t)$.

6.3. Transformation of second order equations into systems of equations

Let us investigate the (λ,μ)-dependence of $Q_i(t)$ $(i=1,\dots,n+2)$. They are determined up to the factor $X=\sum_{(j)} t_j q_{12}^j$. In fact we have the following result.

PROPOSITION 6.3.1. *The matrices Q_i $(i=1,\cdots,n+2)$ are expressed in terms of (λ,μ) and X as follows:*

$$\begin{aligned} q_{11}^i &= M_i(W_i - W), \\ q_{12}^i &= -M_i X, \\ q_{21}^i &= -X^{-1}(W_i - W)\left[M_i(W - W_i) + \theta_i\right], \\ q_{22}^i &= \theta_i - M_i(W_i - W), \end{aligned}$$

where

$$M_i := -\operatorname*{Res}_{x=t_i} \frac{\Lambda(x)}{T(x)} = -\frac{\Lambda(t_i)}{T'(t_i)}, \qquad i=1,\cdots,n+2,$$

$$W_j := \sum_k M^{k,j}\left(\mu_k + \frac{\alpha}{\lambda_k}\right), \qquad j=1,\cdots,n,$$

$$\begin{aligned} M_{n+1}W_{n+1} &:= \sum_j (t_j-1)M_jW_j - \alpha, \\ M_{n+2}W_{n+2} &:= -\sum_j t_jM_jW_j, \\ (\theta_\infty - 1)W &:= \sum_{(j)} W_j(M_jW_j - \theta_j), \end{aligned} \tag{6.3.1}$$

and $\Lambda(x)$, $T(x)$ are defined by

$$\Lambda(x) = \prod_k (x-\lambda_k), \qquad T(x) = \prod_{(i)} (x - t_i).$$

Proof Since

$$q_{12}(x,t) = X\frac{\Lambda(x)}{T(x)} = -\sum_{(i)}\frac{M_iX}{x-t_i},$$

we have

$$q_{12}^i = -M_iX, \quad i = 1,\cdots,n+2.$$

Note that we have

$$\sum_{(j)} M_j = 0 \tag{6.3.2}$$

from the relation $\sum_{(j)} q_{12}^j = 0$, which is a part of the assumption (ii) in Section 6.2.

Next let us determine q_{11}^i ($i = 1,\cdots,n$). Set

$$W := \sum_{(j)} t_j q_{11}^j. \tag{6.3.3}$$

Then, using the assumption (ii) in Section 6.2, we have

$$\sum_{(i)} q_{11}^i = -\alpha. \tag{6.3.4}$$

We write the second equality of (6.2.5) in Proposition 6.2.1 in the following form

$$\sum_i E_{ki}q_{11}^i = -\frac{W}{\lambda_k(\lambda_k-1)} + \mu_k + \frac{\alpha}{\lambda_k},$$

where E_{ki} is given by (4.3.1). Applying Lemma 4.3.1 and using the identity (4.3.13), we get

$$\begin{aligned} q_{11}^i &= M_i\sum_k M^{k,i}\left(\mu_k + \frac{\alpha}{\lambda_k} - \frac{W}{\lambda_k(\lambda_k-1)}\right) \\ &= M_i(W_i - W), \qquad i = 1,\cdots,n, \end{aligned} \tag{6.3.5}$$

where

$$W_i := \sum_k M^{k,i}\left(\mu_k + \frac{\alpha}{\lambda_k}\right),$$

$M^{k,i}$ being given by (4.3.14). q_{11}^{n+1} and q_{11}^{n+2} are determined as follows. First notice that we have the following identities:

$$\sum_{(j)} t_j M_j + 1 = 0, \quad \sum_{(j)} (t_j - 1) M_j + 1 = 0,$$

which are shown by applying the residue theorem to the rational function $x\Lambda(x)/T(x)$. Substitute (6.3.5) into (6.3.3) and (6.3.4), and solve them with respect to q_{11}^{n+1} and q_{11}^{n+2}. Then, if we define W_{n+1} and W_{n+2} by

$$M_{n+1}W_{n+1} = \sum_j (t_j - 1) M_j W_j - \alpha,$$

$$M_{n+2}W_{n+2} = -\sum_j t_j M_j W_j,$$

we see that q_{11}^{n+1} and q_{11}^{n+2} can be written in the form stated in the proposition. Now it is easy to see that q_{21}^i and q_{22}^i are determined by the assumption (i) in Section 6.2.
To complete the proof of the proposition, it is sufficient to express W in terms of λ, μ and X. We have

$$\sum_{(j)} M_j W_j + \alpha = 0, \tag{6.3.6}$$

by the definition of W_{n+1}, W_{n+2}, and we have the Fuchsian relation

$$\sum_{(j)} \theta_j + \theta_\infty + 2\alpha = 1.$$

On the other hand, observing the (2,1)-components of the assumption (ii) in Section 6.2, we have

$$\sum_{(j)} q_{21}^j = 0;$$

substituting the expressions of the q_{21}^j's into the condition, we have

$$\sum_{(i)}(W_i - W)\,[M_i(W - W_i) + \theta_i] = 0.$$

In view of (6.3.2), (6.3.6) and the Fuchs relation, we see that W is expressed as in the proposition. ∎

6.4. Relation between the Garnier system and the Schlesinger system

PROPOSITION 6.4.1. *Assume that $Q_i(t)$ $(i = 1, \cdots, n+2)$ satisfy the Schlesinger system (5.2) and define $(\lambda(t), \mu(t))$ by (6.2.5). Then $X = \sum_{(j)} t_j q_{12}^j$ satisfies the completely integrable Pfaffian equation*

$$d\log X = -\theta_\infty \sum_{(j)} M_j dt_j. \tag{6.4.1}$$

To prove the proposition, we need two lemmas.

LEMMA 6.4.2. *Suppose (λ, μ) satisfies the Garnier system. Then $\sigma := \sum_k \lambda_k$ satisfies the differential equations*

$$D_i\sigma = M_i[2W_i - L_i],$$

where W_i is given by (6.3.1) and

$$L_i := \sum_{(h)}(\theta_h - \delta_{ih})(t_h - 1 + \frac{\delta_{ih}}{M_i}) - 2\alpha - \theta_\infty m^i,$$

$$m^i := \sigma - \sum_h^i t_h.$$

Proof Since $(\lambda_1, \cdots, \lambda_n)$ satisfies the Garnier system, we have

$$D_i\lambda_k = M_i\left[2M^{k,i}\mu_k - M^{k,i}\sum_{(h)}\frac{\theta_h - \delta_{ih}}{\lambda_k - t_h}\right]. \tag{6.4.2}$$

So

$$D_i\sigma = M_i\left[2\sum_k M^{k,i}\mu_k - \sum_{(h)}(\theta_h - \delta_{ih})\sum_k\frac{M^{k,i}}{\lambda_k - t_h}\right].$$

To compute the second term of the right-hand side, we show

$$\sum_k\frac{M^{k,i}}{\lambda_k - t_h} = t_h - 1 + m^i + \frac{\delta_{ih}}{M_i}, \qquad h = 1, \cdots, n+2. \tag{6.4.3}$$

We prove it only when $h = i$, since the other cases are proved in the same way. Consider a rational function

$$R_i(x) = \frac{T(x)}{(x - t_i)^2\Lambda(x)},$$

then we see that

$$\begin{aligned}
\operatorname*{Res}_{x=\lambda_k} R_i(x)\,dx &= \frac{T(\lambda_k)}{(\lambda_k - t_i)^2\Lambda'(\lambda_k)} \\
&= \frac{M^{k,i}}{\lambda_k - t_i}, \qquad k = 1, \cdots, n, \\
\operatorname*{Res}_{x=t_i} R_i(x)\,dx &= \frac{T'(t_i)}{\Lambda(t_i)} \\
&= -\frac{1}{M_i}, \\
\operatorname*{Res}_{x=\infty} R_i(x)\,dx &= -t_i + 1 - m^i.
\end{aligned}$$

Then (6.4.3) is a consequence of the residue theorem. Now noting that W_i can be written as

$$W_i = \sum_k M^{k,i}\mu_k + \alpha(m^i - 1),$$

we see that the lemma is obtained by virtue of (6.4.2), (6.4.3) and the Fuchs relation. ∎

LEMMA 6.4.3. *Suppose that (λ, μ) satisfies the Garnier system. Then the 1-form $\sum_i M_i\, dt_i$ is closed, i.e.,*

$$D_i M_j = D_j M_i \qquad \text{for} \quad i, j = 1, \cdots, n.$$

Proof Since $M_i = -\Lambda(t_i)/T'(t_i)$, we have

$$D_j M_i = M_i \left[-\sum_k \frac{1}{t_i - \lambda_k} D_j \lambda_k + \frac{1}{t_i - t_j} \right], \qquad i \neq j.$$

So we have

$$\begin{aligned}(6.4.4)\ & D_i M_j - D_j M_i \\ &= \sum_k \left(\frac{M_i}{\lambda_k - t_i} D_j \lambda_k - \frac{M_j}{\lambda_k - t_j} D_i \lambda_k \right) - \frac{M_i + M_j}{t_i - t_j} \\ &= (t_i - t_j) M_i M_j \sum_k \frac{T(\lambda_k)}{(\lambda_k - t_i)^2 (\lambda_k - t_j)^2 \Lambda'(\lambda_k)} - \frac{M_i + M_j}{t_i - t_j}.\end{aligned}$$

Here, to show the second equality, we used the equality (4.5.1) in the form

$$\begin{aligned}&\frac{M_i}{\lambda_k - t_i} D_j \lambda_k - \frac{M_j}{\lambda_k - t_j} D_i \lambda_k \\ &= -(t_i - t_j) M_i M_j \frac{T(\lambda_k)}{(\lambda_k - t_i)^2 (\lambda_k - t_j)^2 \Lambda'(\lambda_k)}.\end{aligned}$$

We assert that the right-hand side of (6.4.4) vanishes. In fact, for a rational function

$$R(x) = \frac{T(x)}{(x - t_i)^2 (x - t_j)^2 \Lambda(x)}$$

it is easy to see that

$$\operatorname*{Res}_{x=\lambda_k} R(x)\,dx = \frac{T(\lambda_k)}{(\lambda_k - t_i)^2(\lambda_k - t_j)^2\Lambda'(\lambda_k)},$$

$$\operatorname*{Res}_{x=t_i} R(x)\,dx = \frac{T'(t_i)}{(t_i - t_j)^2\Lambda(t_i)} = \frac{-1}{(t_i - t_j)^2 M_i},$$

$$\operatorname*{Res}_{x=t_j} R(x)\,dx = \frac{T'(t_j)}{(t_i - t_j)^2\Lambda(t_j)} = \frac{-1}{(t_i - t_j)^2 M_j},$$

$$\operatorname*{Res}_{x=\infty} R(x)\,dx = 0;$$

so the residue theorem implies the assertion, and hence the lemma. ∎

Proof of Proposition 6.4.1 Observing the (1,2)-components of both sides of the Schlesinger system (5.2), we have

$$dq_{12}^j = \sum_{(i)} (q_{11}^i q_{12}^j + q_{12}^i q_{22}^j - q_{11}^j q_{12}^i - q_{12}^j q_{22}^i)\, d\log(t_i - t_j), \quad i \neq j.$$

Putting the expressions for $q_{11}^i, q_{12}^i, q_{22}^i, q_{11}^j, q_{12}^j, q_{22}^j$ given in Proposition 6.3.1 into the above equation, we obtain for any $j(\neq i)$,

$$(6.4.5) \quad D_i \log X = \frac{M_i}{t_i - t_j}\left[2(W_i - W_j) + \frac{\theta_j}{M_j} - \frac{\theta_i}{M_i}\right] - D_i \log M_j.$$

We show that the right-hand side of (6.4.5) equals to $-\theta_\infty M_i$. By Lemma 6.4.2, we have

$$\begin{aligned}
2(W_i - W_j) &= (\frac{1}{M_i}D_i - \frac{1}{M_j}D_j)\sigma + (L_i - L_j) \\
&= (t_i - t_j)\left[\frac{1}{M_i}\sum_k \frac{1}{\lambda_k - t_j} D_i\lambda_k + \sum_k \frac{M^{k,i}}{(\lambda_k - t_j)^2}\right] \\
&\quad + \frac{\theta_i - 1}{M_i} - \frac{\theta_j - 1}{M_j} - (1 + \theta_\infty)(t_i - t_j) \\
&= (t_i - t_j)\left[\frac{1}{M_i}\sum_k \frac{1}{\lambda_k - t_j} D_i\lambda_k + 1\right] - \frac{1}{M_j} \\
&\quad + \frac{\theta_i - 1}{M_i} - \frac{\theta_j - 1}{M_j} - (1 + \theta_\infty)(t_i - t_j).
\end{aligned}$$

Here the second equality is assured by

$$L_i - L_j = \frac{\theta_i - 1}{M_i} - \frac{\theta_j - 1}{M_j} - (1 + \theta_\infty)(t_i - t_j)$$

and by (4.5.1) which can be written as

$$\frac{1}{M_i} D_i \lambda_k - \frac{1}{M_j} D_j \lambda_k$$
$$= \frac{t_i - t_j}{M_i} \frac{1}{\lambda_k - t_j} D_i \lambda_k + (t_i - t_j) \frac{M^{k,i}}{(\lambda_k - t_j)^2} = 0, \qquad i \neq j,$$

and the third equality is verified by applying the residue theorem to $T(x)/(x - t_i)(x - t_j)^2 \Lambda(x)$. Thus we have

$$2(W_i - W_j) + \frac{\theta_j}{M_j} - \frac{\theta_i}{M_i} = \frac{t_i - t_j}{M_i} \sum_k \frac{1}{\lambda_k - t_j} D_i \lambda_k - \frac{1}{M_i} - \theta_\infty (t_i - t_j).$$

Since

$$D_i \log M_j = \frac{1}{t_j - t_i} + \sum_k \frac{1}{\lambda_k - t_j} D_i \lambda_k, \quad i \neq j,$$

we see that the right-hand side of (6.4.5) is equal to $-\theta_\infty M_i$. Combining this with Lemma 6.4.3, we arrive at Proposition 6.4.1. ∎

7. The Polynomial Hamiltonian system $\mathcal{H}_n$ associated with $\mathcal{G}_n$

In this section we shall transform $\mathcal{G}_n$, which has movable algebraic branch points (Remark 6.2.3), into a Hamiltonian system enjoying the Painlevé property whose Hamiltonians are polynomials in dependent variables. To do so, we use (6.2.2), Proposition 6.2.1, and Theorem 5.4.

7.1. Transformation of $\mathcal{G}_n$ into the polynomial Hamiltonian system $\mathcal{H}_n$

Define a locally biholomorphic mapping:

$$\phi : U \times B \to \mathbb{C}^n \times B,$$

$$(\lambda, t) \mapsto (q, s)$$

$$U := \{\lambda = (\lambda_i) \in \mathbb{C}^n; \lambda_i \neq \lambda_j \text{ for } i \neq j\}$$

by

$$q_i = -t_i M_i, \quad s_i = \frac{t_i}{t_i - 1}, \qquad i = 1, \cdots, n, \tag{7.1.1}$$

where M_i are symmetric polynomials of the λ_j's defined by (4.3.2). It induces the bundle map $\phi_* : (\lambda, \mu, t) \mapsto (q, p, s)$

$$\begin{array}{ccc} T^*U \times B & \xrightarrow{\phi_*} & T^*\mathbb{C}^n \times B \\ \downarrow & & \downarrow \\ U \times B & \xrightarrow{\phi} & \mathbb{C}^n \times B \end{array}$$

with

$$\mu_k = \sum_i p_i \partial q_i / \partial \lambda_k, \tag{7.1.2}$$

where T^*U (resp. $T^*\mathbb{C}^n$) denotes the cotangent bundle of U (resp. $\mathbb{C}^n$).

The bundle map ϕ_* extends to a symplectic transformation Φ : $(\lambda, \mu, t, K) \to (q, p, s, H)$: we define

$$H_i := -(t_i - 1)^2 \left[K_i + \sum_j p_j \frac{\partial q_j}{\partial t_i} \right], \tag{7.1.3}$$

so that

$$\sum_i (p_i dq_i - H_i ds_i) = \sum_i (\mu_i d\lambda_i - K_i dt_i).$$

Firstly let us express $p_1, \cdots, p_n$ as functions of (λ, μ, t).

LEMMA 7.1.1. *The mapping* $\phi_* : (\lambda, \mu, t) \to (q, p, s)$ *is given by* (7.1.1) *and*

$$p_i = -(t_i - 1) \sum_k \frac{M^{k,i} \mu_k}{\lambda_k(\lambda_k - 1)}, \qquad i = 1, \cdots, n, \tag{7.1.4}$$

where $M^{k,i}$ *is defined by* (4.3.2).

Proof Since $q_i = -t_i M_i$, we have from (7.1.2)

$$\mu_k = \sum_i \frac{q_i p_i}{\lambda_k - t_i}, \qquad k = 1, \cdots, n, \tag{7.1.5}$$

or equivalently

$$E \, {}^t\!\left(\frac{q_1 p_1}{t_1(t_1 - 1)}, \cdots, \frac{q_n p_n}{t_n(t_n - 1)} \right) = {}^t\!\left(\frac{\mu_1}{\lambda_1(\lambda_1 - 1)}, \cdots, \frac{\mu_n}{\lambda_n(\lambda_n - 1)} \right),$$

where $E = (E_{ki})_{k,i=1,\cdots,n}$ is the matrix given by (4.3.1). Applying Lemma 4.3.1 to this linear equation with unknowns $q_1 p_1, \cdots, q_n p_n$, we obtain

$$q_i p_i = M_i t_i (t_i - 1) \sum_k \frac{M^{k,i} \mu_k}{\lambda_k(\lambda_k - 1)}. \quad \blacksquare$$

The main theorem of this section is the following.

THEOREM 7.1.2. *The symplectic transformation* Φ *takes the Garnier system* $\mathcal{G}_n$ *into the system* $\mathcal{H}_n$*:*

$$\begin{cases} dq_i = \sum_j \{H_j, q_i\} \, ds_j, \\ dp_i = \sum_j \{H_j, p_i\} \, ds_j, \end{cases} \qquad i = 1, \cdots, n,$$

with Hamiltonians:

$$H_i := \frac{1}{e(s_i)}\left[\sum_{j,k} E^i_{jk}(s,q)p_jp_k - \sum_j F^i_j(s,q)p_j + \kappa q_i\right],$$

where

$$e(x) = x(x-1),$$

and E^i_{jk}, $F^i_j \in \mathbb{C}(s)[q]$ *are given by*

$$E^i_{jk} := \begin{cases} q_iq_jq_k, & \text{if } i \neq j \neq k \neq i, \\ q_iq_j(q_j - R_{ji}), & \text{if } i \neq j = k, \\ q_iq_k(q_k - R_{ki}), & \text{if } i = j \neq k, \\ q_i(q_i-1)(q_i-s_i) - \sum_l^i S_{il}q_iq_l, & \text{if } i = j = k, \end{cases}$$

$$F^i_j(s,q) := \begin{cases} Aq_iq_j - \theta_iR_{ij}q_j - \theta_jR_{ji}q_i, & \text{if } i \neq j, \\ (\theta_{n+1}-1)q_i(q_i-1) + \theta_{n+2}q_i(q_i-s_i) \\ \quad + \theta_i(q_i-1)(q_i-s_i) \\ \quad + \sum_k^i\{\theta_kq_i(q_i-R_{ik}) - \theta_iS_{ik}q_k\}, & \text{if } i = j \end{cases}$$

with

$$R_{ij} := \frac{s_i(s_j-1)}{s_j-s_i}, \quad S_{ij} := \frac{s_i(s_i-1)}{s_i-s_j},$$

$$A := \sum_{(i)} \theta_i - 1, \quad \kappa = \frac{1}{4}(A^2 - \theta_\infty^2). \tag{7.1.6}$$

Notice that E^i_{jk}, $F^i_j \in \mathbb{C}(s)[q]$ are symmetric with respect to i, j, k:

$$E^i_{jk} = E^j_{ik} = E^i_{kj}, \quad F^i_j = F^j_i \tag{7.1.7}$$

and that

$$R_{ij} + R_{ji} = 1, \quad R_{ij} + S_{ji} = s_i. \tag{7.1.8}$$

7.2. Proof of Theorem 7.1.2.

To prove the theorem, we decompose the transformation Φ into two symplectic transformations Φ_1 and Φ_2:

$$\Phi_1 : (\lambda_i, \mu_i, K_i, t_i) \to (q_i, p_i, K_i', t_i)$$

$$\Phi_2 : (q_i, p_i, K_i', t_i) \to (q_i, p_i, H_i, s_i)$$

so that $\Phi = \Phi_2 \circ \Phi_1$. The Hamiltonians $K_1', \cdots, K_n'$ are determined by

$$\sum_i (\mu_i d\lambda_i - K_i dt_i) = \sum_i (p_i dq_i - K_i' dt_i),$$

and hence by

$$K_i' = K_i + \sum_j p_j \frac{\partial q_j}{\partial t_i}, \qquad i = 1, \cdots, n.$$

We express the Hamiltonians K_i' explicitly in terms of (q, p, t).

PROPOSITION 7.2.1. *The Hamiltonians K_i' $(i = 1, \cdots, n)$ are given as follows:*

$$\begin{aligned}
&- t_i K_i'(q, p, t) \\
&= \left\{ q_i(q_i - 1)(q_i - \frac{t_i}{t_i - 1}) + \sum_k^i \frac{t_i(t_k - 1)}{(t_i - 1)(t_i - t_k)} q_i q_k \right\} p_i^2 \\
&+ \sum_k^i q_i q_k (q_k - \frac{t_k}{t_k - t_i}) p_k^2 + 2 \sum_k^i q_i q_k (q_i - \frac{t_i}{t_i - t_k}) p_i p_k \\
&+ \sum_{1 \le k, l \le n, \ i,k,l \text{ distinct}} q_i q_k q_l p_k p_l \\
&- \Big\{ (\theta_{n+1} - 1) q_i (q_i - 1) + \theta_{n+2} q_i (q_i - \frac{t_i}{t_i - 1}) \\
&\quad + \theta_i (q_i - 1)(q_i - \frac{t_i}{t_i - 1}) \\
&\quad + \sum_k^i \{ \theta_k q_i (q_i - \frac{t_i}{t_i - t_k}) + \theta_i q_k \frac{t_i(t_k - 1)}{(t_i - 1)(t_i - t_k)} \} \Big\} p_i \\
&- \sum_k^i \left\{ A q_i q_k + \frac{\theta_k t_k q_i - \theta_i t_i q_k}{t_i - t_k} \right\} p_k + \kappa q_i,
\end{aligned}$$

where the symbol $\sum_k^i$ *stands for the summation over* $k = 1, \cdots, i-1, i+1, \cdots, n$.

The proof will follow from two lemmas given below. First, let us compute the terms in the expression of K_i given by (4.3.7):

$$M_i \sum_k M^{k,i} \mu_k^2 = M_i \sum_k \sum_{a,b} M^{k,i} \frac{q_a p_a}{\lambda_k - t_a} \frac{q_b p_b}{\lambda_k - t_b}$$

$$= M_i \sum_{a,b} M(i,a,b)(q_a p_a)(q_b p_b),$$

where

$$M(i,a,b) := \sum_k \frac{M^{k,i}}{(\lambda_k - t_a)(\lambda_k - t_b)}$$

$$= \sum_k \frac{T(\lambda_k)}{(\lambda_k - t_i)(\lambda_k - t_a)(\lambda_k - t_b)\Lambda'(\lambda_k)}.$$

Notice that $M(i,a,b)$ are symmetric with respect to i, a, b, i.e.,

$$M(i,a,b) = M(a,i,b) = M(i,b,a).$$

LEMMA 7.2.2. *We have the following identities:*

$$M_i M(i,a,b) = \begin{cases} M_i, & \text{if } i \neq a \neq b \neq i, \\ M_i + \dfrac{1}{t_i - t_a}, & \text{if } a \neq b = i, \\ M_i - \left(D \log \Lambda(x) - D \log \dfrac{T(x)}{x - t_i} \right) \Big|_{x=t_i}, & \\ & \text{if } i = a = b. \end{cases}$$

Proof By considering the rational function in x:

$$Z(i,a,b;x) = \frac{T(x)}{(x - t_i)(x - t_a)(x - t_b)\Lambda(x)},$$

we see immediately from the definition that

$$
\left\{
\begin{aligned}
&M(i,a,b) = \sum_k \operatorname*{Res}_{x=\lambda_k} Z(i,a,b;x)\,dx, \\
&\operatorname*{Res}_{x=\infty} Z(i,a,b;x)\,dx = -1.
\end{aligned}
\right.
\tag{7.2.1}
$$

There are three cases: (1) i, a, b are all distinct, (2) $a \neq b$, $i = b$, (3) $a = b = i$. In case (1), since $Z(i,a,b;x)$ is holomorphic at $x = t_1, \cdots, t_n$, the residue theorem:

$$
\sum_k \operatorname*{Res}_{x=\lambda_k} Z(i,a,b;x)\,dx + \operatorname*{Res}_{x=\infty} Z(i,a,b;x)\,dx = 0
$$

leads to $M(i,a,b) = 1$. In cases (2) and (3), since $x = t_i$ is a pole of $Z(i,a,b;x)$, the residue theorem and (7.2.1) leads to

$$
M(i,a,b) = -\operatorname*{Res}_{x=t_i} Z(i,a,b;x)\,dx + 1.
$$

As for (2), we see

$$
\operatorname*{Res}_{x=t_i} Z(i,a,i;x)\,dx = \frac{T'(t_i)}{\Lambda(t_i)(t_i - t_a)} = \frac{-1}{M_i(t_i - t_a)};
$$

and hence we have the conclusion. As for (3), we have

$$
\begin{aligned}
&\operatorname*{Res}_{x=t_i} Z(i,i,i;x)\,dx \\
&\quad = D\left(\frac{T(x)}{(x-t_i)\Lambda(x)}\right)\Big|_{x=t_i} \\
&\quad = \frac{1}{M_i}\left(D\log\Lambda(x) - D\log\frac{T(x)}{x-t_i}\right)\Big|_{x=t_i}. \quad \blacksquare
\end{aligned}
$$

In order to express $M_i M(i,i,i)$ in terms of $t_1, \cdots, t_n$ and $M_1, \cdots, M_n$, we need the following lemma.

LEMMA 7.2.3.

$$D\log\Lambda(x)|_{x=t_i} = \frac{1}{e(t_i)M_i}\left\{\sum_k^i \frac{e(t_i)M_i + e(t_k)M_k}{t_i - t_k} - 1\right\}.$$

Proof We write

$$\Lambda(x) = x^n + \sigma_1 x^{n-1} + \cdots + \sigma_{n-1}x + \sigma_n,$$

where

$$\sigma_i := (-1)^i \sum_{1\le k_1<\cdots<k_i\le n} \lambda_{k_1}\cdots\lambda_{k_i}.$$

Let us express

$$D\Lambda(x)|_{x=t_i} = nt_i^{n-1} + (n-1)\sigma_1 t_i^{n-2} + \cdots + \sigma_{n-1} \tag{7.2.2}$$

in terms of $t_1,\cdots,t_n$ and $M_1,\cdots,M_n$. Since M_j are given by

$$\Lambda(t_j) = -T'(t_j)M_j,$$

by setting

$$W = \begin{pmatrix} t_1^{n-1} & t_1^{n-2} & \cdots & t_1 & 1 \\ \vdots & \vdots & & \vdots & \vdots \\ t_n^{n-1} & t_n^{n-2} & \cdots & t_n & 1 \end{pmatrix},$$

we have

$$W\ {}^t(\sigma_1,\cdots,\sigma_n) = {}^t(u_1,\cdots,u_n), \tag{7.2.3}$$

where

$$u_j = -T'(t_j)M_j - t_j^n, \qquad j=1,\cdots,n.$$

Let W_{ij} be the (j,i)-cofactor of the matrix W. Then, solving (7.2.3), we get

$$\sigma_k = \sum_j \frac{W_{kj}}{\det W}u_j, \qquad k=1,\cdots,n.$$

Substituting these expressions into (7.2.2), we obtain

$$D\Lambda(x)|_{x=t_i} = nt_i^{n-1} + \frac{1}{\det W}\sum_j L_{ij}u_j,$$

where

$$\begin{aligned} L_{ij} &= \sum_k (n-k)t_i^{n-k-1}W_{kj} \\ &= \left(\frac{\partial}{\partial t_j}\det W\right)|_{t_j=t_i}. \end{aligned}$$

On the other hand, it is easy to see

$$\det W = (-1)^{j-1}\prod_{\ell}^{j}(t_j - t_\ell)\prod_{\alpha<\beta,\neq j}(t_\alpha - t_\beta),$$

$$\left(\frac{\partial}{\partial t_j}\det W\right)\Big|_{t=t_i} = (-1)^{j-1}\prod_{\ell}^{i,j}(t_i - t_\ell)\prod_{\alpha<\beta,\neq j}(t_\alpha - t_\beta).$$

Since $T'(t_i) = e(t_i)\prod_\ell^i (t_i - t_h)$, we have

$$\frac{L_{ij}}{\det W} = \begin{cases} \displaystyle\sum_k^i \frac{1}{t_i - t_k}, & \text{if } j = i, \\ \displaystyle\frac{e(t_j)}{e(t_i)(t_i - t_j)}\frac{T'(t_i)}{T'(t_j)}, & \text{if } j \neq i. \end{cases}$$

It follows that

$$D\Lambda(x)|_{x=t_i} = -\frac{T'(t_i)}{e(t_i)}\sum_k^i \frac{e(t_i)M_i + e(t_k)M_k}{t_i - t_k} + R_i,$$

where

$$R_i = \frac{T'(t_i)}{e(t_i)}\sum_k^i\left(\frac{e(t_k)t_k^n}{(t_k - t_i)T'(t_k)} - \frac{e(t_i)t_i^n}{(t_i - t_k)T'(t_i)}\right) + nt_i^{n-1}.$$

The lemma follows immediately if we show

$$R_i = \frac{T'(t_i)}{e(t_i)}. \tag{7.2.4}$$

Putting

$$R_i(x) := \frac{x(x-1)x^n}{(x-t_i)T(x)},$$

we have

$$\operatorname*{Res}_{x=\infty} R_i(x)\,dx = -1,$$

$$\operatorname*{Res}_{x=t_k} R_i(x)\,dx = \frac{e(t_k)t_k^n}{(t_k-t_i)T'(t_k)}$$

if $k \neq i$, and

$$\begin{aligned}\operatorname*{Res}_{x=t_i} R_i(x)\,dx &= D\{(x-t_i)^2 R_i(x)\}|_{x=t_i}\\ &= \frac{e(t_i)t_i^n}{T'(t_i)}\left[\frac{n}{t_i} - \sum_k^i \frac{1}{t_i-t_k}\right].\end{aligned}$$

Then (7.2.4) is a consequence of the residue theorem:

$$\sum_k \operatorname*{Res}_{x=t_k} R_i(x)\,dx + \operatorname*{Res}_{x=\infty} R_i(x)\,dx = 0. \quad \blacksquare$$

Proof of Proposition 7.2.1 Recall that the Hamiltonians K_i and K_i' are linked as

$$K_i' = K_i + \sum_j p_j \frac{\partial q_j}{\partial t_i}, \qquad i = 1, \cdots, n,$$

and that K_i are given by

$$K_i = M_i \sum_k M^{k,i}\left(\mu_k^2 - \mu_k \sum_{(m)} \frac{\theta_m - \delta_{im}}{\lambda_k - t_m}\right) + \kappa M_i.$$

Since

$$D\log\frac{T(x)}{x-t_j}\Big|_{x=t_j}=\sum_k^j\frac{1}{t_j-t_k}+\frac{1}{t_j}+\frac{1}{t_j-1},$$

we have, by virtue of Lemmas 7.2.2 and 7.2.3,

$$\begin{aligned}
&M_i\sum_k M^{k,i}\mu_k^2\\
&=M_i\sum_{a,b}M(i,a,b)(q_ap_a)(q_bp_b)\\
&=-\frac{q_i}{t_i}\Bigg[\Bigg\{(q_i-1)(q_i-\frac{t_i}{t_i-1})+\sum_k^i\frac{t_i(t_k-1)}{(t_i-1)(t_i-t_k)}q_k\Bigg\}p_i^2\\
&\quad+\sum_a^i q_a(q_a-\frac{t_a}{t_a-t_i})p_a^2+2\sum_a^i q_a(q_i-\frac{t_i}{t_i-t_a})p_ip_a\\
&\quad+\sum_{a,b,a\neq b}^i q_aq_bp_ap_b\Bigg].
\end{aligned}$$

In a similar way we can calculate the terms:

$$M_i\sum_k M^{k,i}\mu_k\frac{\theta_m-\delta_{im}}{\lambda_k-t_m}.$$

Finally, to express $\sum_k p_k\partial q_k/\partial t_i$ in terms of (q,p,t), we have only to notice

$$\frac{\partial q_k}{\partial t_i}=\begin{cases}\dfrac{q_k}{t_k-t_i}, & \text{if } k\neq i\\[2ex] q_i\Big(D\log\Lambda(x)-D\log\dfrac{T(x)}{x-t_i}+\dfrac{1}{t_i}\Big)\Big|_{x=t_i}, & \text{if } k=i,\end{cases}$$

and apply Lemma 7.2.3. ∎

Proof of Theorem 7.1.2 In order that the transformation

$$\Phi_2:(q_i,p_i,K_i',t_i)\to(q_i,p_i,H_i,s_i)$$

be symplectic, the Hamiltonians H_i $(i = 1, \cdots, n)$ of the system $\mathcal{H}_n$ should satisfy $\sum_i K_i' dt_i = \sum_i H_i ds_i$, therefore H_i is given by

$$H_i = -\frac{1}{(s_i - 1)^2} K_i'.$$

Using the explicit expressions for the K_i' $(k = 1, \cdots, n)$ in Proposition 7.2.1, the theorem follows immediately by computations. ∎

REMARK 7.2.4. In case $n = 1$, (i) the Hamiltonian H_1 is

$$\frac{1}{s(s-1)}[q(q-1)(q-s)p^2 - \{\theta_1(q-1)(q-s) \\ + (\theta_2 - 1)q(q-1) + \theta_3 q(q-s)\}p + \kappa q],$$

and the Hamiltonian system is the sixth Painlevé system $\mathcal{H}_{VI}$,

(ii) The transformation $(\lambda, t) \to (q, s)$ defined by (7.1.1) is just the transformation T_3 in Proposition 1.3.2.

7.3. τ-function associated with $\mathcal{H}_n$

THEOREM 7.3.1. *Let $(q(s), p(s))$ be a solution of the system $\mathcal{H}_n$, then the 1-form*

$$\omega := \sum_i H_i(s, q(s), p(s))\, ds_i$$

is closed.

Proof Let $\Gamma = \sum_{i<j} \Gamma_{ij} ds_i \wedge ds_j$ be the fundamental 2-form associated with the Hamiltonian system $\mathcal{H}_n$ (see Chapter 1, §6). Then, by making use of the explicit forms of the Hamiltonians H_i, it is easy to verify that $H_i(s, q, p)$ $(i = 1, \cdots, n)$ satisfy

$$\frac{\partial H_j}{\partial s_i} = \frac{\partial H_i}{\partial s_j}, \qquad i, j = 1, \cdots, n;\ i \neq j, \tag{7.3.1}$$

where $\partial f / \partial s_i$ denotes the partial derivative with respect to s_i of f, regarded as a function in q, p and s. Let $(q(s), p(s))$ be a solution of $\mathcal{H}_n$

and set $H_i(s) = H_i(s, q(s), p(s))$. The integrability condition for $\mathcal{H}_n$, which implies the vanishing of Γ_{ij} along any solution of $\mathcal{H}_n$ (Remark 4.5.8), says that

$$D_j H_i - D_i H_j = \frac{\partial H_j}{\partial s_i} - \frac{\partial H_i}{\partial s_j}. \tag{7.3.2}$$

Combining (7.3.1) and (7.3.2), we obtain

$$D_j H_i - D_i H_j = 0$$

for any $i, j = 1, \cdots, n$; this yields

$$\begin{aligned} d\omega &= -\sum_{i<j} (D_j H_i - D_i H_j)\, ds_i \wedge ds_j \\ &= 0. \quad \blacksquare \end{aligned}$$

REMARK 7.3.2. The above theorem allows us to define, for any solution $(q(s), p(s))$ of $\mathcal{H}_n$, a function $\tau(s)$ called the τ-function by

$$d \log \tau(s) = \omega.$$

$\tau(s)$ is determined up to multiplicative constant factors.

PROPOSITION 7.3.3. *If $(Q_1(t), \cdots, Q_{n+2}(t))$ satisfies the Schlesinger system (5.2), then $(q(s), p(s))$ defined below satisfies the Hamiltonian system $\mathcal{H}_n$.*

$$\begin{cases} q_i = t_i \dfrac{q_{12}^i}{X}, \\[2ex] p_i = \dfrac{X}{t_i} \left(\dfrac{q_{11}^i}{q_{12}^i} + (t_i - 1) \dfrac{q_{11}^{n+1}}{q_{12}^{n+1}} - t_i \dfrac{q_{11}^{n+2}}{q_{12}^{n+2}} \right), \\[2ex] s_i = \dfrac{t_i}{t_i - 1}, \end{cases}$$

where $X = \sum_i t_i q_{12}^i$.

Proof This follows from Propositions 6.2.1 and 6.3.1 together with the residue theorem. In fact, by virtue of (7.1.4) and (6.2.5)

$$\begin{aligned} p_i &= -(t_i - 1)\sum_k \frac{M^{k,i}}{\lambda_k(\lambda_k - 1)}\mu_k \\ &= -(t_i - 1)\sum_{(a)} q_{11}^a M(i,a) \end{aligned}$$

where

$$\begin{aligned} M(i,a) :&= \sum_k \frac{M^{k,i}}{\lambda_k(\lambda_k - 1)(\lambda_k - t_a)} \\ &= \sum_k \frac{T(\lambda_k}{\lambda_k(\lambda_k - 1)(\lambda_k - t_i)(\lambda_k - t_a)\Lambda'(\lambda_k)}. \end{aligned}$$

Consider the rational function in x:

$$Z(i,a;x) := \frac{T(x)}{x(x-1)(x-t_i)(x-t_a)\Lambda(x)}.$$

It is easy to see

$$M(i,a) = \sum_k \operatorname*{Res}_{x=\lambda_k} \; Z(i,a;x)dx,$$

$$\operatorname*{Res}_{x=\infty} \; Z(i,a;x)dx = 0.$$

If $a \neq i, n+1, n+2$, then the poles of $Z(i,a;x)$ are only $x = \lambda_k$. Then we have from the residue theorem $M(i,a) = 0$. Moreover we can show:

$$\begin{aligned} M(i,i) &= -\operatorname*{Res}_{x=t_i} \; Z(i,i;x)dx \\ &= -\frac{1}{t_i(t_i-1)}\frac{T'(t_i)}{\Lambda(t_i)} \\ &= -\frac{1}{t_i(t_i-1)}\frac{X}{q_{12}^i}, \end{aligned}$$

$$M(i,n+1) = -\operatorname*{Res}_{x=t_{n+1}} Z(i,n+1;x)dx$$
$$= -\frac{1}{t_i}\frac{X}{q_{12}^{n+1}},$$

$$M(i,n+2) = -\operatorname*{Res}_{x=t_{n+1}} Z(i,n+1;x)dx$$
$$= -\frac{1}{t_i-1}\frac{X}{q_{12}^{n+2}}.$$

This proves the proposition. ∎

COROLLARY 7.3.4. *The Hamiltonian system $\mathcal{H}_n$ enjoys the Painlevé property. Any solution of $\mathcal{H}_n$ is continued meromorphically on the universal covering space of $B = \mathbb{C}^n \setminus \Xi$.*

Proof This follows from Theorem 5.4 and Proposition 7.3.3. ∎

8. Symmetries of the Garnier system $\mathcal{G}_n$ and of the system $\mathcal{H}_n$

In this section, we investigate a group of symmetries for the Garnier system $\mathcal{G}_n$ and for the system $\mathcal{H}_n$. Roughly speaking, a group of symmetries for $\mathcal{G}_n$ is a group of birational transformations $(\lambda, \mu, t) \mapsto (\lambda', \mu', t')$ which leave $\mathcal{G}_n$ invariant modulo change of parameters.

8.1. Symmetries of $\mathcal{G}_n$

Let us consider a Pfaffian system:

$$(E) \qquad dx_i = \sum_{j=1}^{n} F_{ij}(x, t, \theta)\, dt_j, \qquad i = 1, ..., m,$$

where $F_{ij}(x, t, \theta)$ are rational functions in $(x, t) = (x_1, \ldots, x_m, t_1, \ldots, t_n)$ depending on parameters $\theta \in \mathbb{C}^N$. The system E with a parameter θ is denoted by $E(\theta)$. For a birational transformation $T : (x, t) \to (x', t')$, we denote by $T \cdot E(\theta)$ the system of differential equations in the variables (x', t') obtained from $E(\theta)$ by the transformation T.

DEFINITION 8.1.1. A *group of symmetries* for the system E is a group whose element is a pair $\sigma = (T, \ell)$ of a birational transformation $T : (x, t) \to (x', t')$ and an affine transformation $\ell : \mathbb{C}^N \to \mathbb{C}^N$ such that $T \cdot E(\theta) = E(\ell(\theta))$.

We give a group of symmetries for the Garnier system $\mathcal{G}_n$ and for the corresponding polynomial Hamiltonian system $\mathcal{H}_n$. Let $V = \{\theta = (\theta_1, \ldots, \theta_{n+3}) \in \mathbb{C}^{n+3}\}$ be the space of parameters of the systems $\mathcal{G}_n$ and $\mathcal{H}_n$; and let $\mathcal{G}_n(\theta)$ and $\mathcal{H}_n(\theta)$ be the systems $\mathcal{G}_n$ and $\mathcal{H}_n$ with fixed parameters θ, respectively.

First we treat the Garnier system $\mathcal{G}_n$. Let $T_m : (\lambda, \mu, t) \mapsto (\lambda', \mu', t')$ $(m = 1, \ldots, n, n+2, n+3)$ be the birational transformations given as follows.

$T_m \ (m = 1, \ldots, n)$:

$$\begin{cases} \lambda_i' = \dfrac{t_m - \lambda_i}{t_m - 1}, \\ \mu_i' = -(t_m - 1)\mu_i, \\ t_i' = \begin{cases} \dfrac{t_m - t_i}{t_m - 1}, & (i \neq m, n+1), \\ \dfrac{t_m}{t_m - 1}, & (i = m); \end{cases} \end{cases}$$

T_{n+2} :

$$\begin{cases} \lambda_i' = 1 - \lambda_i, \\ \mu_i' = -\mu_i, \\ t_i' = 1 - t_i; \end{cases}$$

T_{n+3} :

$$\begin{cases} \lambda_i' = \dfrac{\lambda_i}{\lambda_i - 1}, \\ \mu_i' = -(\lambda_i - 1)^2 \mu_i - \alpha(\lambda_i - 1), \\ t_i' = \dfrac{t_i}{t_i - 1}. \end{cases}$$

Let $\ell_m : V \to V$ be the linear transformations defined by the following exchanges of coordinates:

$$\ell_m : \ \theta_m \longleftrightarrow \theta_{n+1}, \ (m = 1, \ldots, n)$$

$$\ell_{n+2} : \ \theta_{n+2} \longleftrightarrow \theta_{n+1},$$

$$\ell_{n+3} : \ \theta_{n+3} \longleftrightarrow \theta_{n+2}.$$

Set $\sigma_m = (T_m, \ell_m) \ (m = 1, \ldots, n, n+2, n+3)$ and let G be the group

$$G = \langle \sigma_1, \ldots, \sigma_n, \sigma_{n+2}, \sigma_{n+3} \rangle$$

generated by the elements $\sigma_1, \ldots, \sigma_n, \sigma_{n+2}, \sigma_{n+3}$. Then we have

THEOREM 8.1.2. *The group G is a group of symmetries for the Garnier system $\mathcal{G}_n$ and is isomorphic to the symmetric group $\mathfrak{S}_{n+3}$ on $n+3$ letters.*

Proof First we sketch the idea of the proof. Let us consider the linear differential equation

$$D_x^2 y + p_1(x,t) D_x y + p_2(x,t) y = 0, \quad D_x = d/dx, \tag{8.1.1}$$

with coefficients given by (6.2.4). The Riemann scheme of the equation reads

$$\begin{pmatrix} x = t_i, & x = t_{n+1} = 0, & x = t_{n+2} = 1, & t = t_{n+3} = \infty, & z = \lambda_i \\ 0 & 0 & 0 & \alpha & 0 \\ \theta_i & \theta_{n+1} & \theta_{n+2} & \alpha + \theta_{n+3} & 2 \end{pmatrix},$$

where

$$\alpha = -\frac{1}{2}\Big(\sum_{i=1}^{n+3} \theta_i - 1\Big).$$

For the equation (8.1.1), we consider a projective transformation of the independent variable ϕ: $\mathbb{P}^1 \ni x \mapsto z \in \mathbb{P}^1$ such that ϕ takes the set of singular points $S = \{t_1, \ldots, t_n, 0, 1, \infty\}$ into a set $S' = \{t'_1, \ldots, t'_n, 0, 1, \infty\}$, and then we consider a change of the unknown $w = \psi(z) y$ which takes, at each singular point in the finite plane $\mathbb{C}$, one of the characteristic exponents into zero. Let

$$D_z^2 w + P_1(z,t) D_z w + P_2(z,t) w = 0, \quad D_z = d/dz, \tag{8.1.2}$$

be the consequent equation under the change of variables $(z, w) = (\phi(x), \psi(z) y)$. This change of variables induces a correspondence

$$\begin{cases} \lambda'_i = \phi(\lambda_i), \\ \mu'_i = \mu'_i(\lambda, \mu, t), \qquad i = 1, \ldots, n, \\ t'_i = t'_i(t), \end{cases} \tag{8.1.3}$$

where $\mu'_1, \ldots, \mu'_n$ are defined by

$$\mu'_i := \operatorname*{Res}_{z=\lambda'_i} P_2(z,t), \qquad i = 1, \ldots, n. \tag{8.1.4}$$

Let us assume the family (8.1.1) is monodromy preserving; then so is the family (8.1.2). By the way, Theorem 4.1 says that the family (8.1.2) is monodromy preserving if and only if λ' and μ' are functions of $t' = (t'_1, \ldots, t'_n)$ solving the Garnier system $\mathcal{G}_n(\theta')$ for some $\theta' \in V$. Hence the transformation $T_i : (\lambda, \mu, t) \to (\lambda', \mu', t')$, given by (8.1.3), takes the Garnier system $\mathcal{G}_n(\theta)$ into $\mathcal{G}_n(\theta')$ for some θ'. We set $\ell(\theta) := \theta'$.

Now we give the details of the above procedure for σ_{n+3}. Consider the projective transformation $\phi : x \mapsto z = x/(x-1)$ which exchanges singular points 1 and ∞ and leaves the point 0 fixed. The Riemann scheme of the linear differential equation thus obtained is

$$\begin{pmatrix} z = t'_i, & z = 0, & z = 1, & z = \infty, & z = \lambda'_i \\ 0 & 0 & \alpha & 0 & 0 \\ \theta_i & \theta_{n+1} & \alpha + \theta_{n+3} & \theta_{n+2} & 2 \end{pmatrix},$$

where

$$\lambda'_i = \frac{\lambda_i}{\lambda_i - 1}, \quad t'_i = \frac{t_i}{t_i - 1}, \qquad i = 1, \ldots, n. \tag{8.1.5}$$

In order to make one of the exponents at $z = 1$ equal to zero, we make the change of the unknown $y \to w$:

$$w = \psi(z) y, \quad \psi(z) = (z-1)^{-\alpha};$$

and get the equation (8.1.2) with coefficients

$$\begin{cases} P_1(z,t) = \dfrac{2\alpha + 2}{z - 1} - \dfrac{p_1(x,t)}{(z-1)^2}, \\ P_2(z,t) = \dfrac{\alpha(\alpha - 1)}{(z-1)^2} + \dfrac{\alpha}{z-1}\left\{\dfrac{2}{z-1} - \dfrac{p_1(x,t)}{(z-1)^2}\right\} \\ \qquad + \dfrac{p_2(x,t)}{(z-1)^4}. \end{cases} \tag{8.1.6}$$

On the other hand, as we have seen in Section 4.1, the coefficients can be written as follows:

$$P_1(z,t) = \sum_{i=1}^{n+1} \frac{1-\theta_i}{z-t'_i} + \frac{1-\theta_\infty}{z-t'_{n+2}} - \sum_k \frac{1}{z-\lambda'_k},$$

$$P_2(z,t) = \frac{\kappa'}{z(z-1)} - \sum_i \frac{t'_i(t'_i-1)K'_i}{z(z-1)(z-t'_i)}$$

$$+\sum_k \frac{\lambda'_k(\lambda'_k-1)\mu'_k}{z(z-1)(z-\lambda'_k)},$$

where

$$\kappa' = \frac{1}{4}\left[(\theta_1 + \cdots \theta_{n+1} + \theta_{n+3} - 1)^2 - \theta_{n+2}{}^2\right].$$

Thus λ', μ', t' are related with λ , μ , t by (8.1.5) and

(8.1.7) $$\mu'_i = -(\lambda_i - 1)^2\mu_i - \alpha(\lambda_i - 1), \qquad i = 1, \ldots, n,$$

and $K'_i(\lambda', \mu', t) := \operatorname*{Res}_{z=t'_i} P_2(z)$ are obtained from $K_i(\lambda, \mu, t)$ by replacing θ by $\ell_{n+3}(\theta)$, and (λ, μ, t) by (λ', μ', t'). Since the monodromy preserving deformation for (8.1.1) induces that for the equation (8.1.2), Theorem 4.1.2 says that the transformation T_{n+3} defined by (8.1.5) and (8.1.7) takes the Garnier system $\mathcal{G}_n(\theta)$ into $\mathcal{G}_n(\ell_{n+3}(\theta))$; this gives the generator σ_{n+3}. The other generators $\sigma_1, \ldots, \sigma_n, \sigma_{n+2}$ of the group G can be obtained in a similar way (see Remark 8.1.3 below). The statement $G \simeq \mathfrak{S}_{n+3}$ is now clear from the construction. ∎

REMARK 8.1.3. For later use, we give explicit forms of the projective transformation $z = \phi(x)$ and the change of unknown $w = \psi(z)y$ for each σ_m. We give also the correspondence $S \to S'$ of singularities of the equations (8.1.1) and (8.1.2).
(σ_m) $(m = 1, \ldots, n)$:

$$\begin{cases} z = \dfrac{t_m - x}{t_m - 1}, \quad w = y, \\ (t_1, \cdots, t_n, 0, 1, \infty) \mapsto (t'_1, \cdots, t'_{m-1}, 0, t'_{m+1}, \cdots, t'_n, t'_m, 1, \infty), \end{cases}$$

(σ_{n+2}):

$$\begin{cases} z = 1 - x, \quad w = y, \\ (t_1, \cdots, t_n, 0, 1, \infty) \mapsto (t'_1, \cdots, t'_n, 1, 0, \infty), \end{cases}$$

(σ_{n+3}):

$$\begin{cases} z = \dfrac{x}{x-1}, \quad w = (z-1)^{-\alpha} y, \\ (t_1, \cdots, t_n, 0, 1, \infty) \mapsto (t'_1, \cdots, t'_n, 0, \infty, 1). \end{cases}$$

8.2. Symmetries of $\mathcal{H}_n$

We shall present a group of symmetries for the Hamiltonian system $\mathcal{H}_n$ which is naturally induced from the group G. For each generator $\sigma_m = (T_m, \ell_m)$ of the group G, we associate a birational symplectic transformation $\hat{T}_m$: $(q, p, s, H) \to (q', p', s', H')$ such that the following diagram commutes:

$$\begin{array}{ccc} (\lambda, \mu, t) & \xrightarrow{T_m} & (\lambda', \mu', t') \\ \downarrow \Phi & & \downarrow \Phi \\ (q, p, s) & \xrightarrow{\hat{T}_m} & (q', p', s') \end{array}$$

where the right and left vertical arrows indicate the symplectic transformations, given in Theorem 7.1.2, which take the Garnier system $\mathcal{G}_n(\theta)$ into the Hamiltonian system $\mathcal{H}_n(\theta)$. Note that the Hamiltonians H_i (resp. H'_i) are polynomials in q_k and p_l (resp. q'_k and p'_l). Let us define the birational transformation $\hat{T}_m$: $(q, p, s) \to (q', p', s')$ as follows:

$\hat{T}_m\,(m=1,\ldots,n):$

$$q_i' = \begin{cases} \dfrac{q_i}{R_{im}} & (i \neq m), \\[2ex] -\dfrac{s_m}{s_m-1}(g_s-1) & (i=m), \end{cases}$$

$$p_i' = \begin{cases} R_{im}(p_i - \dfrac{s_m}{s_i}p_m) & (i \neq m), \\[2ex] -(s_m-1)p_m & (i=m), \end{cases}$$

$$s_i' = \begin{cases} \dfrac{s_m-s_i}{s_m-1} & (i \neq m, n+1), \\[2ex] \dfrac{s_m}{s_m-1} & (i=m), \end{cases}$$

$\hat{T}_{n+2}$:

$$q_i' = \frac{q_i}{s_i},$$

$$p_i' = s_i p_i,$$

$$s_i' = \frac{1}{s_i},$$

$\hat{T}_{n+3}$:

$$q_i' = \frac{q_i}{g_1-1},$$

$$p_i' = (g_1-1)\left(p_i + \alpha - \sum_a q_a p_a\right),$$

$$s_i' = \frac{s_i}{s_i-1},$$

where R_{im} are given by (7.1.5) and

$$g_1 := \sum_a q_a, \quad g_s := \sum_a \frac{q_a}{s_a}.$$

LEMMA 8.2.1. *If a birational transformation $X : (q, p, s) \to (q', p', s')$ satisfies $X \circ \Phi = \Phi \circ T_m$, then we have $X = \hat{T}_m$.*

Proof. We show the lemma for $m = n + 3$. For the other cases the lemma can be proved in a similar way. Using the explicit form of T_{n+3} and Φ, we see that

$$s_i' = \frac{t_i'}{t_i' - 1} = t_i = \frac{s_i}{s_i - 1}.$$

We express $q_1', \ldots, q_n'$ in terms of q and s. Set

$$\Lambda_*(x) := \prod_i (x - \lambda_i'), \quad T_*(x) := \prod_{(i)} (x - t_i'),$$

where $t_{n+1}' = 0$, $t_{n+2}' = 1$. Then we have

$$\begin{aligned} q_i' &= t_i' \frac{\Lambda_*(t_i')}{T_*'(t_i')} \\ &= -t_i \frac{\Lambda(t_i)}{T'(t_i)} \prod_k \frac{1 - t_k}{1 - \lambda_k} \\ &= -q_i \left(\prod_k \frac{1 - \lambda_k}{1 - t_k} \right)^{-1}. \end{aligned}$$

Apply the residue theorem to a rational function $Z(x) := x\Lambda(x)/T(x) = \prod_k (x - \lambda_k)/(x - 1) \prod_k (x - t_k)$; we have

$$\operatorname*{Res}_{x=t_k} Z(x)\,dx = \frac{t_k \Lambda(t_k)}{T'(t_k)} = q_k, \qquad k = 1, \ldots, n,$$

$$\operatorname*{Res}_{x=1} Z(x)\,dx = \prod_k \frac{1 - \lambda_k}{1 - t_k},$$

$$\operatorname*{Res}_{x=\infty} Z(x)\,dx = -1$$

and

$$\prod_k \frac{1 - \lambda_k}{1 - t_k} = 1 - g_1. \tag{8.2.2}$$

Hence we obtain $q_i' = q_i/(g_1 - 1)$ $(i = 1, \dots, n)$ and

$$g_i' = \sum_a q_a' = \frac{g_1}{g_1 - 1}.$$

Next we express $p_1', \dots, p_n'$ in terms of q, p and s. By using the explicit forms of T_{n+3} and of Φ, we see that

$$\begin{aligned} p_i' &= -(t_i' - 1) \sum_k \frac{M_*^{k,i} \mu_k'}{\lambda_k'(\lambda_k' - 1)} \\ &= (t_i' - 1) \sum_k \frac{M_*^{k,i}\{(\lambda_k - 1)^2 \mu_k + \alpha(\lambda_k - 1)\}}{\lambda_k'(\lambda_k' - 1)}, \end{aligned}$$

where

$$M_*^{k,i} := \frac{T_*(\lambda_k')}{(\lambda_k' - t_i')\Lambda_*'(\lambda_k')}.$$

Put

$$A_i := (t_i' - 1) \sum_k \frac{M_*^{k,i} (\lambda_k - 1)^2 \mu_k}{\lambda_k'(\lambda_k' - 1)},$$

$$B_i := (t_i' - 1) \sum_k \frac{M_*^{k,i} (\lambda_k - 1)}{\lambda_k'(\lambda_k' - 1)},$$

then $p_i' = A_i + \alpha B_i$. First we compute B_i:

$$B_i := (t_i' - 1) \sum_k \frac{M_*^{k,i}}{\lambda_k'(\lambda_k' - 1)^2},$$

by applying the residue theorem to

$$Z_*(x) = \frac{\prod_{m \neq i}(x - t_m')}{(x - 1)\Lambda_*(x)}.$$

Exactly by the same computations to those in the proof of Proposition 7.3.3, we have from (8.2.2),

$$\begin{aligned} B_i &= -\prod_k \frac{1-t'_k}{1-\lambda'_k} \\ &= \frac{1}{g'_1 - 1} \\ &= g_1 - 1. \end{aligned}$$

Next we compute $A_1, \ldots, A_n$. By the construction of σ_{n+2} and that of Φ, we see easily that the transformation

$$(q_1, \ldots, q_n, p_1, \ldots, p_n) \mapsto (q'_1, \ldots, q'_n, A_1, \ldots, A_n)$$

coincides with the induced bundle map: $T^*U \to T^*\mathbb{C}^n$ of the mapping $U \to \mathbb{C}^n$ given by

$$(q_1, \ldots, q_n) \mapsto (q'_1, \ldots, q'_n) = \left(\frac{q_1}{g_1 - 1}, \ldots, \frac{q_n}{g_1 - 1}\right),$$

where

$$U = \mathbb{C}^n - \{(q_1, \ldots, q_n);\ g_1 - 1 = 0\}.$$

Hence we have $\sum_k p_k dq_k = \sum_k A_k dq'_k$, which is equivalent, by putting $\vec{A} = {}^t(A_1, \ldots, A_n)$, to

$$\left((g_1 - 1)I_n - \begin{pmatrix} q_1 & \cdots & q_n \\ \vdots & & \vdots \\ q_1 & \cdots & q_n \end{pmatrix} \right) \vec{A} = (g_1 - 1)^2 \begin{pmatrix} p_1 \\ \vdots \\ p_n \end{pmatrix}.$$

Solving this equation, we obtain

$$A_i = (g_1 - 1)\left(p_i - \sum_a q_a p_a\right).$$

Combining the expressions of A_i and B_i, we see that the transformation X coincides with $\hat{T}_{n+3}$. ∎

We set $\hat{\sigma}_m = (\hat{T}_m, \ell_m)$, $(m = 1, \ldots, n, n+2, n+3)$. Then, as a consequence of Theorem 8.1.2 and Lemma 8.2.1, we have

THEOREM 8.2.2.
(i) *The birational transformations $\hat{T}_m$ extend to symplectic transformations which make the diagram (8.2.1) commute.*

(ii) *The group $\hat{G} = \langle \hat{\sigma}_1, \ldots, \hat{\sigma}_n, \hat{\sigma}_{n+2}\hat{\sigma}_{n+3} \rangle$ is a group of symmetries for the system $\mathcal{H}_n$ and is isomorphic to the symmetric group $\mathfrak{S}_{n+3}$ on $n+3$ letters.*

8.3. Prolongation of the system $\mathcal{H}_n$

In this subsection, we prove, as an application of Theorem 8.2.2, a result analogous to Corollary 1.5.3. To this end, we use the following lemma, whose proof will be given at the end of the present section.

LEMMA 8.3.1. *Define $\tau_m = (T'_m, \ell'_m) \in \hat{G}$ by*

$$\tau_m := (\hat{\sigma}_{n+3}\hat{\sigma}_m)\hat{\sigma}_{n+2}(\hat{\sigma}_{n+3}\hat{\sigma}_m)^{-1}, \quad m = 1, \cdots, n.$$

Then $T'_m : (q, p, s) \to (q', p', s')$ is given by

$$q'_i = \begin{cases} -\dfrac{q_i}{q_m} & (i \neq m), \\[2ex] \dfrac{1}{q_m} & (i = m), \end{cases}$$

$$p'_i = \begin{cases} -q_m p_i & (i \neq m), \\[2ex] -q_m \left(\alpha + \sum_a q_a p_a \right) & (i = m), \end{cases}$$

$$s'_i = \begin{cases} \dfrac{s_i}{s_m} & (i \neq m), \\[2ex] \dfrac{1}{s_m} & (i = m). \end{cases}$$

Let us define the affine bundle Σ_α of rank n over $\mathbb{P}^n$ as follows. Let $z := (z_0, \dots, z_n)$ be homogeneous coordinates of $\mathbb{P}^n$; let $(U_i, \mathbf{z}^i)$ be the i-th affine charts of $\mathbb{P}^n$:

$$U_i := \{z \in \mathbb{P}^n | z_i \neq 0\},$$

$$\mathbf{z}^i := (z_0^i, \dots, z_{i-1}^i, z_{i+1}^i, \dots, z_n^i), \quad z_j^i = \frac{z_j}{z_i}.$$

We identify two points $(\mathbf{z}^i, \zeta^i) := (\mathbf{z}^i, \zeta_0^i, \dots, \zeta_{i-1}^i, \zeta_{i+1}^i, \dots, \zeta_n^i) \in U_i \times \mathbb{C}^n$ and $(\mathbf{z}^j, \zeta^j) := (\mathbf{z}^j, \zeta_0^j, \dots, \zeta_{j-1}^j, \zeta_{j+1}^j, \dots, \zeta_n^j) \in U_j \times \mathbb{C}^n$ when $\mathbf{z}^i$ and $\mathbf{z}^j$ represent the same point in $\mathbb{P}^n$ and

$$\zeta_k^j = \begin{cases} z_j^i \zeta_k^i & (k \neq i, j), \\ z_j^i \left(-\alpha - \sum_{a \neq i} z_a^i \zeta_a^i \right) & (k = i). \end{cases} \tag{8.3.1}$$

Denote by $\sim$ the equivalence relation and define the affine bundle Σ_α over $\mathbb{P}^n$ by

$$\Sigma_\alpha := \left(\bigcup_{i=0}^n U_i \times \mathbb{C}^n \right) / \sim .$$

We define the transition maps

$$\Phi_m : U_0 \times \mathbb{C}^n \ni (\mathbf{z}^0, \zeta^0) \mapsto (\mathbf{z}^m, \zeta^m) \in U_m \times \mathbb{C}^n,$$

$m = 1, \dots, n$ by (8.3.1).

For each $m = 1, \dots, n$, let $H_i(q,p,s)$ and $H_i^m(q,p,s)$ $(i = 1, \dots, n)$ be the Hamiltonians of $\mathcal{H}_n(\theta)$ and $\mathcal{H}_n(\ell'_m(\theta))$, respectively. Let R_m be a birational map defined by

$$\begin{aligned} R_m : (q,p,s) \mapsto & (q', p', s^m) \\ = & \Big(-q_m, -q_2, \dots, -q_{m-1}, q_1, -q_{m+1}, \dots, -q_n, \\ & - p_m, -p_2, \dots, -p_{m-1}, p_1, -p_{m+1}, \dots, -p_n, \\ & \frac{s_1}{s_m}, \dots, \frac{s_{m-1}}{s_m}, \frac{1}{s_m}, \frac{s_{m+1}}{s_m}, \dots, \frac{s_n}{s_m} \Big). \end{aligned}$$

If we define functions $K_i^m(q', p', s^m)$ by

$$R_m{}^* \left(\sum K_i^m(q', p', s^m) ds_i^m \right) = \sum H_i(q, p, s) ds_i,$$

then the map R_m extends to the symplectic transformation

$$(q, p, s, H^m) \to (R_m(q, p, s), K^m).$$

The composition $R_m \circ T'_m$ is shown to be

$$R_m \circ T'_m : (q, p, s) \mapsto (\Phi_m(q, p), s)$$

and is extended to the symplectic transformation

$$(q, p, s, H) \mapsto (\Phi_m(q, p), s, K^m).$$

Therefore the system $\mathcal{H}_n$ in $U_0 \times \mathbb{C}^n \times B$ is transformed into the Hamiltonian system on $U_m \times \mathbb{C}^n \times B$ with Hamiltonians $K^m = (K_j^m)_{j=1,\cdots,n}$. We call the collection of the Hamiltonian systems on $U_m \times \mathbb{C}^n \times B$ ($m = 0, \ldots, n$) obtained above, the Hamiltonian system on $\Sigma_\alpha \times B$. The fact that K_j^m ($j = 1, \ldots, n$) are polynomials in $(\mathbf{z}^m, \zeta^m)$ leads to the result:

PROPOSITION 8.3.2. *The Hamiltonian system $\mathcal{H}_n$ extends to a Hamiltonian system on $\Sigma_\alpha \times B$ with Hamiltonians holomorphic on $\Sigma_\alpha \times B$.*

REMARK 8.3.3. In the case $n = 1$, the above proposition is equivalent to Corollary 1.5.3.

Proof of Lemma 8.3.1. Firstly we note that τ_m is associated with the linear transformation caused by the exchange of coordinates:

$$\hat{\ell}_m : \theta_m \longleftrightarrow \theta_{n+3}$$

To find the explicit form of the transformation T'_m, we make use of the method adopted in the proof of Theorem 8.1.2. In this case, the transformation from (8.1.1) to (8.1.2) is given by:

$$z = \frac{(t_m - 1)x}{t_m - x}, \quad w = (z - 1 + t_m)^{-\alpha} y.$$

It follows that

$$t'_i = \begin{cases} \dfrac{t_m - 1}{t_m - t_i} \cdot t_i & (i \neq m), \\ 1 - t_m & (i = m), \end{cases}$$

$$\lambda'_k = \frac{t_m - 1}{t_m - \lambda_k} \qquad (k = 1, \cdots, n).$$

Then we have , for $i \neq m$,

$$\begin{aligned} q'_i &= t'_i \frac{\Lambda_*(t'_i)}{T'_*(t'_i)} \\ &= -\frac{t_i}{t_m} \frac{\Lambda(t_i)}{\Lambda(t_m)} \frac{T'(t_m)}{T'(t_i)} \\ &= -\frac{q_i}{q_m}, \\ s'_i &= \frac{s_i}{s_m} \end{aligned}$$

and

$$q'_m = \frac{1}{q_m}, \quad s'_m = \frac{1}{s_m}.$$

On the other hand, since, in this case,

$$\begin{aligned} P_2(z,t) = \frac{e(t'_m)^2}{(z - t'_m)^4} p_2(x,t) + \frac{\alpha(\alpha - 1)}{(z - t'_m)^2} \\ \frac{\alpha}{z - t'_m} \left\{ \frac{2}{z - t'_m} + \frac{e(t_m)}{(z - t'_m)^2} p_1(x,t) \right\}, \end{aligned}$$

we obtain

$$\begin{aligned} \mu'_k := & \operatorname*{Res}_{z=\lambda'_k} P_2(z,t) \\ &= -\frac{\alpha}{\lambda'_k - t'_m} + \frac{e(t'_m)}{(\lambda'_k - t'_m)^2} \mu_k \\ &= \frac{\lambda_k - t_m}{e(t_m)} (\alpha + (\lambda_k - t_m)\mu_k), \end{aligned}$$

where $P_2(z,t)$ is a coefficient of the equation (8.1.2) and $e(t) = t(t-1)$. Then, by definition,

$$
\begin{aligned}
p_i' &= -(t_i'-1)\sum_k \frac{M_*^{k,i}}{\lambda_k'(\lambda_k'-1)}\mu_k' \\
&= A_i + B_i,
\end{aligned}
$$

where

$$
A_i = (t_i'-1)\alpha \sum_k \frac{M_*^{k,i}}{\lambda_k'(\lambda_k'-1)(\lambda_k'-t_m')},
$$

$$
B_i = -(t_i'-1)e(t_m')\sum_k \frac{M_*^{k,i}}{\lambda_k'(\lambda_k'-1)(\lambda_k'-t_m')^2}\mu_k.
$$

Now recall the proof of Lemma 8.2.1 and make again use of similar computations to those in the proof of Proposition 7.3.3. For example, we find:

$$
B_m = (t_m'-1)\sum_a^m q_a p_a (t_m'-t_a')N_*(m,a) + (t_m'-1)q_m p_m M_*(m,m),
$$

where

$$
N_*(m,a) = \sum_k \frac{M_*^{k,m}}{\lambda_k'(\lambda_k'-1)(\lambda_k'-t_m')(\lambda_k'-t_a')},
$$

$$
M_*(m,m) = \sum_k \frac{M_*^{k,m}}{\lambda_k'(\lambda_k'-1)(\lambda_k'-t_m')}.
$$

By means of the residue theorem for rational functions,

$$
\frac{T_*(x)}{x(x-1)(x-t_m')^2(x-t_a')\Lambda_*(x)},
$$

$$
\frac{T_*(x)}{x(x-1)(x-t_m')^2\Lambda_*(x)},
$$

we obtain finally

$$(t'_m - t'_a)N_*(m,a) = -\frac{1}{q'_m} = -q_m,$$

$$M_*(m,m) = \frac{1}{q'_m} = q_m.$$

The proof can be completed by computations. ∎

9. Particular solutions of the Hamiltonian system $\mathcal{H}_n$

We saw in Section 1.7 that if the parameters in the Painlevé equation P_{VI} satisfy certain conditions, then P_{VI} has particular solutions expressed in terms of the Gauss hypergeometric function. As a generalization, we shall see in this section that, under certain conditions on parameters, the Hamiltonian system $\mathcal{H}_n$ has solutions expressed in terms of Lauricella's hypergeometric functions in n variables.

9.1. The Lauricella hypergeometric series F_D

We present a generalization of the hypergeometric function due to (Appell-) Lauricella.

Let $I := \{m = (m_1, \dots, m_n); m_1, \dots, m_n \in \mathbb{N}_0\}$; and for $m \in I$, let $|m| := m_1 + \cdots + m_n$. For $x = (x_1, \dots, x_n)$ and $m \in I$, we set $x^m := x_1^{m_1} \cdots x_n^{m_n}$. Let us define the *Lauricella hypergeometric series* in n variables:

$$F_D(\alpha, \beta_1, \dots, \beta_n, \gamma; x) := \sum_{m \in I} \frac{(\alpha)_{|m|}(\beta_1)_{m_1} \cdots (\beta_n)_{m_n}}{(\gamma)_{|m|}(1)_{m_1} \cdots (1)_{m_n}} x^m,$$

where α, $\beta_1, \dots, \beta_n$, γ are complex constants such that $\gamma \neq -1, -2, \dots$, and

$$(\alpha)_k := \alpha(\alpha+1)\cdots(\alpha+k-1).$$

PROPOSITION 9.1.1. *The power series $F_D(\alpha, \beta_1, \dots, \beta_n, \gamma; x)$ converges in the polydisc $\Delta = \{x \in \mathbb{C}^n; |x_1| < 1, \dots, |x_n| < 1\}$.*

Proof. Set

$$C_m := \frac{(\alpha)_{|m|}(\beta_1)_{m_1} \cdots (\beta_n)_{m_n}}{(\gamma)_{|m|}(1)_{m_1} \cdots (1)_{m_n}}. \tag{9.1.1}$$

It is easily seen that there exists a positive constant C such that

$$|C_m| < C \quad \text{for all } m \in I$$

and hence that $F_D(\alpha, \beta_1, \ldots, \beta_n, \gamma; x)$ is majorized by the series

$$\sum_{m_1,\ldots,m_n=0}^{\infty} C x_1^{m_1} \cdots x_n^{m_n} = \frac{C}{(1-x_1)\cdots(1-x_n)},$$

which converges in the unit polydisc Δ. Hence $F_D(\alpha, \beta_1, \cdots, \beta_n, \gamma; x)$ converges uniformly on any compact set in Δ. ∎

PROPOSITION 9.1.2. *If α and γ are not negative integers, the series $F_D(\alpha, \beta_1, \ldots, \beta_n, \gamma; x)$ satisfies the system of differential equations:*

$$L_i u = 0, \quad M_{ij} u = 0, \qquad i, j = 1, \ldots, n, \tag{9.1.2}$$

where L_i and M_{ij} are linear differential operators defined by

$$L_i := \left(\gamma - 1 + \sum_i \delta_i\right)\delta_i - x_i\left(\alpha + \sum_i \delta_i\right)(\beta_i + \delta_i), \tag{9.1.3}$$

$$M_{ij} := (x_i - x_j)\delta_i\delta_j - \beta_j x_j \delta_i + \beta_i x_i \delta_j, \tag{9.1.4}$$

and $\delta_i := x_i \partial/\partial x_i$.

Proof. Let $e_i = (0, \ldots, 0, 1, 0, \ldots, 0) \in I$ be the multi-index with 1 at the i-th position. Notice that the operators $\delta_1, \ldots, \delta_n$ commute with each other:

$$[\delta_i, \delta_j] = 0, \qquad i, j = 1, \ldots, n, \tag{9.1.5}$$

and x^m ($m \in I$) are simultaneous eigenfunctions of the operators $\delta_1, \ldots, \delta_n$:

$$\delta_i x^m = m_i x^m, \qquad i = 1, \ldots, n. \tag{9.1.6}$$

In particular we have

$$(\delta_1 + \cdots + \delta_n)x^m = (m_1 + \cdots + m_n)x^m.$$

By the definition of the coefficients C_m, we have

$$\frac{C_{m+e_i} x^{m+e_i}}{C_m x^m} = \frac{(\alpha + m_1 + \cdots + m_n)(\beta_i + m_i)}{(\gamma + m_1 + \cdots + m_n)(1 + m_i)} x_i,$$

i.e.,

$$(\gamma + m_1 + \cdots + m_n)(1 + m_i)C_{m+e_i}x^{m+e_i}$$
$$- x_i(\alpha + m_1 + \cdots + m_n)(\beta_i + m_i)C_m x^m = 0.$$

Using (9.1.5) and (9.1.6), this is equivalent to

$$\left(\gamma - 1 + \sum_i \delta_i\right)\delta_i C_{m+e_i}x^{m+e_i}$$
$$- x_i\left(\alpha + \sum_i \delta_i\right)\cdot(\beta_i + \delta_i)C_m x^m = 0.$$

Thus we have

$$L_i \cdot F_D(\alpha, \beta_1, \ldots, \beta_n, \gamma; x) = 0, \qquad i = 1, \ldots, n.$$

To deduce the second equation of (9.1.2), we consider operators $\delta_i L_j - \delta_j L_i$ which annihilate F_D. Then it is easy to see that

$$\begin{aligned} &\delta_i L_j - \delta_j L_i \\ &= \left(\alpha + 1 + \sum_i \delta_i\right)\{(x_i - x_j)\delta_i\delta_j + \beta_i x_j \delta_i - \beta_j x_i \delta_j\}. \end{aligned} \tag{9.1.7}$$

Since $(\alpha + 1 + \sum_i \delta_i)u(x) = 0$ for $u \in \mathbb{C}\{x\}$ implies $u = 0$ by virtue of the assumption on α, it follows from (9.1.7) that the series F_D satisfies the second equation of (9.1.2). ∎

COROLLARY 9.1.3. *The hypergeometric series F_D satisfies the system of differential equations*

$$P_i u = 0, \quad Q_{ij} u = 0, \qquad i, j = 1, \ldots, n, \tag{9.1.8}$$

where

$$P_i := x_i(1 - x_i)D_i^2 + (1 - x_i)\sum_j x_j D_i D_j \tag{9.1.9}$$
$$+ \{\gamma - (\alpha + \beta_i + 1)x_i\}D_i - \beta_i \sum_j x_j D_j - \alpha\beta_i,$$

$$Q_{ij} := (x_i - x_j)D_i D_j + \beta_i D_j - \beta_j D_i \tag{9.1.10}$$

and $D_i := \partial/\partial x_i$.

Proof. Using the commutation relations:

$$[\delta_i, x_i] = x_i, \quad [\delta_i, D_i] = -D_i, \qquad i = 1, \ldots, n,$$

we see that $L_i = x_i P_i$, where

$$P_i = \left(\gamma + \sum_i \delta_i\right) D_i - \left(\alpha + \sum_i \delta_i\right)(\beta_i + \delta_i).$$

Expressing P_i in terms of $D_1, \ldots, D_n$, we obtain the first equations of (9.1.8). It is immediate that the second equation is derived from the second equation of (9.1.2). ∎

The system (9.1.2) as well as (9.1.8) is called the *Lauricella hypergeometric differential equation* and is denoted by $E_D(\alpha, \beta_1, \ldots, \beta_n, \gamma)$. Let

$$\Xi := \bigcup_{i \neq j} \Xi_{ij}, \quad \Xi_{ij} := \{x \in \mathbb{C}^n;\ x_i = x_j\}, \quad i, j = 1, \ldots, n+2; i \neq j,$$

$$B := \mathbb{C}^n \setminus \Xi,$$

where we set $x_{n+1} = 0, x_{n+2} = 1$.

PROPOSITION 9.1.4.
(i) *For any point* $P \in B$, *there are* $n+1$ *linearly independent holomorphic solutions of* $E_D(\alpha, \beta_1, \ldots, \beta_n, \gamma)$ *at* P;
(ii) $F_D(\alpha, \beta_1, \ldots, \beta_n, \gamma; x)$ *is the unique holomorphic solution* $u(s)$ *of the system* (9.1.2) *such that* $u(0) = 1$.

Proof. (i): Set $\vec{u} := {}^t(u_0, u_1, \ldots, u_n) := {}^t(u, \delta_1 u, \ldots, \delta_n u)$, then the equations (9.1.2) are written as

$$D_i u_0 = \frac{u_i}{x_i},$$

$$D_i u_i = \frac{-\alpha\beta_i}{x_i - 1} u_0 + \{(1-\gamma-\sum_k^i \beta_k)\frac{1}{x_i}$$
$$+ \frac{\gamma-\alpha-\beta_i-1}{x_i-1} + \frac{-\beta_k}{x_i-x_k}\}u_i$$
$$+\sum_k^i \beta_i\{\frac{1}{x_i-x_k} - \frac{1}{x_i-1}\}u_k,$$

$$D_i u_j = \beta_j(-\frac{1}{x_i} + \frac{1}{x_i-x_j})u_i - \frac{\beta_i}{x_i-x_j}u_j \quad (i \neq j).$$

Hence we have the Pfaffian system

$$d\vec{u} = \left(\sum_{1\le i<j\le n+2} A_{ij} d\log\varphi_{ij}\right)\vec{u},$$

where $\varphi_{ij} := x_i - x_j$ and

$$A_{ij} = \begin{array}{c} \\ i \\ j \end{array}\!\!\begin{array}{c} \begin{array}{cc} i & j \end{array} \\ \begin{pmatrix} \beta_i & -\beta_j \\ \beta_i & -\beta_j \end{pmatrix} \end{array},$$

$$A_{i,n+1} = i\begin{pmatrix} 0 & \dots & 0 & \overset{i}{1} & 0 & \dots & 0 \\ & & & -\beta_1 & & & \\ & & & \vdots & & & \\ & & & -\beta_{i-1} & & & \\ & & & 1-\gamma-\beta_i & & & \\ & & & -\beta_{i+1} & & & \\ & & & \vdots & & & \\ & & & -\beta_n & & & \end{pmatrix},$$

$$A_{i,n+2} = i \begin{pmatrix} & & i & \\ -\alpha\beta_i & -\beta_i \ldots -\beta_i & \gamma-\alpha-\beta_i-1 & -\beta_i \ldots -\beta_i \\ & & & \end{pmatrix}.$$

By the identity

$$d\log\varphi_{ij} \wedge d\log\varphi_{jk} = d\log\varphi_{jk} \wedge d\log\varphi_{ki} + d\log\varphi_{ki} \wedge d\log\varphi_{ji},$$

the integrability condition $\Omega \wedge \Omega = 0$ reduces to

$$[A_{ij}, A_{jk} + A_{ik}] = [A_{ij} + A_{jk}, A_{ik}] = 0 \quad \text{if} \quad i < j < k$$

$$[A_{ij}, A_{kl}] = 0 \quad \text{if} \quad i, j, k, l \text{ are distinct},$$

which can be easily checked.
(ii): Suppose that the series $u(x) = \sum_{m \in I} C_m x^m$, $C_0 = 1$ satisfies formally the equations $L_i \cdot u = 0$ $(i = 1, \cdots, n)$. Substitute $u(x)$ into the equations; we get recurrence formulae for $\{C_m\}_{m \in I}$:

$$C_{m+e_i} = \frac{(\alpha + m_1 + \cdots + m_n)(\beta_i + m_i)}{(\gamma + m_1 + \cdots + m_n)(1 + m_i)} C_m, \qquad i = 1, \ldots, n.$$

Determining successively C_m by these recursive formulas, with the initial condition $C_0 = 1$, we obtain

$$C_m = \frac{(\alpha)_{|m|}(\beta_1)_{m_1} \cdots (\beta_n)_{m_n}}{(\gamma)_{|m|}(1)_{m_1} \ldots (1)_{m_n}}. \quad \blacksquare$$

COROLLARY 9.1.5. *The holomorphic function $F_D(\alpha, \beta_1, \ldots, \beta_n, \gamma; x)$ at $0 \in \mathbb{C}^n$ can be continued analytically to a holomorphic function on the universal covering space of $B = \mathbb{C}^n \setminus \Xi$.*

Any solution of $E_D(\alpha, \beta_1, \ldots, \beta_n, \gamma)$ is called a *hypergeometric function of Lauricella.*

9.2. Particular solutions of $\mathcal{H}_n$

We show that, when the parameters θ belong to some hyperplanes in the affine space V, the system $\mathcal{H}_n(\theta)$ admits solutions which can be expressed by logarithmic derivatives of hypergeometric functions of Lauricella. In fact we have

THEOREM 9.2.1. *Suppose that the parameters $\theta = (\theta_1, \dots, \theta_{n+2}, \theta_\infty)$ satisfy*

$$\sum_{(i)} \theta_i + \theta_\infty - 1 = 0, \tag{9.2.1.$\sharp$}$$

or

$$\sum_{(i)} \theta_i - \theta_\infty - 1 = 0. \tag{9.2.1.$\flat$}$$

Then the system $\mathcal{H}_n(\theta)$ has solutions $(q(s), p(s))$ given by

$$\begin{cases} q_i(s) = A^{-1} e(s_i) D_i \left(\log((1 - s_i)^{\theta_i} u(\theta; s))\right), \\ p_i(s) = 0, \end{cases}$$

$i = 1, \dots, n$, where $D_i = \partial / \partial s_i$,

$$e(x) = x(x-1), \quad A = \sum_{(i)} \theta_i - 1$$

and $u(\theta; s)$ is an arbitrary solution of $E_D(L(\theta))$,

$$L(\theta) := (1 - \theta_{n+2}, \theta_1, \cdots, \theta_n, A - \theta_{n+2} + 1).$$

Proof. The condition (9.2.1.$\sharp$ or $\flat$) for the parameters θ implies $\kappa = 0$, κ being the parameter of the Hamiltonians $H_1, \dots, H_n$ for the system $\mathcal{H}_n(\theta)$. Then the system $\mathcal{H}_n(\theta)$ can be written as follows:

$$\begin{cases} e(s_i) D_i q_j = 2 \sum_k E^i_{jk} p_k - F^i_j, \\ e(s_i) D_i p_j = - \sum_{k,l} \dfrac{\partial E^i_{kl}}{\partial q_j} p_k p_l + \sum_k \dfrac{\partial F^i_k}{\partial q_j} p_k. \end{cases} \tag{9.2.2}$$

Let $p_1 = \cdots = p_n = 0$, then the second equation of (9.2.2) is satisfied. The first equation of (9.2.2) reduces to the following system for $q_1, \ldots, q_n$:

$$e(s_i)D_i q_j = -F^i_j(s,q), \qquad i,j = 1,\ldots,n. \tag{9.2.3}$$

Since the system $\mathcal{H}_n(\theta)$ is completely integrable, so is the system (9.2.3). Since the terms $F^i_j(s,q)$ in the right-hand side of (9.2.3) are polynomials of degree two in $q_1, \ldots, q_n$, the system (9.2.3) can be considered as a several-variables version of the classical Riccati equation. Recall here that the Riccati equation can be linearized by a suitable change of unknown (cf. Proposition 1.1.3). Imitating the classical case, let us linearize (9.2.3) by introducing a new unknown. To this end we first see that, for any given solution $q(s)$ of (9.2.3), the 1-form

$$\Omega := \sum_i \frac{q_i(s)}{e(s_i)} ds_i$$

is closed. In fact, we have

$$\begin{aligned} d\Omega &= \sum_i \frac{dq_i(s) \wedge ds_i}{e(s_i)} \\ &= \sum_{i,j} \frac{1}{e(s_i)} D_j q_i \, ds_j \wedge ds_i \\ &= \sum_{i<j} \frac{1}{e(s_i)e(s_j)} (F^i_j - F^j_i) ds_i \wedge ds_j \\ &= 0, \end{aligned}$$

because of the symmetry (7.1.7) of the rational functions F^i_j. Since Ω is closed, for any solution $q(s)$ of (9.2.3) there is a function $u(s)$ such that

$$\Omega = A^{-1} d \log \left(\prod_k (s_k - 1)^{\theta_k} u(s) \right),$$

i.e.,

$$q_i(s) = A^{-1} e(s_i) \left\{ \frac{\theta_i}{s_i - 1} + \frac{D_i u}{u} \right\}, \qquad i = 1, \ldots, n. \tag{9.2.4}$$

Note that $u(s)$ is determined up to multiplicative constants. Substituting (9.2.4) into the system (9.2.3), we get a system of linear differentrial equations:

$$\begin{aligned}
&s_i(s_i-1)D_i^2u \\
=&(s_i-1)\sum_k^i s_k\left[\frac{\theta_k}{s_k-s_i}D_iu+\frac{\theta_i}{s_i-s_k}D_ku\right] \\
&+\left[\sum_{(m)}^{n+2}\theta_m-(2+\theta_i-\theta_{n+2})s_i\right]D_iu \\
&-\theta_i\sum_k^i s_kD_ku-(1-\theta_{n+2})\theta_iu,
\end{aligned}$$

$$(s_i-s_j)D_iD_ju=\theta_jD_iu-\theta_iD_ju=0, \qquad i,j=1,\ldots,n;$$

which is just the system $E_D(1-\theta_{n+2},\theta_1,\ldots,\theta_n,A-\theta_{n+2}+1)$. ∎

4 Studies on Singularities of Non-linear Differential Equations

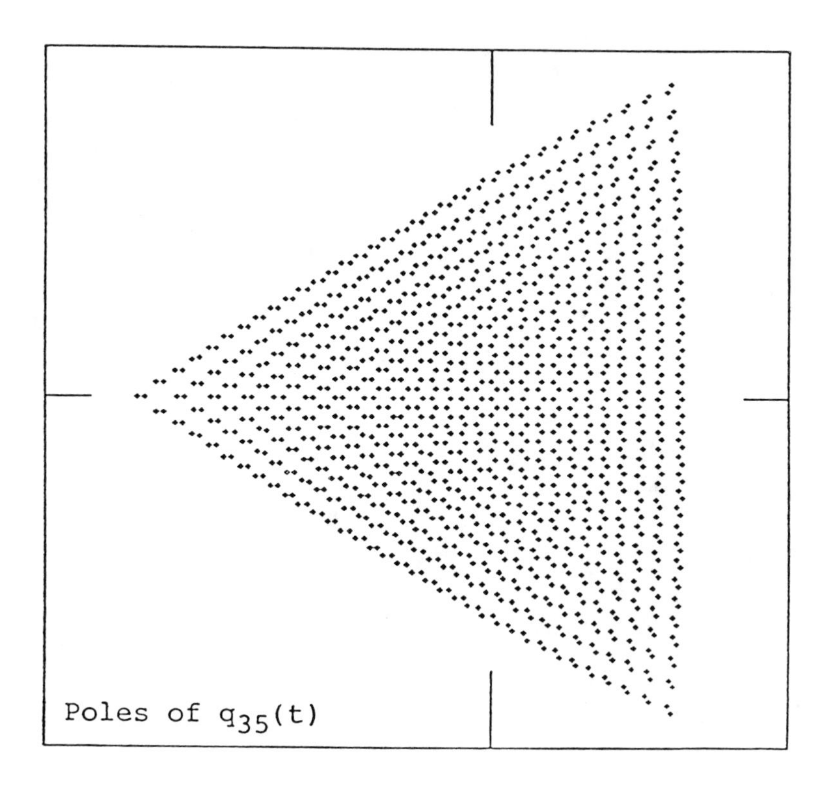

Consider a system of nonlinear ordinary differential equations of the form

$$(0.0.1) \qquad y_i' = g_i(x, y_1, \cdots, y_n) \quad (i = 1, \cdots, n)$$

in n unknowns $y_1, \cdots, y_n$, where the g_i's are meromorphic in $x, y_1, \cdots, y_n$ with poles along $x = a$. We say that equation (0.0.1) has a *singular point of regular type* at $x = a$ if every g_i has at most a simple pole at $x = a$, and *of irregular type* otherwise.

For an n-th order differential equation

$$y^{(n)} = g(x, y, y', \cdots, y^{(n-1)}),$$

where g is a meromorphic function having poles along $x = a$, the singular point $x = a$ is said to be *of regular type* if it is equivalent to a system of n differential equations of the first order for which $x = a$ is a singular point of regular type.

In this chapter, we give some methods of constructing solutions of nonlinear differential equations near singular points of regular type. The main purpose of Section 1 is to establish a classical fundamental theorem (Theorem 1.3.1) under the so-called Poincaré condition. We also discuss the case $n = 1$ in detail. In Section 2, we study the fixed singular points of regular type of Painlevé equations. Unfortunately, Theorem 1.3.1 is not applicable to Painlevé equations because Poincaré's condition is completely violated. In order to treat Painlevé equations, we first transform them into a normal form and next give a local expression of its solutions at the singular point (Theorem 2.2.1).

We use, in this chapter, the following notation:

1)

$$\mathbf{0} = (0, \cdots, 0) \in \mathbb{Z}^n,$$

$$E[1] = (1, 0, \cdots, 0) \in \mathbb{Z}^n,$$

$$\vdots$$

$$E[n] = (0, \cdots, 0, 1) \in \mathbb{Z}^n.$$

For a column vector $X = {}^t(x_1, \cdots, x_n) \in \mathbb{C}^n$ and for a multi-index

$$K = (k(1), \cdots, k(n)) \in \mathbb{N}_0{}^n,$$

we write

$$X^K = x_1^{k(1)} \cdots x_n^{k(n)},$$

$$|K| = k(1) + \cdots + k(n).$$

2) For two indices $K = (k(1), \cdots, k(n))$ and $L = (\ell(1), \cdots, \ell(n))$, we write

$$K \succ L$$

if and only if

$$k(1) + \cdots + k(n) > \ell(1) + \cdots + \ell(n),$$

or

$$k(1) + \cdots + k(n) = \ell(1) + \cdots + \ell(n),$$
$$k(i) = \ell(i) \ (i < j), \quad k(j) > \ell(j)$$

for some $1 \leq j \leq n$. We write

$$K \succeq L$$

if and only if $K \succ L$ or $K = L$, and $K \prec L$ if $L \succ K$. We use similar notations on $(\mathrm{N}_0)^{n+1}$. Thus

$$\begin{aligned}(0, \mathbf{0}) &\prec (0, E[n]) \prec (0, E[n-1]) \prec \cdots \prec (0, E[1]) \prec (1, \mathbf{0}) \\ &\prec (0, 2E[n]) \prec (0, E[n-1] + E[n]) \prec \cdots \\ &\qquad \prec (0, 2E[1]) \prec \cdots \prec (1, E[1]) \prec (2, \mathbf{0}) \\ &\prec (0, 3E[n]) \prec \cdots\cdots \prec (3, \mathbf{0}) \\ &\prec \cdots\cdots .\end{aligned}$$

3) For a column vector $X = {}^t(x_1, \cdots, x_n) \in \mathbb{C}^n$ and an $n \times n$ matrix $A = (a_{ij})_{1 \leq i,j \leq n}$, we define $\|X\|$ and $\|A\|$ by

$$\|X\| = \max_i |x_i|,$$

$$\|A\| = \max_i \sum_{j=1}^{n} |a_{ij}|.$$

Note, for $X, Y \in \mathbb{C}^n$, $c \in \mathbb{C}$ and an $n \times n$ matrix A, that

$$X = 0 \text{ if and only if } \|X\| = 0, \tag{0.0.2}$$

$$\|cX\| = |c|\|X\|, \tag{0.0.3}$$

$$\|X + Y\| \leq \|X\| + \|Y\|, \tag{0.0.4}$$

$$\|AX\| \leq \|A\|\|X\|. \tag{0.0.5}$$

In Section 1, vectors, multi-indices and matrices will be denoted by capital letters, while their components will be denoted by the corresponding small letters.

1. Singularities of Regular Type

In this section we consider a system of the form

$$xy_i' = f_i(x, y_1, \cdots, y_n) \quad (i = 1, \cdots, n), \tag{1.0.1}$$

where the $f_i(x, y_1, \cdots, y_n)$ are holomorphic at $x = 0, y_1 = 0, \cdots, y_n = 0$ and satisfy

$$f_i(0, \cdots, 0) = 0 \quad (i = 1, \cdots, n).$$

In our terminology, (1.0.1) has a singular point of regular type at $x = 0$.

1.1. Holomorphic solutions

First we give a solution of (1.0.1) which is holomorphic at $x = 0$. Equation (1.0.1) can be written in the vectorial form:

$$xY' = F(x, Y), \tag{1.1.1}$$

where $Y = {}^t(y_1, \cdots, y_n) \in \mathbb{C}^n$ is the unknown vector and $F(x, Y) = {}^t(f_1(x, Y), \cdots, f_n(x, Y))$ is a vector-valued function. We expand $F(x, Y)$ into a convergent power series in x and Y:

$$F(x, Y) = \sum_{j+|K|\geq 1} A_{jK} x^j Y^K \tag{1.1.2}$$

which is bounded in the domain:

$$|x| < r_0, \ \|Y\| < r_1 \quad (r_0 > 0, \ r_1 > 0).$$

Here K denotes the multi-index $(k(1),\cdots,k(n))\in(\mathbb{N}\cup\{0\})^n$ and A_{jK} the column vector ${}^t(a_{jK}^{(1)},\cdots,a_{jK}^{(n)})\in\mathbb{C}^n$.

Suppose that the series

$$Y=\sum_{j\geq 1}C_jx^j,\quad C_j={}^t(c_j^{(1)},\cdots,c_j^{(n)})\in\mathbb{C}^n \tag{1.1.3}$$

satisfies equation (1.1.1) formally. Then we have

$$\sum_{j\geq 1}jC_jx^j=\sum_{j\geq 1}AC_jx^j$$
$$+\sideset{}{'}\sum_{(j,K)}A_{jK}x^j\prod_{\ell=1}^{n}\Bigl(\sum_{p\geq 1}c_p^{(\ell)}x^p\Bigr)^{k(\ell)},$$

where $\sum'_{(j,K)}$ stands for the summation over $(j,K)\succeq(1,\mathbf{0})$, that is,

$$j+|K|\geq 1\text{ and }(j,K)\neq(0,E[\ell])\quad(1\leq\ell\leq n),$$

and A denotes the matrix:

$$A=\begin{pmatrix}\dfrac{\partial f_1}{\partial y_1}(0,\mathbf{0})\cdots\dfrac{\partial f_1}{\partial y_n}(0,\mathbf{0})\\ \vdots\qquad\qquad\vdots\\ \dfrac{\partial f_n}{\partial y_1}(0,\mathbf{0})\cdots\dfrac{\partial f_n}{\partial y_n}(0,\mathbf{0})\end{pmatrix}.$$

We have

$$(jI-A)C_j=V_j,\quad j\geq 1, \tag{1.1.4}$$

where

$$V_j=\sideset{}{'}\sum_{(m,K)}\ \sum_{m+p_1+\cdots+p_{|K|}=j}A_{mK}c_{p_1}^{(1)}\cdots c_{p_{k(1)}}^{(1)}$$
$$\times\cdots c_{p_{k(1)+\cdots+k(n-1)+1}}^{(n)}\cdots c_{p_{|K|}}^{(n)}. \tag{1.1.5}$$

Notice that each component of V_j is a polynomial with positive coefficients in $a_{mK}^{(\ell)}$'s and $c_p^{(\ell)}$'s, where m, K, ℓ and p satisfy

$$(m,K)\succeq(1,\mathbf{0}),\ 1\leq m+|K|\leq j,\ 1\leq p\leq j-1,\ 1\leq\ell\leq n.$$

Then we have

PROPOSITION 1.1.1. *If none of the eigenvalues of A is a positive integer, then equation* (1.1.1) *admits a solution of the form*

$$Y = \sum_{j \geq 1} C_j x^j$$

convergent at $x = 0$. Furthermore there is a unique solution holomorphic at $x = 0$ satisfying $Y(0) = \mathbf{0}$.

Proof. One can see by (1.1.4) that the coefficients C_j's are determined uniquely, because, for each $j \geq 1$, V_j is a function of $C_1, \cdots, C_{j-1}$ and $jI - A \in GL(n, \mathbb{C})$. It is sufficient to show the convergence of the formal series. There exist positive constants M and ε such that

$$\|A_{jK}\| \leq M r_0^{-j} r_1^{-|K|} \quad \text{for} \quad (j, K) \succeq (1, \mathbf{0}), \tag{1.1.6}$$

$$\|(jI - A)^{-1}\| \leq \varepsilon^{-1} \quad \text{for} \quad j \geq 1. \tag{1.1.7}$$

The Cauchy inequality yields inequality (1.1.6). Inequality (1.1.7) is derived as follows: There exists a non-singular matrix P such that $P^{-1}AP$ is a Jordan canonical form

$$P^{-1}AP = \Lambda = \begin{pmatrix} \Lambda_1 & & 0 \\ & \ddots & \\ 0 & & \Lambda_m \end{pmatrix}$$

where Λ_i $(i = 1, \cdots, m)$ is a Jordan block of size $s(i)$ written in the form

$$\Lambda_i = \begin{pmatrix} \lambda_i & & 0 \\ & \ddots & \\ 0 & & \lambda_i \end{pmatrix} \quad \text{or} \quad \Lambda_i = \begin{pmatrix} \lambda_i & 1 & & 0 \\ & \ddots & \ddots & \\ & & \ddots & 1 \\ 0 & & & \lambda_i \end{pmatrix}.$$

By the assumption on A, for every positive integer j, the matrices $jI_{s(i)} - \Lambda_i$ are nonsingular and we have

$$\begin{aligned} (jI - A)^{-1} &= P(jI - \Lambda)^{-1} P^{-1} \\ &= P \begin{pmatrix} (jI_{s(1)} - \Lambda_1)^{-1} & & \\ & \ddots & \\ & & (jI_{s(m)} - \Lambda_m)^{-1} \end{pmatrix} P^{-1}. \end{aligned}$$

Note that, if $\alpha \neq 0$,

$$\begin{pmatrix} \alpha & 1 & & 0 \\ & \ddots & \ddots & \\ & & \ddots & 1 \\ 0 & & & \alpha \end{pmatrix}^{-1} = \frac{1}{\alpha}I - \frac{1}{\alpha^2}\Delta + \frac{1}{\alpha^3}\Delta^2 - \cdots + (-1)^k \frac{1}{\alpha^k}\Delta^k,$$

where k is the size of the matrix and Δ is the following k by k matrix:

$$\begin{pmatrix} 0 & 1 & & 0 \\ & \ddots & \ddots & \\ & & \ddots & 1 \\ 0 & & & 0 \end{pmatrix}.$$

By the assumption on A, there exists a positive constant M such that

$$|j - \lambda_i|^{-1} \leq M$$

for integers $j \geq 1$ and $i = 1, \ldots, m$. Using this estimate, we have

$$\begin{aligned} \|(jI - \Lambda)^{-1}\| &= \max_{1 \leq i \leq m} \|(jI_{s(i)} - \Lambda_i)^{-1}\| \\ &\leq (M + M^2 + \cdots + M^\sigma) \quad (\sigma = \max_{1 \leq i \leq m} s(i)), \end{aligned}$$

and hence

$$\|(jI - A)^{-1}\| \leq \|P\|\|P^{-1}\|(M + M^2 + \cdots + M^\sigma).$$

Thus we have inequality (1.1.7).

Now consider the following algebraic equation

$$\begin{aligned} \varepsilon\eta &= \sum_{(j,K)}{}' \frac{M}{{r_0}^j {r_1}^{|K|}} x^j \eta^{|K|} \\ &= M\left[\left(1 - \frac{x}{r_0}\right)^{-1}\left(1 - \frac{\eta}{r_1}\right)^{-n} - 1 - \frac{n\eta}{r_1}\right]. \end{aligned} \tag{1.1.8}$$

From the implicit function theorem, it follows that (1.1.8) has a unique holomorphic solution $\eta = \eta(x)$ with $\eta(0) = 0$ in the neighborhood of $x = 0$. Let $\sum_{j\geq 1} \gamma_j x^j$ be the Taylor expansion of $\eta(x)$. We see that the coefficients γ_j are recursively determined by

$$\gamma_j = \varepsilon^{-1} \sum_{(m,K)}' \sum_{m+p_1+\cdots+p_{|K|}=j} \frac{M}{{r_0}^m {r_1}^{|K|}} \gamma_{p_1} \cdots \gamma_{p_{|K|}}. \tag{1.1.9}$$

On the other hand, from (1.1.4), (1.1.5), (1.1.6) and (1.1.7), we have

$$\begin{aligned} &\|C_j\| \\ &\leq \|(jI - A)^{-1}\| \sum_{(m,K)}' \sum_{m+p_1+\cdots+p_{|K|}=j} \|A_{mK}\| \|C_{p_1}\| \cdots \|C_{p_{|K|}}\| \\ &\leq \varepsilon^{-1} \sum_{(m,K)}' \sum_{m+p_1+\cdots+p_{|K|}=j} \frac{M}{{r_0}^m {r_1}^{|K|}} \|C_{p_1}\| \cdots \|C_{p_{|K|}}\|. \end{aligned} \tag{1.1.10}$$

By (1.1.9) and (1.1.10), we obtain

$$\|C_j\| \leq \gamma_j \quad \text{for} \quad j \geq 1,$$

which proves the convergence of $\sum_{j\geq 1} C_j x^j$. ∎

1.2. One-dimensional case

1.2.1. Formal transformations

In this section we construct a family of solutions of equation (1.1.1) when $n = 1$. Let us write it in the form

$$xy' = f(x, y), \tag{1.2.1}$$

where

$$f(x, y) = a_{10}x + \lambda y + \sum_{i+k\geq 2} a_{ik} x^i y^k,$$

for $|x| < r_0, |y| < r_1$. First we shall find a formal transformation $y = p(x, z)$ which changes equation (1.2.1) into an equation of simpler form.

Consider the transformation

$$T_1: \quad y = z + px,$$

where p is a parameter.

PROPOSITION 1.2.1. *Equation* (1.2.1) *is transformed by* T_1 *into*

$$xz' = g(x, z) = \sum_{j+k\geq 1} b_{jk}x^j z^k, \tag{1.2.2}$$

where

$$b_{10} = a_{10} + (\lambda - 1)p, \quad b_{01} = a_{01} = \lambda.$$

Proof. Equation (1.2.1) is transformed by T_1 into

$$\begin{aligned} xz' &= f(x, z + px) - px \\ &= (a_{10} + \lambda p - p)x + \lambda z + \sum_{j+k\geq 2} a_{jk}x^j (z + px)^k, \end{aligned}$$

which implies the proposition. ∎

Next we consider the transformation

$$T_m: \quad y = z + \sum_{j=0}^{m} p_j x^j z^{m-j} \quad (m \geq 2),$$

where p_j are parameters.

PROPOSITION 1.2.2. *Equation* (1.2.1) *is formally transformed by* T_m ($m \geq 2$) *into*

$$xz' = h(x, z) = \sum_{j+k\geq 1} c_{jk}x^j z^k, \tag{1.2.3}$$

where

$$\begin{aligned} c_{jk} &= a_{jk}, & j + k \leq m - 1, \\ c_{jk} &= a_{jk} - [(k-1)\lambda + j]p_j - (k+1)a_{10}p_{j-1}, & j + k = m, \end{aligned}$$

(*we set* $p_{-1} = 0$).

Proof. The formal inverse of T_m can be written as

$$z = y - \sum_{j=0}^{m} p_j x^j y^{m-j} + [x, y]_m \tag{1.2.4}$$

where $[x, y]_m$ denotes a linear combination of terms whose total degree (with respect to x and y) is greater than m. Differentiating (1.2.4), we have

$$\begin{aligned} xz' &= xy' - \sum_{j=0}^{m} jp_j x^j y^{m-j} - \sum_{j=0}^{m} (m-j) p_j x^j y^{m-j-1} (xy') + \cdots \\ &= f(x, y) - \sum_{j=0}^{m} jp_j x^j y^{m-j} - \sum_{j=0}^{m} (m-j) p_j x^j y^{m-j-1} f(x, y) + \cdots . \end{aligned}$$

Substituting T_m into the right-hand side, we have

$$\begin{aligned} xz' = f(x, z) &+ \sum_{j=0}^{m} \lambda p_j x^j z^{m-j} - \sum_{j=0}^{m} jp_j x^j z^{m-j} \\ &- \sum_{j=0}^{m} (m-j) a_{10} p_j x^{j+1} z^{m-j-1} - \sum_{j=0}^{m} (m-j) \lambda p_j x^j z^{m-j} + \cdots \\ = &\sum_{j+k \leq m-1} a_{jk} x^j z^k - \sum_{j=0}^{m} [((m-j-1)\lambda + j) p_j - a_{j,m-j}] x^j z^{m-j} \\ &- \sum_{j=0}^{m} (m-j+1) a_{10} p_{j-1} x^j z^{m-j} + \cdots \quad (p_{-1} = 0). \end{aligned}$$

Hence we have

$$\begin{aligned} c_{jk} &= a_{jk}, & j+k &\leq m-1, \\ c_{jk} &= a_{jk} - [(k-1)\lambda + j] p_j - (k+1) a_{10} p_{j-1}, & j+k &= m. \quad \blacksquare \end{aligned}$$

We shall construct a formal transformation by composing infinitely many transformations $T_1, T_2, \cdots, T_m, \cdots$, where the parameters of each T_m will be suitably chosen.

Let (E_1) be the equation obtained from (1.2.1) by the transformation T_1. Replacing the variable z in (E_1) by y, we apply the transformation T_2 to (E_1) and obtain equation (E_2). In this way, applying the transformations $T_1, T_2, \cdots, T_m$ to (1.2.1) successively, we obtain the equation which is written as

$$(E_m) \qquad xz' = h_m(x, z) = \sum_{j+k\geq 1} c_{jk}^{(m)} x^j z^k,$$

where $c_{jk}^{(m)}$ are constants.

Case 1. $\lambda \in \mathbb{C} - \mathbb{Q} \cap (-\infty, 0] - \mathbb{N}$
If we put $p = a_{10}/(1 - \lambda)$ in the transformation T_1, then

$$c_{10}^{(1)} = 0, \quad c_{01}^{(1)} = \lambda$$

in (E_1). Next putting $p_0 = c_{02}^{(1)}/\lambda$, $p_1 = c_{11}^{(1)}$, $p_2 = c_{20}^{(1)}/(2 - \lambda)$ in the transformation T_2, we have

$$c_{01}^{(2)} = \lambda,$$

$$c_{jk}^{(2)} = 0, \quad j + k \leq 2, \quad (j, k) \neq (0, 1)$$

in (E_2). Since $(k - 1)\lambda + j \neq 0$ (for $(j, k) \neq (0, 1)$), we can choose $T_3, \cdots, T_m$ so that

$$c_{01}^{(m)} = \lambda,$$
$$c_{jk}^{(m)} = 0, \quad j + k \leq m, \quad (j, k) \neq (0, 1)$$

in (E_m) . Therefore, composing the transformations given above, we obtain a formal transformation

$$y = \sum_{j+\ell\geq 1} p_{j\ell} x^j z^\ell \quad (p_{01} = 1)$$

by which equation (1.2.1) is transformed into

$$(1.2.5) \qquad xz' = \lambda z.$$

Case 2. $\lambda \in \mathbb{N}$

As long as $(j,k) \neq (0,1), (\lambda, 0)$, we can choose T_m so that $c_{jk}^{(m)} = 0$ for $j+k \leq m$. When $\lambda = 1$, we put $p = 0$ in T_1 and when $\lambda \neq 1$, we put $p_\lambda = 0$ in T_λ. Thus we obtain a formal transformation $y = \sum_{j+\ell \geq 1} p_{j\ell} x^j z^\ell$ $(p_{01} = 1, \;\; p_{\lambda 0} = 0)$ by which equation (1.2.1) is transformed into

$$xz' = \lambda z + bx^\lambda, \tag{1.2.6}$$

where b is a complex constant.

Case 3. $\lambda = 0$

In this case, by Proposition 1.1.1, equation (1.2.1) admits a convergent power series solution $\varphi(x) = \sum_{j \geq 1} c_j x^j$. By the transformation $y = v + \varphi(x)$, equation (1.2.1) is transformed into the equation

$$xv' = f_0(x,v) = \sum_{i+k \geq 1, k \geq 1} a'_{ik} x^i v^k. \tag{1.2.7}$$

In place of (1.2.1), we start from equation (1.2.7). By Proposition 1.2.2, the transformation T_m $(m \geq 2)$ takes equation (1.2.7) into equation (1.2.3) with

$$\begin{aligned} c_{ik} &= a'_{ik}, & i+k &\leq m-1, \\ c_{ik} &= a'_{ik} - ip_i, & i+k &= m. \end{aligned}$$

Hence if we put $p_i = c_{ik}^{(m-1)}/i$ $(i \geq 1)$ and $p_0 = 0$ in T_m $(m \geq 2)$, then by the transformation $T_m \cdots T_2$, we get equation (E_m) with $c_{ik}^{(m)} = 0$ $(i \geq 1)$. Thus we obtain a formal transformation

$$v = \sum_{j+k \geq 1} p_{jk}^{(1)} x^j u^k, \quad p_{01}^{(1)} = 1, \quad p_{0k}^{(1)} = 0 \quad (k \geq 2),$$

by which equation (1.2.7) is changed into

$$xu' = u^{s+1} \sum_{k \geq 0} b_k u^k \quad (b_0 = b \neq 0), \tag{1.2.8}$$

where s is a positive integer. Now we make a change of the unknown:

$$u = z(1 + r_n z^n) \tag{1.2.9}$$

where n is a positive integer and r_n is a constant. Since $z = u(1 - r_n u^n + \cdots)$, if we apply this transformation to equation (1.2.8), it becomes

$$\begin{aligned} xz' &= xu'(1 - r_n(n+1)u^n + \cdots) \\ &= (1 - r_n(n+1)u^n + \cdots)u^{s+1} \sum_{k\geq 0} b_k u^k \\ &= (1 - r_n(n+1)z^n(1 + r_n z^n)^n + \cdots)z^{s+1} \\ &\quad \times (1 + r_n z^n)^{s+1} \sum_{k\geq 0} b_k z^k (1 + r_n z^n)^k \\ &= z^{s+1}\Bigg[\sum_{k\geq 0} b_k(z^k + r_n k z^{k+n} + \cdots) \\ &\quad + r_n(s-n)z^n \sum_{k\geq 0} b_k(z^k + r_n k z^{k+n} + \cdots) + \cdots\Bigg] \\ &= z^{s+1}\Bigg[\sum_{k\geq 0} b_k z^k - b r_n(n-s)z^n + \cdots\Bigg] \\ &= z^{s+1} \sum_{k\geq 0} c_k^{(n)} z^k, \end{aligned}$$

where

$$\begin{aligned} c_k^{(n)} &= b_k, \qquad && 0 \leq k \leq n-1, \\ c_n^{(n)} &= b_n - b r_n(n-s), && k = n. \end{aligned}$$

When $n \neq s$, we choose r_n so that $c_n^{(n)} = 0$, $c_k^{(n)} = b_k$ $(0 \leq k \leq n-1)$. Composing the transformations of the form (1.2.9) for $n \neq s$, we obtain the transformation

$$u = z + \sum_{k\geq 2} r_k' z^k$$

by which (1.2.8) is changed into an equation of the form

$$xz' = z^{s+1}(b + b'z^s), \tag{1.2.10}$$

where b and b' are complex constants, and s is a positive integer. Thus we have constructed a formal transformation

$$y = \sum_{j+\ell\geq 1} p_{j\ell} x^j z^\ell, \quad p_{01} = 1, \quad p_{0k} = 0\ (k \geq 2)$$

which changes equation (1.2.1) into equation (1.2.10).

Case 4. $\lambda = -\mu/\nu, \quad \mu, \nu \in \mathbb{N}, \quad (\mu, \nu) = 1.$
Notice that $(k-1)\lambda + j = 0$ if and only if

$$j = h\mu, \quad k = h\nu + 1, \quad h = 0, 1, 2, \cdots.$$

Hence there exists a transformation

$$y = \sum_{j+k \geq 1} p_{jk}^{(1)} x^j u^k \tag{1.2.11}$$

with $p_{01}^{(1)} = 1$ and $p_{h\mu, h\nu+1}^{(1)} = 0$, by which (1.2.1) is changed into an equation of the form

$$xu' = -\frac{\mu}{\nu} u + u \sum_{h \geq 1} b_h (x^\mu u^\nu)^h. \tag{1.2.12}$$

If we put $w = x^\mu u^\nu$, this equation is written as

$$xw' = \nu w \sum_{h \geq 1} b_h w^h = \nu w^{s+1} \sum_{h \geq 0} b_{s+h} w^h \quad (b_s \neq 0), \tag{1.2.13}$$

for some positive integer s. In the same way as in Case 3, we can construct a transformation

$$w = \zeta + \sum_{k \geq 2} r_k'' \zeta^k, \tag{1.2.14}$$

by which (1.2.13) is changed into

$$x\zeta' = \zeta^{s+1}(b + b'\zeta^s). \tag{1.2.15}$$

Note that

$$\begin{aligned} u &= x^{-\mu/\nu} \left(\sum_{k \geq 1} r_k'' \zeta^k \right)^{1/\nu} \quad (r_1'' = 1) \\ &= x^{-\mu/\nu} \zeta^{1/\nu} \sum_{k \geq 0} r_k''' \zeta^k \quad (r_0''' = 1). \end{aligned}$$

We put $z = x^{-\mu/\nu}\zeta^{1/\nu}$ ($\zeta = x^\mu z^\nu$). Then we see that equation (1.2.15) is changed into

$$xz' = z\left(\lambda + b\nu^{-1}(x^\mu z^\nu)^s + b'\nu^{-1}(x^\mu z^\nu)^{2s}\right), \tag{1.2.16}$$

by the transformation

$$\begin{aligned} y &= \sum_{j+\ell\geq 1} p_{j\ell}^{(1)} x^j \left(z \sum_{k\geq 0} r_k''' x^{\mu k} z^{\nu k}\right)^\ell \\ &= \sum_{j+\ell\geq 1} p_{j\ell} x^j z^\ell \quad (p_{01} = 1). \end{aligned}$$

Thus we have shown

THEOREM 1.2.3. *There exists a formal transformation of the form*

$$y = \sum_{j+\ell\geq 1} p_{j\ell} x^j z^\ell \quad (p_{01} = 1), \tag{1.2.17}$$

by which equation (1.2.1) is formally changed into an equation of the following form

$$xz' = \lambda z, \qquad \text{if } \lambda \in \mathbb{C} - \mathbb{Q} \cap (-\infty, 0] - \mathbb{N}, \tag{1.2.5}$$

$$xz' = \lambda z + bx^\lambda, \qquad \text{if } \lambda \in \mathbb{N}, \tag{1.2.6}$$

$$xz' = z^{s+1}(b + b'z^s), \qquad \text{if } \lambda = 0, \tag{1.2.10}$$

$$xz' = z\left(\lambda + b\nu^{-1}(x^\mu z^\nu)^s + b'\nu^{-1}(x^\mu z^\nu)^{2s}\right), \tag{1.2.16}$$

$$\text{if } \lambda = -\mu/\nu,\ \mu, \nu \in \mathbb{N}.$$

Here b and b' are complex constants, and s is a positive integer.

It is easy to see that equations (1.2.5) and (1.2.6) have solutions

$$z = Cx^{\lambda} \quad (C \in \mathbb{C})$$

and

$$z = x^{\lambda}(b\log x + C) \quad (C \in \mathbb{C})$$

respectively. Equation (1.2.10) can be solved as follows. If $b = b' = 0$, then $z = C$, and if $b \neq 0, b' = 0$, then $z = (C - sb\log x)^{-1/s}$. In case $bb' \neq 0$, we put

$$\zeta = (b/b')z^{-s}, \quad x = \xi^{-b'/sb^2}.$$

Then equation (1.2.10) becomes

$$\xi\frac{d\zeta}{d\xi} = 1 + \frac{1}{\zeta}.$$

From this equation, we have

$$\zeta - \log(\zeta + 1) = \log\xi + C.$$

Hence equation (1.2.10) admits a solution

$$z = \left(\frac{b'}{b}\mathfrak{A}A\big(C - \frac{sb^2}{b'}\log x\big)\right)^{-1/s},$$

where $\zeta = \mathfrak{A}A(t)$ is an inverse function of $\zeta - \log(\zeta+1) = t$ (the function $\mathfrak{A}A$ is studied in [HKM]).
Summing up, equation (1.2.16) admits a solution of the form

$$\begin{aligned}
&z = Cx^{\lambda}, && \text{if } b = b' = 0,\\
&z = x^{\lambda}(C - sb\log x)^{-1/s\nu}, && \text{if } b' = 0,\\
&z = x^{\lambda}\big(\tfrac{b'}{b}\mathfrak{A}A(C - \tfrac{sb^2}{b'}\log x)\big)^{-1/s\nu}, && \text{if } bb' \neq 0.
\end{aligned}$$

1.2.2. Convergence of formal transformation

In this subsection, we show the convergence of the formal transformation obtained above, in case $\lambda \in \mathbb{C} - (-\infty, 0]$.

THEOREM 1.2.4. *If $\lambda \in \mathbb{C} - (-\infty, 0]$, the transformation*

$$y = \sum_{j+\ell \geq 1} p_{j\ell} x^j z^\ell \quad (p_{01} = 1) \tag{1.2.17}$$

constructed in Section 1.2.1 converges for $|x| < r, |z| < r'$, where r and r' are sufficiently small positive constants.

As an immediate consequence of Theorem 1.2.4, we have

THEOREM 1.2.5. (1) *If $\lambda \in \mathbb{C} - (-\infty, 0] - \mathbb{N}$, then equation (1.2.1) admits a solution of the form*

$$y = \sum_{j+\ell \geq 1} p_{j\ell} x^j (Cx^\lambda)^\ell \quad (p_{01} = 1) \tag{1.2.18}$$

which converges in the domain

$$|x| < r, \quad |Cx^\lambda| < r'.$$

More precisely, the domain can be described in terms of $t := \log x$ as follows:

$$\Re(t) < -M, \quad \Re(\lambda t) < -M$$

for some large M; (these conditions are compatible since $\lambda \notin (-\infty, 0]$).
(2) *If $\lambda \in \mathbb{N}$, then equation (1.2.1) admits a solution of the form*

$$y = \sum_{j+\ell \geq 1} p_{j\ell} x^j (x^\lambda (C + b \log x))^\ell \quad (p_{01} = 1) \tag{1.2.19}$$

which converges in the domain:

$$|x| < r, \quad |x^\lambda (C + b \log x)| < r'.$$

Proof of Theorem 1.2.4. First we consider the case $\lambda \in \mathbb{C} - (-\infty, 0] - \mathbb{N}$. Substituting transformation (1.2.17) into the left and the right-hand sides of equation (1.2.1), we have

$$xy' = \sum (j + \lambda\ell) p_{j\ell} x^j z^\ell$$

and

$$f(x, \sum p_{j\ell} x^j z^\ell) = \lambda \sum p_{j\ell} x^j z^\ell + \sum_{(i,k)}{}' a_{ik} x^i \Big(\sum p_{j\ell} x^j z^\ell\Big)^k$$

where $\sum'_{(i,k)}$ denotes the summation for $(i,k) \succ (0,1)$, namely for $i+k \geq 1, (i,k) \neq (0,1)$. Hence we have

$$\sum_{j+\ell\geq 1} (j+(\ell-1)\lambda) p_{j\ell} x^j z^\ell = \sum_{(i,k)}{}' a_{ik} x^i \Big(\sum_{j+\ell\geq 1} p_{j\ell} x^j z^\ell\Big)^k, \tag{1.2.20}$$

from which we obtain

$$(j+(\ell-1)\lambda) p_{j\ell} = V_{j\ell} \quad \text{for} \quad (j,\ell) \succeq (1,0), \tag{1.2.21}$$

where

$$V_{j\ell} = \sum_{(i,k)}{}' \sum_{(\alpha,\beta;(i,k),(j,\ell))} a_{ik} p_{\alpha_1\beta_1} \cdots p_{\alpha_k\beta_k}. \tag{1.2.22}$$

Here $\sum_{(\alpha,\beta;(i,k),(j,\ell))}$ stands for the summation for $(\alpha_1, \ldots, \alpha_k)$ and $(\beta_1, \ldots, \beta_k)$ satisfying

$$i + \alpha_1 + \ldots + \alpha_k = j$$
$$\beta_1 + \ldots + \beta_k = \ell.$$

Notice that $V_{j\ell}$ is a polynomial with positive coefficients in a finite number of the a_{ik}'s and in $p_{\alpha\beta}$ $((\alpha,\beta) \prec (j,\ell))$. If we put $p_{01} = 1$, then $p_{j\ell}$'s for $(j,\ell) \succ (0,1)$ can be determined uniquely. There exists a sufficiently small positive constant ε such that

$$|j + (\ell-1)\lambda| > \varepsilon$$

for $(j,\ell) \succ (0,1)$ (see Figure 1.2.1), and a positive constant M such that $|a_{ik}| \leq M r_0^{-i} r_1^{-k}$. We put

$$P_{01} = 1$$

and recursively

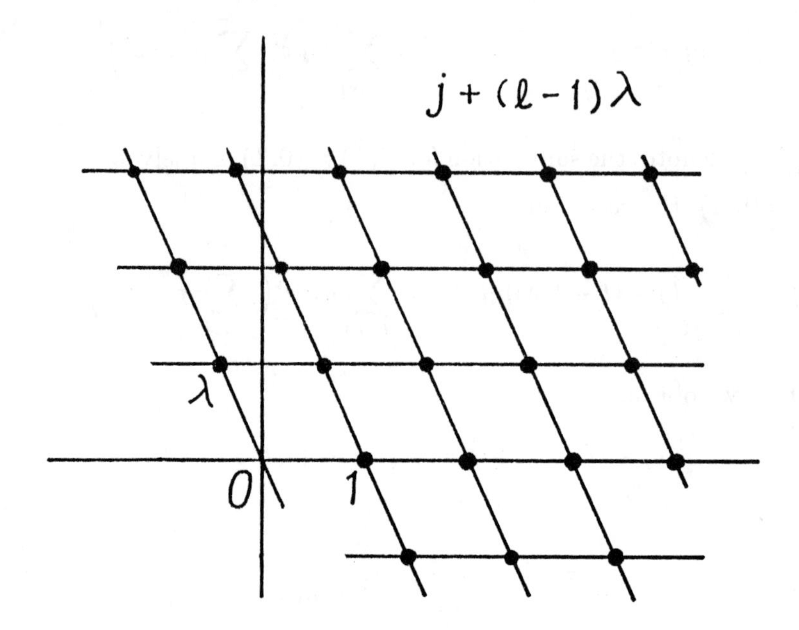

Figure 1.2.1

$$P_{j\ell} = \varepsilon^{-1} \sum_{(i,k)}{}' \sum_{(\alpha,\beta;(i,k),(j,\ell))} \frac{M}{r_0^i r_1^k} P_{\alpha_1\beta_1} \dots P_{\alpha_k\beta_k}$$

for $(j,\ell) \succ (0,1)$. Then we have

$$|p_{j\ell}| \leq P_{j\ell} \quad j+\ell \geq 1.$$

The formal power series

$$Y(x,z) = \sum_{j+\ell\geq 1} P_{j\ell} x^j z^\ell$$

satisfies the following algebraic equation

$$\begin{aligned} Y &= z + \varepsilon^{-1} \sum_{(i,k)}{}' \frac{M}{r_0^i r_1^k} x^i Y^k \\ &= z + \varepsilon^{-1} M \left(\left(1 - \frac{x}{r_0}\right)^{-1} \left(1 - \frac{Y}{r_1}\right)^{-1} - 1 - \frac{Y}{r_1} \right). \end{aligned}$$

The implicit function theorem says that this equation has a unique holomorphic solution $Y = Y(x, z)$ with $Y(0,0) = 0$ in the neighbourhood of $x = z = 0$, which implies the convergence of $\sum_{j+\ell \geq 1} P_{j\ell} x^j z^\ell$. Hence $\sum_{j+\ell \geq 1} p_{j\ell} x^j z^\ell$ also converges.

Next we consider the case $\lambda \in \mathbb{N}$. Substituting (1.2.17) into the left and the right-hand sides of equation (1.2.1), we have

$$xy' = \sum_{j+\ell \geq 1} (j + \lambda\ell) p_{j\ell} x^j z^\ell + b x^\lambda z^{-1} \sum_{j+\ell \geq 1} \ell p_{j\ell} x^j z^\ell,$$

$$f(x, \sum p_{j\ell} x^j z^\ell) = \lambda \sum_{j+\ell \geq 1} p_{j\ell} x^j z^\ell + \sum_{(i,k)}{}' a_{ik} x^i \Big(\sum_{j+\ell \geq 1} p_{j\ell} x^j z^\ell \Big)^k,$$

and hence

$$\begin{aligned} &\sum_{j+\ell \geq 1} (j + (\ell - 1)\lambda) p_{j\ell} x^j z^\ell \\ &= -b x^\lambda z^{-1} \sum_{j+\ell \geq 1} \ell p_{j\ell} x^j z^\ell + \sum_{(i,k)}{}' a_{ik} x^i \Big(\sum_{j+\ell \geq 1} p_{j\ell} x^j z^\ell \Big)^k. \end{aligned} \tag{1.2.23}$$

Thus we have

$$(j + (\ell - 1)\lambda) p_{j\ell} = -b(\ell + 1) p_{j-\lambda, \ell+1} + V_{j\ell} \tag{1.2.24}$$

for $(j, \ell) \succ (0, 1), (j, \ell) \neq (\lambda, 0)$. Here $V_{j\ell}$ is the polynomial given in (1.2.22). The coefficients p_{01} and $p_{\lambda 0}$ can not be determined uniquely. If we put $p_{01} = 1$ and $p_{\lambda 0} = 0$, then the $p_{j\ell}$'s for $(j, \ell) \succ (0, 1), \neq (\lambda, 0)$ can be determined uniquely.

Note that there exists a positive constant B such that

$$\frac{|b|(\ell+1)}{|j+(\ell-1)\lambda|} < B, \quad \frac{1}{|j+(\ell-1)\lambda|} < B$$

for $(j,\ell) \succ (0,1), \neq (\lambda,0)$. If we put

$$P_{01} = 1$$

and

$$P_{j\ell} = B\Big(P_{j-\lambda,\ell+1} + {\sum_{(i,k)}}' \sum_{(\alpha,\beta;(i,k),(j,\ell))} \frac{M}{r_0^i r_1^k} P_{\alpha_1\beta_1}\dots P_{\alpha_k\beta_k}\Big)$$

for $(j,\ell) \succ (0,1)$, then

$$|p_{j\ell}| \le P_{j\ell}.$$

In order to prove the convergence of the series $\sum P_{j\ell}x^j z^\ell$ in x, z, we introduce the quantities $\{Q_{j,j+\ell}\}$ by

$$Q_{01} = 1,$$

$$Q_{j,j+\ell} = B\Big(Q_{j-\lambda,j-\lambda+\ell+1} + {\sum_{(i,k)}}' \sum_{(\alpha,\beta;(i,k),(j,\ell))} \frac{M}{r_0^i r_1^k} Q_{\alpha_1,\alpha_1+\beta_1}\dots Q_{\alpha_k,\alpha_k+\beta_k}\Big), \tag{1.2.25}$$

and consider the power series

$$\sum_{j+\ell\ge 1} Q_{j,j+\ell}\xi^j z^{j+\ell} \quad \Big(= \sum_{j+\ell\ge 1} P_{j\ell}\xi^j z^{j+\ell}\Big) \tag{1.2.26}$$

in ξ and z. Notice that this series is obtained from the series $Y(x,z)$ by the substitution $x = \xi z$.

Consider the algebraic equation

$$\begin{aligned} Y &= z + B\Big\{\xi^\lambda z^{\lambda-1}Y + {\sum_{(i,k)}}' \frac{M}{r_0^i r_1^k}\xi^i z^i Y^k\Big\} \\ &= z + B\xi^\lambda z^{\lambda-1}Y + BM\Big(\Big(1-\frac{\xi z}{r_0}\Big)^{-1}\Big(1-\frac{Y}{r_1}\Big)^{-1} - 1 - \frac{Y}{r_1}\Big). \end{aligned}$$

By the implicit function theorem, this equation has a unique holomorphic solution $Y = Y(\xi, z)$ with $Y(0,0) = 0$ in the neighbourhood of $\xi = z = 0$. Let

$$Y(\xi, z) = \sum_{j+h\geq 1} Y_{jh}\xi^j z^h$$

be the Taylor expansion of $Y(\xi, z)$. Then $\{Y_{jh}\}$ are determined by

$$\begin{aligned} Y_{01} &= 1, \\ Y_{jh} &= B\Big(Y_{j-\lambda,h-\lambda+1} \\ &\qquad + \sum_{(i,k)}{}' \sum_{(\alpha,\gamma;(i,k),(j,h-i))} \frac{M}{r_0^i r_1^k} Y_{\alpha_1,\gamma_1} \ldots Y_{\alpha_k,\gamma_k}\Big) \end{aligned}$$

for $(j,h) \succ (0,1)$. Therefore, we have the following inequalities among the coefficients $\{Y_{jh},\ j \leq h\}$

$$\begin{aligned} Y_{01} &= 1, \\ Y_{jh} &\geq B\Big(Y_{j-\lambda,h-\lambda+1} \\ &\qquad + \sum_{(i,k)}{}' \sum_{(\alpha,\gamma;(i,k),(j,h-i))}{}'' \frac{M}{r_0^i r_1^k} Y_{\alpha_1,\gamma_1} \ldots Y_{\alpha_k,\gamma_k}\Big) \end{aligned} \tag{1.2.27}$$

for $j \leq h$, $(j,h) \succ (0,1)$. Here $\sum''_{(\alpha,\gamma;(i,k),(j,h-i))}$ stands for the summation for $(\alpha_1, \ldots, \alpha_k)$ and $(\gamma_1, \ldots, \gamma_k)$ satisfying

$$\begin{aligned} i + \alpha_1 + \ldots + \alpha_k &= j, \\ \gamma_1 + \ldots + \gamma_k &= h - i, \\ \alpha_1 \leq \gamma_1, \ldots, \alpha_k &\leq \gamma_k. \end{aligned}$$

By (1.2.25) and (1.2.27), we obtain

$$0 \leq P_{j\ell} = Q_{j,j+\ell} \leq Y_{j,j+\ell} \quad \text{for } j + \ell \geq 1.$$

Since $\sum_{j+h\geq 1} Y_{jh}\xi^j z^h$ converges, (1.2.26) also converges. Thus

$$|p_{j\ell}| \leq P_{j\ell} \leq \tilde{M} r^{-j-(j+\ell)} = \tilde{M}(r^2)^{-j} r^{-\ell}$$

for some positive constants r and $\tilde{M}$, which implies that the right-hand side of (1.2.17) converges for $|x| < r^2, |z| < r$. This completes the proof of the theorem. ∎

Note. In case $\lambda \in (-\infty, 0]$, while transformation (1.2.17) may not converge; it gives an asymptotic expansion of an analytic transformation (see [HKM]).

1.3. n-dimensional case

In this section, we treat the n-dimensional equation

$$xY' = F(x, Y), \qquad (Y = {}^t(y_1, \ldots, y_n)), \tag{1.1.1}$$

where

$$F(x, Y) = \sum_{j+|K|\geq 1} A_{jK} x^j Y^K \tag{1.1.2}$$

converges for $|x| < r_0$, $\|Y\| < r_1$. We may assume that the Wronskian matrix

$$\Lambda = \begin{pmatrix} \dfrac{\partial f_1}{\partial y_1}(0, \mathbf{0}) \cdots \dfrac{\partial f_1}{\partial y_n}(0, \mathbf{0}) \\ \vdots \qquad\qquad \vdots \\ \dfrac{\partial f_n}{\partial y_1}(0, \mathbf{0}) \cdots \dfrac{\partial f_n}{\partial y_n}(0, \mathbf{0}) \end{pmatrix}$$

is of the Jordan canonical form:

$$\Lambda = \begin{pmatrix} \lambda_1 & & & \\ \delta_2 & \ddots & & \\ & \ddots & \ddots & \\ & & \delta_n & \lambda_n \end{pmatrix} \qquad (\delta_\nu = 0 \text{ or } 1). \tag{1.3.1}$$

We will obtain a solution of equation (1.1.1) in the following way:

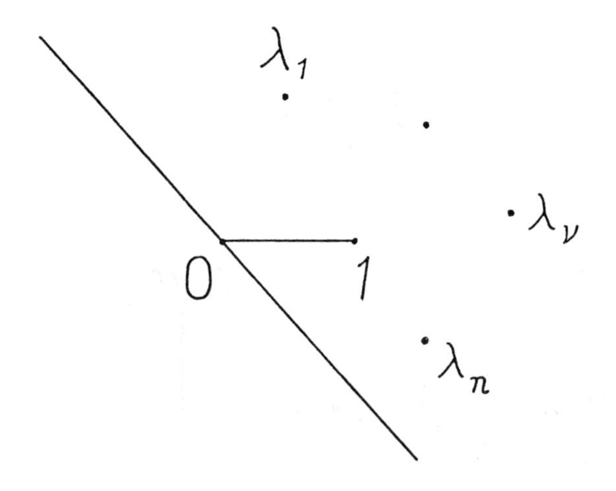

Figure 1.3.1

(a) We construct a transformation $Y = \Phi(x, Z)$ that reduces the equation (1.1.1) into an equation of simpler form.
(b) We substitute a solution Z of the reduced equation into $\Phi(x, Z)$, which leads to solution of (1.1.1)

One expects that, under suitable conditions, equation (1.1.1) can be reduced to

$$xZ' = \Lambda Z.$$

In fact, as is shown below, it is true under the following assumptions on $\lambda_1, \ldots, \lambda_n$.

(A1) For every $(j, L) \succeq (1, \mathbf{0})$; $(L = (\ell(1), \cdots, \ell(n))$ and $\nu = 1, \ldots, n$,

$$j + \lambda_1 \ell(1) + \cdots + \lambda_n \ell(n) - \lambda_\nu \neq 0.$$

(A2: Poincaré's condition) The convex hull of the points $1, \lambda_1, \cdots, \lambda_n$ in the complex plane does not contain the origin, in other words, we can draw a line l through the origin so that $1, \lambda_1, \cdots, \lambda_n$ are situated on one side of l. Equivalently, there exist complex numbers t such that $e^t, e^{\lambda_1 t}, \ldots, e^{\lambda_n t}$ are all small simultaneously.

THEOREM 1.3.1. *Under assumptions* (A1) *and* (A2), *equation* (1.1.1) *admits a solution of the form*

$$Y = \Phi(x, x^\Lambda C) \tag{1.3.2}$$

with the following properties:

1) *The series*

$$\Phi(x, Z) = \sum_{j+|L|\geq 1} P_{jL} x^j Z^L, \tag{1.3.3}$$

$$P_{jL} \in \mathbb{C}^n, Z = \begin{pmatrix} z_1 \\ \vdots \\ z_n \end{pmatrix}, \quad P_{0E[\nu]} = e_\nu = \begin{pmatrix} 0 \\ \vdots \\ 0 \\ 1 \\ 0 \\ \vdots \\ 0 \end{pmatrix} (\nu$$

converges for $|x| < r$, $\|Z\| < r$, *where* r *is a sufficiently small positive constant.*

2) $Z = x^\Lambda C$ *is a solution of the equation*

$$xZ' = \Lambda Z, \tag{1.3.4}$$

where $C = {}^t(c_1, \ldots, c_n) \in \mathbb{C}^n$ *is an arbitrary constant vector.*

REMARK 1.3.1. If we write Λ in the form $\Lambda = \Lambda_1 \oplus \ldots \oplus \Lambda_m$ with

$$\Lambda_i = \begin{pmatrix} \lambda_i & & & \\ 1 & \ddots & & \\ & \ddots & \ddots & \\ & & 1 & \lambda_i \end{pmatrix},$$

then z^Λ is decomposed as

$$x^\Lambda = x^{\Lambda_1} \oplus \cdots \oplus x^{\Lambda_m},$$

where each x^{Λ_i} is given by

$$x^{\Lambda_i} = x^{\lambda_i} \begin{pmatrix} 1 & & & \\ \log x & \ddots & & \\ \vdots & \ddots & \ddots & \\ (\log x)^{m_i-1}/(m_i-1)! & \cdots & \log x & 1 \end{pmatrix}.$$

Here m_i is the size of the matrix Λ_i.

To prove Theorem 1.3.1, we first show the existence of the formal series $\Phi(x, Z)$ such that $\Phi(x, x^{\Lambda}C)$ satisfies equation (1.1.1).

Existence of formal series $\Phi(x, Z)$. Assume that $Y = \Phi(x, x^{\Lambda}C)$ satisfies equation (1.1.1). We put $Z(x) = x^{\Lambda}C$. Substituting $\Phi(x, Z(x))$ into equation (1.1.1), and using $xZ'(x) = \Lambda Z(x)$, we have

$$\begin{aligned} & x\frac{d}{dx}\Phi(x, Z(x)) \\ &= \sum jP_{jL}x^j Z(x)^L \\ &\quad + \sum P_{jL}x^j \Big(xz_1'(x)\ell(1)Z(x)^{L-E[1]} + \cdots + xz_n'(x)\ell(n)Z(x)^{L-E[n]}\Big) \\ &= \sum jP_{jL}x^j Z(x)^L + \sum P_{jL}x^j(\lambda_1\ell(1) + \cdots + \lambda_n\ell(n))Z(x)^L \\ &\quad + \sum P_{jL}x^j \Big(\delta_2\ell(2)Z(x)^{E[1]-E[2]} + \cdots \\ &\qquad\qquad + \delta_n\ell(n)Z(x)^{E[n-1]-E[n]}\Big) Z(x)^L \\ &= \sum \Lambda P_{jL}x^j Z(x)^L + \sum_{(i,K)}{}' A_{iK}x^i \Big(\sum P_{jL}x^j Z(x)^L\Big)^K, \end{aligned}$$

where $\sum'_{(i,K)}$ denotes the summation for $(i, K) \succeq (1, \mathbf{0})$. Since $Z(x) = x^{\Lambda}C$ contains arbitrary constants $c_1, \ldots, c_n$, we have the following identity

$$
\begin{aligned}
&\sum_{j+|L|\geq 1} ((j+\lambda_1\ell(1)+\cdots+\lambda_n\ell(n))I_n-\Lambda)P_{jL}x^jZ^L \\
&= -\sum_{j+|L|\geq 1}(\delta_2\ell(2)Z^{E[1]-E[2]}+\cdots \\
&\quad +\delta_n\ell(n)Z^{E[n-1]-E[n]})P_{jL}x^jZ^L \\
&\quad +\sum_{(i,K)}{}' A_{iK}x^i\Big(\sum_{j+|L|\geq 1}P_{jL}x^jZ^L\Big)^K,
\end{aligned}
\tag{1.3.5}
$$

where x and $Z = {}^t(z_1,\ldots,z_n)$ are regarded as indeterminates. Conversely, if P_{jL}'s are taken so that identity (1.3.5) holds, then $\Phi(x,x^\Lambda C)$ satisfies equation (1.1.1) formally.

Let us see that the P_{jL}'s can be determined so that (1.3.5) holds. First we see the term x^jZ^L in (1.3.5) for $(j,L)\prec(1,\mathbf{0})$. For $(j,L)=(0,E[n])$, we have

$$
\left[\begin{pmatrix} \lambda_n-\lambda_1 & & & \\ & \ddots & & \\ & & \lambda_n-\lambda_{n-1} & \\ & & & 0\end{pmatrix}-\Delta\right]P_{0E[n]}=0,
$$

where $\Delta=\Lambda-\mathrm{diag}[\lambda_1,\ldots,\lambda_n]$. For $(0,E[n-1]),\ldots,(0,E[1])$, we have

$$
\left[\begin{pmatrix} \lambda_{n-1}-\lambda_1 & & & \\ & \ddots & & \\ & & 0 & \\ & & & \lambda_{n-1}-\lambda_n\end{pmatrix}-\Delta\right]P_{0E[n-1]}=-\delta_nP_{0E[n]},
$$

$$
\vdots
$$

$$
\left[\begin{pmatrix} 0 & & & \\ & \lambda_1-\lambda_2 & & \\ & & \ddots & \\ & & & \lambda_1-\lambda_n\end{pmatrix}-\Delta\right]P_{0E[1]}=-\delta_2P_{0E[2]}.
$$

These linear equations are solved by

$$P_{0E[n]} = e_n = \begin{pmatrix} 0 \\ \vdots \\ 0 \\ 1 \end{pmatrix}, \cdots, P_{0E[\nu]} = e_\nu = \begin{pmatrix} 0 \\ \vdots \\ 1 \\ \vdots \\ 0 \end{pmatrix} (\nu, \tag{1.3.6}$$

$$\cdots, P_{0E[1]} = e_1 = \begin{pmatrix} 1 \\ 0 \\ \vdots \\ 0 \end{pmatrix}.$$

Next for $(j, L) \succeq (1, \mathbf{0})$, we have

$$\begin{aligned} &[(j + \lambda_1 \ell(1) + \cdots + \lambda_n \ell(n)) I_n - \Lambda] P_{jL} \\ = &- (\delta_2 \ell(2) P_{j,L-E[1]+E[2]} + \cdots + \delta_n \ell(n) P_{j,L-E[n-1]+E[n]}) \\ &+ V_{jL}, \end{aligned} \tag{1.3.7}$$

where

$$\begin{aligned} V_{jL} = \sum_{(i,K)}{}' \sum_{(\alpha,\Gamma;(i,K),(j,L))} & A_{iK} p^{(1)}_{\alpha_1 \Gamma_1} \cdots p^{(1)}_{\alpha_{k(1)} \Gamma_{k(1)}} \\ & \times \cdots p^{(n)}_{\alpha_{k(1)+\cdots+k(n-1)+1} \Gamma_{k(1)+\cdots+k(n-1)+1}} \cdots p^{(n)}_{\alpha_{|K|} \Gamma_{|K|}} \end{aligned} \tag{1.3.8}$$

and $P_{jL} = {}^t(p^{(1)}_{jL}, \cdots, p^{(n)}_{jL})$. Here $\sum_{(\alpha,\Gamma;(i,K),(j,L))}$ stands for the summation for $((\alpha_1, \cdots, \alpha_{|K|}), (\Gamma_1, \cdots, \Gamma_{|K|}))$ satisfying

$$i + \alpha_1 + \cdots + \alpha_{|K|} = j,$$
$$\Gamma_1 + \cdots + \Gamma_{|K|} = L.$$

Notice that each component of V_{jL} is a polynomial with positive coefficients in a finite number of the $a^{(\nu)}_{iK}$'s (components of A_{iK}) and in $p^{(\nu)}_{\alpha\Gamma}$'s for $(\alpha, \Gamma) \prec (j, L)$ $(1 \le \nu \le n)$. By assumption (A1), P_{jL} can be determined uniquely for every $(j, L) \succeq (1, \mathbf{0})$ so that identity (1.3.5) holds. Thus we have proved the existence of the formal solution $\Phi(x, x^{\Lambda}C)$ of equation (1.1.1).

Convergence of $\Phi(x, Z)$. We show that $\Phi(x, Z)$ converges near $(x, Z) = (0, \mathbf{0})$. By assumptions (A1) and (A2), there exists a positive constant $\varepsilon < 1$ such that

$$(1.3.9) \qquad (1 + |L|)\|((j + \lambda_1 \ell(1) + \cdots + \lambda_n \ell(n))I_n - \Lambda)^{-1}\| < \varepsilon^{-1},$$

for $(j, L) \succeq (1, \mathbf{0})$. This inequality is proved as follows. Set

$$A = (j + \lambda_1 \ell(1) + \cdots + \lambda_n \ell(n))I_n - \Lambda$$

$$= \begin{pmatrix} a_1 & & & \\ -\delta_2 & \ddots & & \\ & \ddots & \ddots & \\ & & -\delta_n & a_n \end{pmatrix}.$$

Let $\mu_0, \mu_1, \cdots, \mu_n$ be the distances between the line l and the points $1, \lambda_1, \cdots, \lambda_n$, respectively(cf. (A2)). Define μ and m by

$$\mu = \min(\mu_0, \mu_1, \cdots, \mu_n),$$

$$m = \max\left(1 + \frac{2}{\mu} \max_i |\lambda_i|, \frac{2n}{\mu}\right).$$

We first consider the case $j + |L| \ge m$. From (A2) and the assumption $j + |L| \ge m$, we have

$$\begin{aligned} |a_i| &= |(j + \lambda_1 \ell(1) + \cdots + \lambda_n \ell(n)) - \lambda_i| \\ &\ge |j + \lambda_1 \ell(1) + \cdots + \lambda_n \ell(n)| - |\lambda_i| \\ &\ge (\mu_0 j + \mu_1 \ell(1) + \cdots + \mu_n \ell(n)) - |\lambda_i| \\ &\ge (j + |L|)\mu - |\lambda_i| \\ &\ge \frac{1}{2}(j + |L|)\mu + \frac{m}{2}\mu - |\lambda_i| \\ &\ge (j + 1 + |L|)\frac{\mu}{2} \\ &\ge n, \end{aligned}$$

which implies

$$|a_i|^{-1} \leq n^{-1}.$$

Let $b_{i\nu}$ be the (i,ν)-th component of A^{-1}, then $b_{i\nu} = 0$ for $i < \nu$, $b_{ii} = a_i{}^{-1}$ and

$$b_{i\nu} = \frac{\delta_{\nu+1} \cdots \delta_i}{a_\nu a_{\nu+1} \cdots a_i} \quad \text{for } i > \nu.$$

Hence we have

$$\begin{aligned}\sum_{\nu=1}^{n} |b_{i\nu}| &\leq \frac{1}{|a_i|} \left\{ \frac{1}{|a_1| \cdots |a_{i-1}|} + \cdots + \frac{1}{|a_{i-1}|} + 1 \right\} \\ &\leq \frac{2}{|a_i|} \leq \frac{4}{(1+|L|)\mu}.\end{aligned}$$

Thus we have

$$(1+|L|)\|A^{-1}\| \leq \frac{4}{\mu}$$

for $j + |L| \geq m$. If we put

$$\varepsilon^{-1} = \max\left(\max_{j+|L|<m,(j,L)\succeq(1,\mathbf{0})} (1+|L|)\|A^{-1}\|, \frac{4}{\mu} \right) > 1,$$

inequality (1.3.9) holds.

By Cauchy's inequality, we have

$$\|A_{iK}\| \leq \frac{M}{r_0{}^i r_1{}^{|K|}} \qquad \text{for} \quad i + |K| \geq 1,$$

where M is a positive constant independent of (i, K). It follows from

(1.3.7), (1.3.8) and (1.3.9) that

$$\begin{aligned}\|P_{jL}\| &\le \|((j+\lambda_1\ell(1)+\cdots+\lambda_n\ell(n))I_n-\Lambda)^{-1}\| \\ &\quad\times\Big\{\ell(2)\|P_{j,L-E[1]+E[2]}\|+\cdots+\ell(n)\|P_{j,L-E[n-1]+E[n]}\| \\ &\quad+\sum_{(i,K)}{}' \sum_{(\alpha,\Gamma;(i,K),(j,L))} \|A_{iK}\|\|P_{\alpha_1\Gamma_1}\|\cdots\|P_{\alpha_{|K|}\Gamma_{|K|}}\|\Big\} \\ &\le \varepsilon^{-1}\Big\{\|P_{j,L-E[1]+E[2]}\|+\cdots+\|P_{j,L-E[n-1]+E[n]}\| \\ &\quad+\sum_{(i,K)}{}' \sum_{(\alpha,\Gamma;(i,K),(j,L))} \frac{M}{r_0{}^i r_1{}^{|K|}}\|P_{\alpha_1\Gamma_1}\|\cdots\|P_{\alpha_{|K|}\Gamma_{|K|}}\|\Big\}\end{aligned}$$

for $(j,L)\succeq(1,\mathbf{0})$. Suggested by these inequalities, we define positive constants γ_{jL} by

$$\gamma_{0E[n]}=1,$$

$$\begin{aligned}\gamma_{jL} &= \varepsilon^{-1}\Big\{\gamma_{j,L-E[1]+E[2]}+\cdots+\gamma_{j,L-E[n-1]+E[n]} \\ &\quad+\sum_{(i,K)}{}' \sum_{(\alpha,\Gamma;(i,K),(j,L))} \frac{M}{r_0{}^i r_1{}^{|K|}}\gamma_{\alpha_1\Gamma_1}\cdots\gamma_{\alpha_{|K|}\Gamma_{|K|}}\Big\}\end{aligned} \tag{1.3.10}$$

for $(j,L)\succ(0,E[n])$. Since $\varepsilon^{-1}>1$, we have

$$\begin{aligned}\gamma_{0E[\nu]} &\ge \varepsilon^{-1}\gamma_{0E[\nu+1]}\ge\varepsilon^{-2}\gamma_{0E[\nu+2]}\ge\cdots \\ &\cdots\ge\varepsilon^{-(n-\nu)}\gamma_{0E[n]}>1=\|P_{0E[\nu]}\|, \qquad \nu=1,\dots,n-1;\end{aligned}$$

and hence

$$\|P_{jL}\|\le\gamma_{jL}\quad \text{for}\ \ (j,L)\succ(0,\mathbf{0}).$$

Therefore it is sufficient to prove the convergence of the series

$$\sum_{j+|L|\ge1}\gamma_{jL}x^jZ^L.$$

If the linear part Λ of the equation (1.1.1) is diagonal, we can prove the convergence of $\sum P_{jL}x^jZ^L$ by the method used in the proof of Proposition 1.1.1. However, since the linear part Λ in general has off-diagonal parts, we make use of the following trick.
We consider the series

$$\sum_{j+|L|\geq 1} \gamma_{jL}x^jT^{W[L]} \quad (T = {}^t(t_1,\cdots,t_n) \in \mathbb{C}^n),$$

where

$$W[L] = \left(\ell(1),\ldots,\sum_{s=1}^{\nu}\ell(s),\ldots,\sum_{s=1}^{n}\ell(s)\right).$$

Notice that $\sum \gamma_{jL}x^jT^{W[L]}$ is the series obtained from $\sum \gamma_{jL}x^jZ^L$ by the substitution

$$z_1 = t_1t_2\cdots t_n, \;\ldots,\; z_\nu = t_\nu t_{\nu+1}\cdots t_n, \;\ldots,\; z_n = t_n.$$

Define $\phi_{jW[L]}$ by

$$\sum_{j+|L|\geq 1} \gamma_{jL}x^jT^{W[L]} = \sum_{j+|L|\geq 1} \phi_{jW[L]}x^jT^{W[L]};$$

then from (1.3.10), we have
(1.3.11)

$$\phi_{0E[n]} = 1$$

$$\phi_{jW[L]} = \varepsilon^{-1}\Big\{\phi_{j,W[L-E[1]+E[2]]} + \cdots + \phi_{j,W[L-E[n-1]+E[n]]}$$

$$+ \sum_{(i,K)}{}' \sum_{(\alpha,\Gamma;(i,K),(j,L))} \frac{M}{{r_0}^i{r_1}^{|K|}}\phi_{\alpha_1W[\Gamma_1]}\cdots\phi_{\alpha_{|K|}W[\Gamma_{|K|}]}\Big\}.$$

Consider the following algebraic equation

$$\psi = t_n + \varepsilon^{-1}\Big\{(t_1 + \cdots + t_{n-1})\psi + \sum_{(i,K)}{}' \frac{M}{r_0^i{r_1}^{|K|}}x^i\psi^{|K|}\Big\}$$

$$= t_n + \varepsilon^{-1}\Big\{(t_1 + \cdots + t_{n-1})\psi$$

$$+ M\left[\left(1-\frac{x}{r_0}\right)^{-1}\left(1-\frac{\psi}{r_1}\right)^{-n} - 1 - \frac{n\psi}{r_1}\right]\Big\}.$$

By the implicit function theorem, this equation has a unique holomorphic solution $\psi = \psi(x,T)$ with $\psi(0,\mathbf{0}) = 0$ in the neighbourhood of $(x,T) = (0,\mathbf{0})$. Let us expand $\psi(x,T)$ into the convergent power series

$$\psi(x,T) = \sum_{j+|W|\geq 1} \psi_{jW} x^j T^W.$$

The coefficients ψ_{jW} are recursively determined by

$$\begin{aligned}
&\psi_{0E[n]} = 1 \\
&\psi_{jW} = \varepsilon^{-1}\Big\{\psi_{j,W-E[1]} + \cdots + \psi_{j,W-E[n-1]} \\
&\qquad + \sum_{(i,K)}' \sum_{(\alpha,H;(i,K),(j,W))} \frac{M}{r_0{}^i r_1{}^{|K|}} \psi_{\alpha_1 H_1} \cdots \psi_{\alpha_{|K|} H_{|K|}}\Big\}
\end{aligned} \tag{1.3.12}$$

for $(j,W) \succ (0,E[n])$. Since

$$\begin{aligned}
&W[L_1] + W[L_2] = W[L_1 + L_2] \\
&W[E[\nu] - E[\nu-1]] = -E[\nu-1], \quad \nu = 2,\ldots,n,
\end{aligned}$$

we have

$$\begin{aligned}
&\psi_{0E[n]} = 1 \\
&\psi_{jW[L]} \geq \varepsilon^{-1}\Big\{\psi_{j,W[L-E[1]+E[2]]} + \cdots \\
&\qquad + \psi_{j,W[L-E[n-1]+E[n]]} \\
&\qquad + \sum_{(i,K)}' \sum_{(\alpha,\Gamma;(i,K),(j,L))} \frac{M}{r_0{}^i r_1{}^{|K|}} \psi_{\alpha_1 W[\Gamma_1]} \cdots \psi_{\alpha_{|K|} W[\Gamma_{|K|}]}\Big\}
\end{aligned} \tag{1.3.13}$$

for special multi-indices $(j,W[L])$ with $(j,L) \succ (0,E[n])$. By (1.3.11) and (1.3.13), we have

$$\begin{aligned}
&\gamma_{0E[n]} = \phi_{0E[n]} = 1, \\
&\gamma_{jL} = \phi_{jW[L]} \leq \psi_{jW[L]} \quad \text{for } (j,L) \succ (0,E[n]).
\end{aligned}$$

Therefore, the convergence of the series $\psi(x,T) = \sum_{j+|W|\geq 1} \psi_{jW} x^j T^W$ implies that of

$$\sum_{j+|L|\geq 1} \gamma_{jL} x^j T^{W[L]} \quad \text{for} \quad |x|, \|T\| < r',$$

where r' is some positive constant. Hence, for an arbitrarily fixed r $(0 < r < r')$, there exists a constant $M' > 0$ such that

$$\begin{aligned} 0 \leq \gamma_{jL} &\leq M' r^{-j-|W[L]|} \\ &= M' r^{-j-n\ell(1)-(n-1)\ell(2)-\cdots-\ell(n)} \\ &= M' r^{-j} (r^n)^{-\ell(1)} (r^{n-1})^{-\ell(2)} \cdots r^{-\ell(n)}, \end{aligned}$$

which proves that $\sum \gamma_{jL} x^j Z^L$ converges for $|x| < r, |z_1| < r^n, |z_2| < r^{n-1}, \cdots, |z_n| < r$. Thus we have completed the proof of Theorem 1.3.1.

2. Fixed Singular Points of Regular Type of Painlevé Equations

Solutions of Painlevé equations (P_J) $(J = I, \cdots, VI)$ have fixed singular points as well as movable singular points which are known to be poles. In this section, we study the behavior of solutions of Painlevé equations near the fixed singular points of regular type. The singular points of irregular type are not studied in this book (cf. [Shim.2], [Tkn.3] and [YosS.1]). We notice that the fixed singular points $t = 0, 1, \infty$ of (P_{VI}) and $t = 0$ of (P_{III}) and (P_V) are of regular type, and the fixed singular point $t = \infty$ of each (P_J) $(J = I, II, III, IV, V)$ is of irregular type.

We can see that Poincaré's condition is not satisfied at the fixed singular points of Painlevé equations; we can not apply Theorem 1.3.1 to Painlevé equations. In order to overcome this difficulty, we proceed as follows: We first transform Painlevé equations into equations of the following form

$$x(xu')' = F_0(x, e^{-u}, xe^u) + F_1(x, e^{-u}, xe^u)(xu') + F_2(x, e^{-u}, xe^u)(xu')^2,$$

where $F_i(x, \xi, \eta)$ $(i = 0, 1, 2)$ are holomorphic at $(x, \xi, \eta) = (0, 0, 0)$ and $F_i(0, 0, 0) = 0$ $(i = 0, 1, 2)$. We next construct solutions of the equation at $x = 0$. The above equation is called the *normal form* of Painlevé equations at fixed singular points of regular type.

The fixed singular points of Painlevé equations can also be studied by the use of the polynomial Hamiltonians obtained in Chapter 3 ([Tkn.2], [KimH.2]).

2.1. Transformation into the normal form

Consider the fifth Painlevé equation (P_V) :

$$\lambda'' = \left(\frac{1}{2\lambda} + \frac{1}{\lambda - 1}\right)\lambda'^2 - \frac{\lambda'}{t} + \frac{(\lambda - 1)^2}{t^2}\left(\alpha\lambda + \frac{\beta}{\lambda}\right) + \gamma\frac{\lambda}{t} + \delta\frac{\lambda(\lambda + 1)}{\lambda - 1}.$$

In order to eliminate the term λ'^2, we change the unknowns by putting

$$\lambda = \tanh^2(u/2) = \left(\frac{1 - e^{-u}}{1 + e^{-u}}\right)^2,$$

which satisfies

$$\lambda_{uu} = \left(\frac{1}{2\lambda} + \frac{1}{\lambda - 1}\right)\lambda_u^2,$$

where $\lambda_u = d\lambda/du$ and $\lambda_{uu} = d^2\lambda/du^2$. Substituting

$$\lambda' = u'\lambda_u, \quad \lambda'' = u''\lambda_u + u'^2\lambda_{uu}$$

into equation (P_V) , and using the above relation among λ_{uu}, λ_u and λ, we obtain

$$t(tu')' = \frac{(\lambda-1)^2}{\lambda_u}\left(\alpha\lambda + \frac{\beta}{\lambda}\right) + \gamma t\frac{\lambda}{\lambda_u} + \delta t^2\frac{\lambda(\lambda+1)}{\lambda_u(\lambda-1)}.$$

Substitute $\lambda = \tanh^2(u/2)$, $\lambda_u = \tanh(u/2)\cdot\cosh^{-2}(u/2)$ into the right-hand side of this equation. Then equation (P_V) is changed into

$$\begin{aligned} t(tu')' &= \left(\alpha\tanh\frac{u}{2} + \beta\tanh^{-3}\frac{u}{2}\right)\cosh^{-2}\frac{u}{2} \\ &\quad + \frac{\gamma}{2}t\sinh u - \frac{\delta}{4}t^2\sinh 2u. \end{aligned} \tag{2.1.1}$$

Notice that equation (2.1.1) can be written as

$$\begin{aligned} t(tu')' &= \left(\alpha\frac{1-e^{-u}}{1+e^{-u}} + \beta\left(\frac{1+e^{-u}}{1-e^{-u}}\right)^3\right)\frac{4e^{-u}}{(1+e^{-u})^2} \\ &\quad + \frac{\gamma}{4}(te^u - te^{-u}) - \frac{\delta}{8}((te^u)^2 - t^2(e^{-u})^2), \end{aligned}$$

which is in normal form with

$$\begin{aligned} F_0(t,\xi,\eta) &= \left(\alpha\frac{1-\xi}{1+\xi} + \beta\left(\frac{1+\xi}{1-\xi}\right)^3\right)\frac{4\xi}{(1+\xi)^2} \\ &\quad + \frac{\gamma}{4}(\eta - t\xi) - \frac{\delta}{8}(\eta^2 - t^2\xi^2), \\ F_1(t,\xi,\eta) &= F_2(t,\xi,\eta) \equiv 0. \end{aligned}$$

Similar treatment is possible around the origin for the two Painlevé equations

$$\lambda'' = \frac{\lambda'^2}{\lambda} - \frac{\lambda'}{t} + \frac{1}{t}(\alpha\lambda^2 + \beta) + \gamma\lambda^3 + \frac{\delta}{\lambda}, \tag{P_{III}}$$

$$(\mathrm{P_{VI}})\quad \lambda'' = \frac{1}{2}\left(\frac{1}{\lambda}+\frac{1}{\lambda-1}+\frac{1}{\lambda-t}\right)\lambda'^2 - \left(\frac{1}{t}+\frac{1}{t-1}+\frac{1}{\lambda-t}\right)\lambda'$$
$$+\frac{\lambda(\lambda-1)(\lambda-t)}{t^2(t-1)^2}\left(\alpha+\beta\frac{t}{\lambda^2}+\gamma\frac{t-1}{(\lambda-1)^2}+\delta\frac{t(t-1)}{(\lambda-t)^2}\right).$$

For equation $(\mathrm{P_{III}})$ we put

$$\lambda = e^{-v},$$

which satisfies

$$\lambda_{vv} = \lambda_v^2/\lambda.$$

Substituting

$$\lambda' = v'\lambda_v = -v'e^{-v},\ \lambda'' = v''\lambda_v + v'^2\lambda_{vv} = -v''e^{-v} + v'^2e^{-v}$$

into equation $(\mathrm{P_{III}})$, we have

$$t(tv')' = -\alpha te^{-v} - \beta te^{v} - \gamma(te^{-v})^2 - \delta(te^{v})^2,$$

which is in normal form with

$$F_0(t,\xi,\eta) = -\alpha t\xi - \beta\eta - \gamma t^2\xi^2 - \delta\eta^2.$$

In order to get an equation in normal form satisfying

$$\frac{\partial F_0}{\partial \xi}(0,0,0) \neq 0,\ \frac{\partial F_0}{\partial \eta}(0,0,0) \neq 0,$$

we make a further change of variables

$$v = u + \log t,\ t^2 = x.$$

Then equation $(\mathrm{P_{III}})$ becomes

$$x(xu_x)_x = -\frac{\alpha}{4}e^{-u} - \frac{\beta}{4}xe^{u} - \frac{\gamma}{4}(e^{-u})^2 - \frac{\delta}{4}(xe^{u})^2,$$

which is in normal form with

$$F_0(x,\xi,\eta) = -\frac{\alpha}{4}\xi - \frac{\beta}{4}\eta - \frac{\gamma}{4}\xi^2 - \frac{\delta}{4}\eta^2,$$
$$F_1(x,\xi,\eta) = F_2(x,\xi,\eta) \equiv 0.$$

This normal form will be useful in the construction of solutions of equation (P_{III}) ; whereas only bounded solutions of (P_{III}) are obtained from the former normal form, we can get, from the latter, unbounded solutions of (P_{III}) as well (cf. Theorem 2.4.1 and its proof).

Next we consider equation (P_{VI}) . Note that the λ'^2-term is approximated by

$$\frac{1}{2}\left(\frac{2}{\lambda}+\frac{1}{\lambda-1}\right)\lambda'^2$$

near the singular point $t=0$. We put

$$\lambda=\cosh^{-2}(u/2)=\frac{4e^{-u}}{(1+e^{-u})^2},$$

which is a solution of

$$\lambda_{uu}=\frac{1}{2}\left(\frac{2}{\lambda}+\frac{1}{\lambda-1}\right)\lambda_u^2.$$

Then equation (P_{VI}) becomes

$$\begin{aligned} &t(tu')' \\ &=\frac{(1-tC(u)^2)S(u)C(u)}{(t-1)^2}\left[\frac{\alpha}{C(u)^4}+\beta t+\frac{\gamma(t-1)}{S(u)^4}+\frac{\delta t(t-1)}{(1-tC(u)^2)^2}\right] \\ &-\left(\frac{t}{t-1}+\frac{tC(u)^2}{1-tC(u)^2}\right)tu'-\frac{tS(u)C(u)}{2(1-tC(u)^2)}(tu')^2, \end{aligned}$$

where

$$C(u)=\cosh(u/2),\quad S(u)=\sinh(u/2):=\frac{e^{u/2}-e^{-u/2}}{2}.$$

Near $t = 0$, this is a normal form with

$$F_0(t,\xi,\eta) = \frac{4-2t-t\xi-\eta}{(t-1)^2}\Big[\frac{\alpha\xi(1-\xi)}{(1+\xi)^3} + \frac{\beta}{16}(\eta - t\xi)$$
$$+ \frac{\gamma(t-1)\xi(1+\xi)}{(1-\xi)^3} + \frac{\delta(t-1)(\eta-t\xi)}{(4-2t-t\xi-\eta)^2}\Big],$$

$$F_1(t,\xi,\eta) = -\frac{t}{t-1} - \frac{2t+t\xi+\eta}{4-2t-t\xi-\eta},$$

$$F_2(t,\xi,\eta) = \frac{t\xi-\eta}{2(4-2t-t\xi-\eta)}.$$

Since equation (P_{VI}) contains the term $\lambda'^2/(\lambda - t)$, we can not get an equation in normal form satisfying $F_1(x,\xi,\eta) = F_2(x,\xi,\eta) \equiv 0$.

Summing up we have

PROPOSITION 2.1.1. *Transform the variables as follows*

$$\begin{aligned} &\lambda = t^{-1}e^{-u}, && t^2 = x \quad \textit{for } (P_{III}) \\ &\lambda = \tanh^2(u/2), && t = x \quad \textit{for } (P_V) \\ &\lambda = \cosh^{-2}(u/2), && t = x \quad \textit{for } (P_{VI}). \end{aligned}$$

Then, near $t = 0$, each equation is changed into an equation in normal form:

$$\begin{aligned} x(xu')' &= F_0(x, e^{-u}, xe^u) \\ &\quad + F_1(x, e^{-u}, xe^u)(xu') + F_2(x, e^{-u}, xe^u)(xu')^2, \end{aligned} \tag{2.1.2}$$

with the following properties:

(1) *The $F_i(x,\xi,\eta)$'s are holomorphic for*

$$|x| < r_0, \ |\xi| < r_1, \ |\eta| < r_1,$$

where r_0 and r_1 are sufficiently small positive constants, and

(2) $F_i(0,0,0) = 0 \quad (i = 0, 1, 2)$.

REMARK 2.1.2. For equations (P_{III}) and (P_V), we have $F_1 = F_2 = 0$.

REMARK 2.1.3. The transformation $y = \tanh^2(u/2)$ is used for equation (P_V) in [LM].

REMARK 2.1.4. For equation (P_{VI}), we do not consider the singular points $t = 1$ and $t = \infty$ because they are transformed into $t = 0$ by simple transformations (see Proposition 1.3.2 in Chapter 3).

2.2. Solutions of equations in normal form

The main part of equation (2.1.2) is $x(xu')' = 0$, which admits a solution

$$u_0(x) = -\omega \log x + \kappa,$$

where ω and κ are integration constants. We are going to prove that equation (2.1.2) admits a solution close to $u_0(x)$.

THEOREM 2.2.1. *Let*

$$\omega \in \mathbb{C} - (-\infty, 0] - [1, +\infty) \tag{2.2.1}$$

be a constant and let κ be an arbitrary complex constant. Then equation (2.1.2) admits a holomorphic solution such that

$$\begin{aligned} u(\omega, \kappa; x) &= -\omega \log x + \kappa + \mathcal{O}(|x| + |e^{-\kappa} x^{\omega}| + |e^{\kappa} x^{1-\omega}|), \\ xu'(\omega, \kappa; x) &= -\omega + \mathcal{O}(|x| + |e^{-\kappa} x^{\omega}| + |e^{\kappa} x^{1-\omega}|) \end{aligned} \tag{2.2.2}$$

in the domain

$$\mathcal{D}(r) = \{x \in \mathcal{R}_0; |x| < r, |e^{-\kappa} x^{\omega}| < r, |e^{\kappa} x^{1-\omega}| < r\}, \tag{2.2.3}$$

where r is a sufficiently small positive constant depending on ω, and $\mathcal{R}_0$ is the universal covering space of $\mathbb{C} - \{0\}$. $\mathcal{O}$ denotes Landau's symbol.

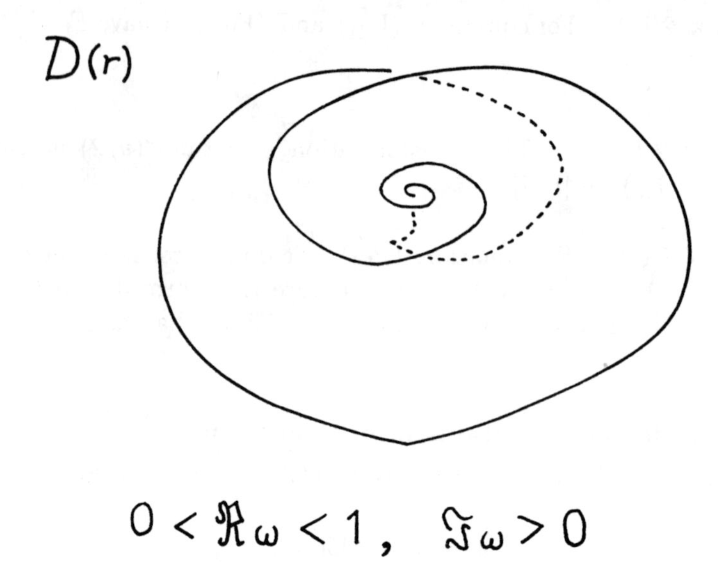

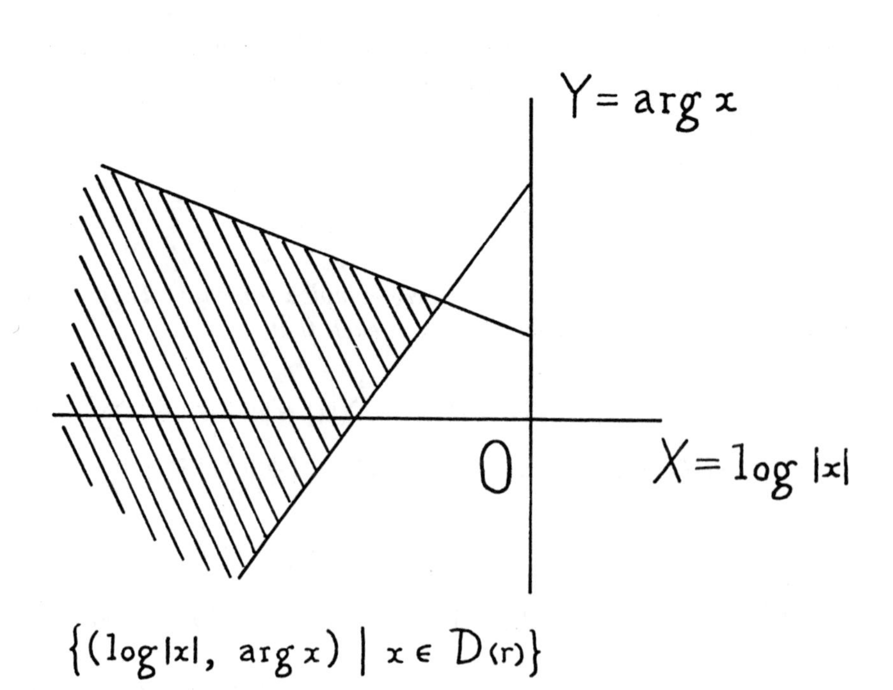

Figure 2.2.1

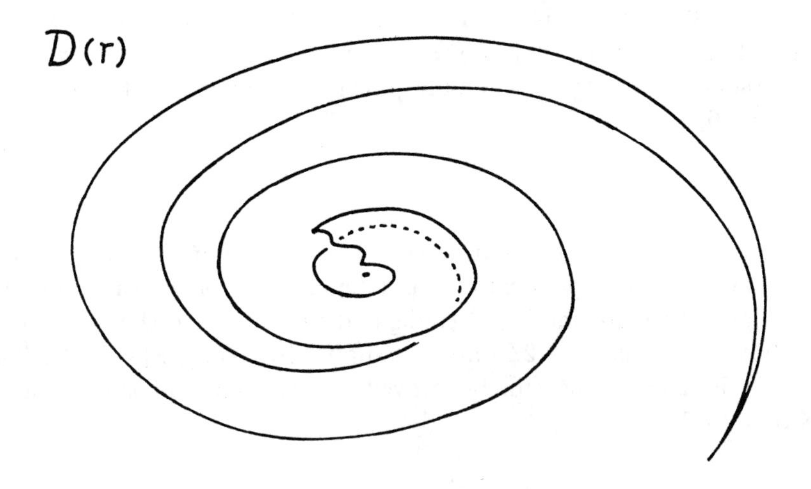

$\Re\omega < 0, \ \Im\omega > 0$

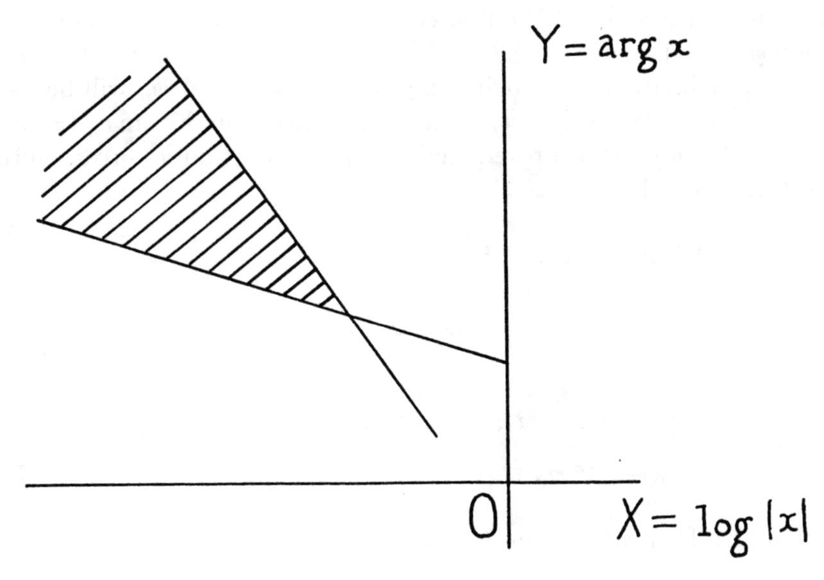

$\{(\log|x|, \arg x) \mid x \in D(r)\}$

Figure 2.2.2

REMARK 2.2.2. It is known that the $\mathcal{O}$-terms in (2.2.2) can be expressed as a convergent power series in x, $e^{-\kappa}x^{\omega}$ and $e^{\kappa}x^{1-\omega}$, and that the solution $u(\omega, \kappa; x)$ is holomorphic in ω and κ in a certain domain (see [Shim.6]).

REMARK 2.2.3. The dominating constants corresponding to the Landau symbol $\mathcal{O}$ are independent of κ and depend only on ω. Namely there exists a positive constant $M(\omega)$ (independent of κ) such that the moduli of both $\mathcal{O}$-terms in (2.2.2) are dominated by $M(\omega)(|x| + |e^{-\kappa}x^{\omega}| + |e^{\kappa}x^{1-\kappa}|)$ in $\mathcal{D}(r)$. This will be proved in the course of the proof of Theorem 2.2.1.

Our proof will proceed as follows. New unknown v will be introduced. The equation for v will be given by (2.3.1). Estimates of the coefficients Φ and Ψ of equation (2.3.1) will be given in Lemmas 2.3.2 and 2.3.3. A system (2.3.5) of integral equations with the unknowns (v, w) will be introduced. The first component $v(x)$ of the holomorphic solution $(v(x), w(x))$ solves (2.3.1). System (2.3.5) will be solved by successive approximation. Approximating sequences v_n and w_n will be introduced, the identity $xv_n' = w_n$ and the existence of limits $\lim v_n$ $(=: v)$ and $\lim w_n$ $(=: w)$ will be proved, and estimates for v and w corresponding to (2.2.2) will be obtained.

2.3. Proof of Theorem 2.2.1

We put

$$u = -\omega \log x + \kappa + v.$$

Since

$$xu' = -\omega + xv', \qquad x(xu')' = x(xv')',$$

$$e^{-u} = e^{-\kappa}x^{\omega}e^{-v}, \qquad xe^{u} = e^{\kappa}x^{1-\omega}e^{v},$$

equation (2.1.2) is changed into

$$\begin{aligned} x(xv')' = {} & F_0(x, e^{-\kappa}x^{\omega}e^{-v}, e^{\kappa}x^{1-\omega}e^{v}) \\ & + F_1(x, e^{-\kappa}x^{\omega}e^{-v}, e^{\kappa}x^{1-\omega}e^{v})(-\omega + xv') \\ & + F_2(x, e^{-\kappa}x^{\omega}e^{-v}, e^{\kappa}x^{1-\omega}e^{v})(-\omega + xv')^2, \end{aligned}$$

which can be written in the form

$$\begin{aligned} x(xv')' &= G_0(x, e^{-\kappa}x^{\omega}e^{-v}, e^{\kappa}x^{1-\omega}e^{v}) \\ &\quad + G_1(x, e^{-\kappa}x^{\omega}e^{-v}, e^{\kappa}x^{1-\omega}e^{v})(xv') \\ &\quad + G_2(x, e^{-\kappa}x^{\omega}e^{-v}, e^{\kappa}x^{1-\omega}e^{v})(xv')^2 \end{aligned}$$

with

$$G_0 := F_0 - \omega F_1 + \omega^2 F_2, \quad G_1 := F_1 - 2\omega F_2 \text{ and } G_2 := F_2.$$

Define Φ and Ψ by

$$\begin{aligned} \Phi(x, \xi, \eta) &:= G_0(x, \xi, \eta), \\ \Psi(x, \xi, \eta, v, w) &:= G_0(x, \xi e^{-v}, \eta e^{v}) - G_0(x, \xi, \eta) \\ &\quad + G_1(x, \xi e^{-v}, \eta e^{v})w + G_2(x, \xi e^{-v}, \eta e^{v})w^2, \end{aligned}$$

then we have

PROPOSITION 2.3.1. *By the transformation*

$$u = -\omega \log x + \kappa + v$$

equation (2.1.2) *is changed into the equation:*

$$\text{(2.3.1)} \quad x(xv')' = \Phi(x, e^{-\kappa}x^{\omega}, e^{\kappa}x^{1-\omega}) + \Psi(x, e^{-\kappa}x^{\omega}, e^{\kappa}x^{1-\omega}, v, xv')$$

which has the following properties:

(1) $\Phi(x, \xi, \eta)$ *is holomorphic in*

$$|x| < r', \quad |\xi| < r', \quad |\eta| < r'$$

and $\Psi(x, \xi, \eta, v, w)$ *is holomorphic in*

$$|x| < r', \quad |\xi| < r', \quad |\eta| < r', \quad |v| < r', \quad |w| < r',$$

where r' *is a sufficiently small positive constant,*

(2) $\Phi(0, 0, 0) = 0$

and

(3) $\Psi(x,\xi,\eta,0,0) = \Psi(0,0,0,v,w) = 0.$

Proof. By the definition of Φ and Ψ, we have

$$G_0(x,\xi e^{-v},\eta e^{v}) + G_1(x,\xi e^{-v},\eta e^{v})w + G_2(x,\xi e^{-v},\eta e^{v})w^2 \\ = \Phi(x,\xi,\eta) + \Psi(x,\xi,\eta,v,w).$$

If we put

$$\xi = e^{-\kappa}x^{\omega}, \quad \eta = e^{\kappa}x^{1-\omega}, \quad w = xv',$$

it is easy to see that equation (2.1.2) can be written in the form (2.3.1). By Proposition 2.1.1,(1), the functions $G_0(x,\xi,\eta)$ and $G_i(x,\xi e^{-v},\eta e^{v})$ $(i = 0,1,2)$ are holomorphic at $(x,\xi,\eta,v) = (0,0,0,0)$, and hence $\Phi(x,\xi,\eta)$ and $\Psi(x,\xi,\eta,v,w)$ are holomorphic near $(x,\xi,\eta,v,w) = (0,0,0,0,0)$. Furthermore, by Proposition 2.1.1, (2) we have

$$\Phi(0,0,0) = G_0(0,0,0) \\ = F_0(0,0,0) - \omega F_1(0,0,0) + \omega^2 F_2(0,0,0) = 0,$$

$$\Psi(x,\xi,\eta,0,0) = G_0(x,\xi,\eta) - G_0(x,\xi,\eta) = 0,$$

and

$$\Psi(0,0,0,v,w) = G_1(0,0,0)w + G_2(0,0,0)w^2 \\ = (F_1(0,0,0) - 2\omega F_2(0,0,0))w + F_2(0,0,0)w^2 = 0,$$

which prove the proposition. ∎

The function Ψ has the following properties.

LEMMA 2.3.2. *In the domain*

$$|x| < r', \quad |\xi| < r', \quad |\eta| < r', \quad |v_i| < r', \quad |w_i| < r' \quad (i = 1,2),$$

we have

$$\begin{aligned} &\Psi(x,\xi,\eta,v_2,w_2) - \Psi(x,\xi,\eta,v_1,w_1) \\ &= (v_2 - v_1)\Psi_1(x,\xi,\eta,v_1,v_2,w_1,w_2) \\ &\quad + (w_2 - w_1)\Psi_2(x,\xi,\eta,v_1,v_2,w_1,w_2), \end{aligned} \tag{2.3.2}$$

where Ψ_1 *and* Ψ_2 *are holomorphic in the domain, and satisfy*

$$\Psi_i(0,0,0,v_1,v_2,w_1,w_2)=0 \quad (i=1,2). \tag{2.3.3}$$

Proof.

$$\begin{aligned}
&\Psi(x,\xi,\eta,v_2,w_2)-\Psi(x,\xi,\eta,v_1,w_1)\\
&=\int_0^1 \frac{\partial}{\partial\theta}\Psi(x,\xi,\eta,V(\theta,v_1,v_2),V(\theta,w_1,w_2))d\theta\\
&=(v_2-v_1)\int_0^1 \Psi_v(x,\xi,\eta,V(\theta,v_1,v_2),V(\theta,w_1,w_2))d\theta\\
&\quad+(w_2-w_1)\int_0^1 \Psi_w(x,\xi,\eta,V(\theta,v_1,v_2),V(\theta,w_1,w_2))d\theta,
\end{aligned}$$

where $V(\theta,\alpha,\beta)=\beta\theta+\alpha(1-\theta)$. We put

$$\begin{aligned}
&\Psi_1(x,\xi,\eta,v_1,v_2,w_1,w_2)\\
&=\int_0^1 \Psi_v(x,\xi,\eta,V(\theta,v_1,v_2),V(\theta,w_1,w_2))d\theta
\end{aligned}$$

and

$$\begin{aligned}
&\Psi_2(x,\xi,\eta,v_1,v_2,w_1,w_2)\\
&=\int_0^1 \Psi_w(x,\xi,\eta,V(\theta,v_1,v_2),V(\theta,w_1,w_2))d\theta.
\end{aligned}$$

By Proposition 2.3.1, these functions are holomorphic in the domain

$$|x|<r', \quad |\xi|<r', \quad |\eta|<r', \quad |v_i|<r', \quad |w_i|<r' \quad (i=1,2).$$

Note that

$$\Psi_v(x,\xi,\eta,v,w)=\frac{1}{2\pi\sqrt{-1}}\int_{|t-v|=\varepsilon}\frac{\Psi(x,\xi,\eta,t,w)}{(t-v)^2}dt$$

for

$$|x| < r', \ |\xi| < r', \ |\eta| < r', \ |v| < r', \ |w| < r',$$

where ε is a sufficiently small positive constant depending on r'. From this formula and Proposition 2.3.1, (3) we have

$$\Psi_v(0,0,0,v,w) = 0$$

for $|v| < r', \ |w| < r'$. Therefore we have

$$\Psi_1(0,0,0,v_1,v_2,w_1,w_2) = 0$$

and similarly

$$\Psi_2(0,0,0,v_1,v_2,w_1,w_2) = 0$$

in the domain $|v_i| < r', \ |w_i| < r' \ \ (i = 1,2)$. These imply the lemma. ∎

LEMMA 2.3.3. *If*

$$|x| < \frac{r'}{2}, \ \ |e^{-\kappa}x^{\omega}|, |e^{\kappa}x^{1-\omega}| < \frac{r'}{2}, \ \ |v_i|, |w_i| < \frac{r'}{2} \quad (i = 1,2),$$

then there exists a positive constant c such that

$$|\Phi(x, e^{-\kappa}x^{\omega}, e^{\kappa}x^{1-\omega})| \leq c(|x| + |e^{-\kappa}x^{\omega}| + |e^{\kappa}x^{1-\omega}|),$$

$$|\Psi(x, e^{-\kappa}x^{\omega}, e^{\kappa}x^{1-\omega}, v_2, w_2) - \Psi(x, e^{-\kappa}x^{\omega}, e^{\kappa}x^{1-\omega}, v_1, w_1)|$$

$$\leq c(|x| + |e^{-\kappa}x^{\omega}| + |e^{\kappa}x^{1-\omega}|)(|v_2 - v_1| + |w_2 - w_1|).$$

Proof. Since $\Psi_i(0,0,0,v_1,v_2,w_1,w_2) = 0 \ (i = 1,2)$, we have

$$\begin{aligned}
&\Psi_i(x,\xi,\eta,v_1,v_2,w_1,w_2) \\
&= \int_0^1 \frac{\partial}{\partial\theta}\Psi_i(\theta x, \theta\xi, \theta\eta, v_1, v_2, w_1, w_2)d\theta \\
&= x\int_0^1 \Psi_{ix}(\theta x, \theta\xi, \theta\eta, v_1, v_2, w_1, w_2)d\theta \\
&\quad + \xi\int_0^1 \Psi_{i\xi}(\theta x, \theta\xi, \theta\eta, v_1, v_2, w_1, w_2)d\theta \\
&\quad + \eta\int_0^1 \Psi_{i\eta}(\theta x, \theta\xi, \theta\eta, v_1, v_2, w_1, w_2)d\theta
\end{aligned}$$

in the domain

$$|x| < r', \ |\xi| < r', \ |\eta| < r', \ |v_j| < r', \ |w_j| < r' \quad (j = 1, 2).$$

By Proposition 2.3.2,

$$\int_0^1 \Psi_{ix}(\cdots)d\theta, \quad \int_0^1 \Psi_{i\xi}(\cdots)d\theta, \quad \int_0^1 \Psi_{i\eta}(\cdots)d\theta$$

are holomorphic and bounded in the domain

$$|x| < \frac{r'}{2}, \ |\xi| < \frac{r'}{2}, \ |\eta| < \frac{r'}{2}, \ |v_i| < \frac{r'}{2}, \ |w_i| < \frac{r'}{2}.$$

Hence we have

$$\begin{aligned} (2.3.4) \qquad & |\Psi_i(x, e^{-\kappa}x^{\omega}, e^{\kappa}x^{1-\omega}, v_1, v_2, w_1, w_2)| \\ & \leq c(|x| + |e^{-\kappa}x^{\omega}| + |e^{\kappa}x^{1-\omega}|) \ \ (i = 1, 2), \end{aligned}$$

for a positive constant c if $(x, e^{-\kappa}x^{\omega}, e^{\kappa}x^{1-\omega}, v_1, v_2, w_1, w_2)$ is in the domain. By this inequality and (2.3.2), we have the second inequality. Using $\Phi(0,0,0) = 0$, we can prove the first inequality in a similar way as in the proof of (2.3.4). ∎

We consider the system of integral equations

$$(2.3.5) \qquad \begin{aligned} w(x) = & \int_{\Gamma(x)} t^{-1}\{\Phi(t, e^{-\kappa}t^{\omega}, e^{\kappa}t^{1-\omega}) \\ & + \Psi(t, e^{-\kappa}t^{\omega}, e^{\kappa}t^{1-\omega}, v(t), w(t))\}dt, \\ v(x) = & \int_{\Gamma(x)} s^{-1} \int_{\Gamma(s)} t^{-1}\{\Phi(t, e^{-\kappa}t^{\omega}, e^{\kappa}t^{1-\omega}) \\ & + \Psi(t, e^{-\kappa}t^{\omega}, e^{\kappa}t^{1-\omega}, v(t), w(t))\}dtds \end{aligned}$$

with the unknowns v and w, where the path of integration $\Gamma(x)$ is given as follows:

1) If $\Im\omega \neq 0$, then

$$\Gamma(x) = \left\{ t \in \mathcal{R}_0; |t^\omega| = |x^\omega| \left| \frac{t}{x} \right|^{1/2}, \ 0 < |t| < |x| \right\}.$$

2) If $0 < \omega < 1$, then

$$\Gamma(x) = \{t \in \mathcal{R}_0; t = \tau \exp(\sqrt{-1} \arg x), \ 0 < \tau < |x|\}.$$

Note that, if the right-hand sides of (2.3.5) are holomorphic and if they can be differentiated under the sign of integration, then the first component $v(x)$ of a system $(v(x), w(x))$ of holomorphic solutions of the integral equation is a solution of the differential equation (2.3.1).

Put $t = \tau e^{\sqrt{-1}\theta} \in \Gamma(x)$. In case $\Im\omega \neq 0$, since

$$\begin{aligned} \left| \frac{t^\omega}{x^\omega} \right| &= \exp\left(\Re\left(\omega \log \frac{t}{x} \right) \right) \\ &= \exp\left(\Re\omega \log \frac{\tau}{|x|} - \Im\omega(\theta - \arg x) \right) \end{aligned}$$

and since the path $\Gamma(x)$ is defined by

$$\left| \frac{t^\omega}{x^\omega} \right| = \left| \frac{t}{x} \right|^{1/2}, \tag{2.3.6}$$

we have

$$\exp\left(\Re\omega \log \frac{\tau}{|x|} - \Im\omega(\theta - \arg x) \right) = \exp\left(\frac{1}{2} \log \frac{\tau}{|x|} \right).$$

on $\Gamma(x)$. Hence the path $\Gamma(x)$ is given by

$$\begin{aligned} &t = \tau \exp\left(\left(\arg x + \frac{\Re\omega - 1/2}{\Im\omega} \log \frac{\tau}{|x|} \right) \sqrt{-1} \right), \quad \text{if} \quad \Im\omega \neq 0, \\ &t = \tau \exp(\sqrt{-1} \arg x), \quad \text{if} \quad 0 < \omega < 1, \end{aligned} \tag{2.3.7}$$

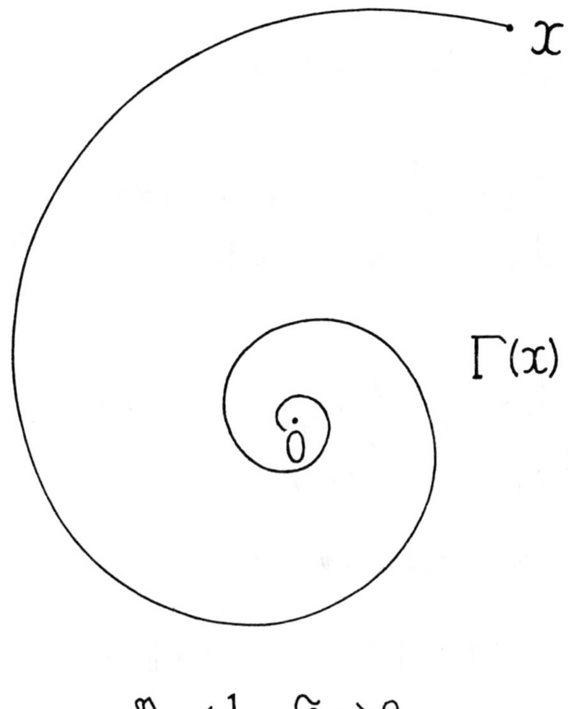
x
Γ(x)
0
ℜω < 1/2 , ℑω > 0

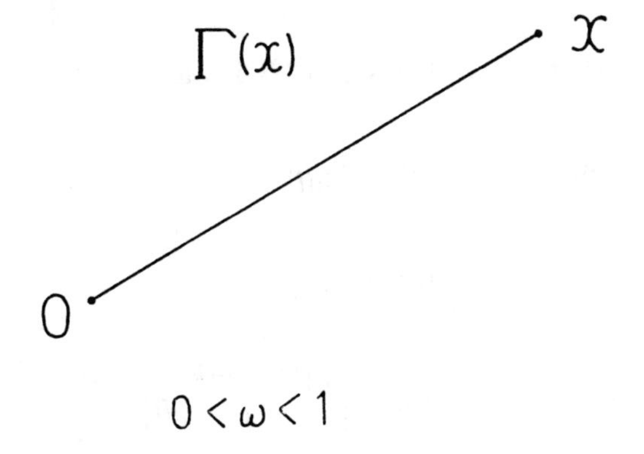
x
Γ(x)
0
0 < ω < 1

where $0 < \tau < |x|$. So we have

$$|dt| \leq L(\omega) d\tau, \tag{2.3.8}$$

where

$$L(\omega) = \begin{cases} \sqrt{1 + \left(\frac{\Re\omega - 1/2}{\Im\omega}\right)^2}, & \text{if } \Im\omega \neq 0, \\ 1, & \text{if } 0 < \omega < 1, \end{cases}$$

because

$$|dt| = \begin{cases} \left|1 + \frac{\Re\omega - 1/2}{\Im\omega}\sqrt{-1}\right| d\tau, & \text{if } \Im\omega \neq 0, \\ d\tau, & \text{if } 0 < \omega < 1. \end{cases}$$

LEMMA 2.3.4. *If ω satisfies (2.2.1), we have*

$$\int_{\Gamma(x)} |t|^{-1}(|t| + |e^{-\kappa}t^{\omega}| + |e^{\kappa}t^{1-\omega}|)^n |dt|$$
$$\leq \frac{K(\omega)}{n}(|x| + |e^{-\kappa}x^{\omega}| + |e^{\kappa}x^{1-\omega}|)^n,$$

for every $n \in \mathbb{N}$, where $K(\omega)$ is a positive constant (≥ 1) depending on ω.

Proof. Assume that $\Im\omega \neq 0$. Using (2.3.6) and (2.3.8), we have

$$\int_{\Gamma(x)} |t|^{-1}|t|^i|e^{-\kappa}t^{\omega}|^j|e^{\kappa}t^{1-\omega}|^k|dt|$$
$$= \int_{\Gamma(x)} |t|^{-1+i+k}|e^{-\kappa}|^{j-k}|t^{\omega}|^{j-k}|dt|$$
$$= \int_{\Gamma(x)} |t|^{-1+i+k}|e^{-\kappa}|^{j-k}\left(|x^{\omega}|\left|\frac{t}{x}\right|^{1/2}\right)^{j-k}|dt|$$
$$= \int_{\Gamma(x)} |e^{-\kappa}x^{\omega-1/2}|^{j-k}|t|^{-1+i+(j+k)/2}|dt|$$

$$\leq |e^{-\kappa} x^{\omega - 1/2}|^{j-k} \int_0^{|x|} \tau^{-1+i+(j+k)/2} L(\omega) d\tau \qquad (\tau = |t|)$$

$$= |e^{-\kappa} x^{\omega - 1/2}|^{j-k} \frac{|x|^{i+(j+k)/2}}{i + (j+k)/2} L(\omega)$$

$$\leq \frac{2L(\omega)}{n} |x|^i |e^{-\kappa} x^{\omega}|^j |e^{\kappa} x^{1-\omega}|^k,$$

for arbitrary nonnegative integers i, j, k satisfying $i + j + k = n$. Hence putting $K(\omega) = 2L(\omega)$, we have the assertion.
When $0 < \omega < 1$, we have

$$\int_{\Gamma(x)} |t|^{-1} |t|^i |e^{-\kappa} t^{\omega}|^j |e^{\kappa} t^{1-\omega}|^k |dt|$$

$$\leq |e^{-\kappa}|^{j-k} \int_0^{|x|} \tau^{-1+i+\omega j+(1-\omega)k} L(\omega) d\tau$$

$$= |e^{-\kappa}|^{j-k} \frac{|x|^{i+\omega j+(1-\omega)k}}{i + \omega j + (1-\omega)k} L(\omega)$$

$$\leq \frac{L(\omega)}{n \min\{\omega, 1-\omega\}} |x|^i |e^{-\kappa} x^{\omega}|^j |e^{\kappa} x^{1-\omega}|^k,$$

for arbitrary nonnegative integers $i,\ j,\ k$ satisfying $i+j+k = n$. Hence putting $K(\omega) = L(\omega)/\min\{\omega, 1-\omega\}$, we have the assertion. ∎

We show that equation (2.3.5) has a system of solutions $(v(x), w(x))$ satisfying

$$|v(x)| = \mathcal{O}(|x| + |e^{-\kappa} x^{\omega}| + |e^{\kappa} x^{1-\omega}|)$$

and

$$|w(x)| = \mathcal{O}(|x| + |e^{-\kappa} x^{\omega}| + |e^{\kappa} x^{1-\omega}|)$$

in $\mathcal{D}(r)$. We prove this by the method of successive approximation.

Let $\{v_n(x)\}_{n\geq 0}$ and $\{w_n(x)\}_{n\geq 0}$ be sequences defined by

$$v_0(x) = w_0(x) = 0$$

and

$$\begin{aligned} w_n(x) &= \varphi_1(x) + I_1(v_{n-1}, w_{n-1}; x), \\ v_n(x) &= \varphi(x) + I_0(v_{n-1}, w_{n-1}; x), \end{aligned} \tag{2.3.9}$$

where

$$\varphi_1(x) = \int_{\Gamma(x)} t^{-1}\Phi(t, e^{-\kappa}t^{\omega}, e^{\kappa}t^{1-\omega})dt,$$

$$\varphi(x) = \int_{\Gamma(x)} t^{-1}\varphi_1(t)dt,$$

$$I_1(v, w; x) = \int_{\Gamma(x)} t^{-1}\Psi(t, e^{-\kappa}t^{\omega}, e^{\kappa}t^{1-\omega}, v(t), w(t))dt$$

and

$$I_0(v, w; x) = \int_{\Gamma(x)} t^{-1}I_1(v, w; t)dt.$$

Put

$$V_n(x) = v_n(x) - v_{n-1}(x),$$
$$W_n(x) = w_n(x) - w_{n-1}(x) \qquad (n = 1, 2, \cdots).$$

We need the following lemma.

LEMMA 2.3.5. *For a sufficiently small positive constant* r ($<$ $\min\{r'/2, 1/3\}$), *and for every positive integer* n, *the functions* $v_n(x)$, $w_n(x)$, $V_n(x)$ *and* $W_n(x)$ *are holomorphic in* $\mathcal{D}(r)$ *and satisfy*

$$xv_n'(x) = w_n(x), \quad xV_n'(x) = W_n(x).$$

Moreover we have

(2.3.10) $|v_n(x)|, \quad |w_n(x)| < r'/3,$
(2.3.11) $|V_n(x)| \leq (2c)^n K(\omega)^{2n}(|x| + |e^{-\kappa}x^{\omega}| + |e^{\kappa}x^{1-\omega}|)^n (n!)^{-1},$
(2.3.12) $|W_n(x)| \leq (2c)^n K(\omega)^{2n}(|x| + |e^{-\kappa}x^{\omega}| + |e^{\kappa}x^{1-\omega}|)^n (n!)^{-1}$

in $\mathcal{D}(r)$, *where* c *and* $K(\omega)$ *are the constants given in Lemmas 2.3.3 and 2.3.4, respectively.*

Assuming this lemma for a moment, we finish the proof of Theorem 2.2.1.

By (2.3.11) and (2.3.12), the series $\sum V_k(x)$ and $\sum W_k(x)$ converge uniformly in the domain $\mathcal{D}(r)$; let us denote the limits by $v(x)$ and $w(x)$, respectively:

$$\begin{aligned} v(x) &:= \sum_{k=1}^{\infty} V_k(x) \\ &= \lim_{n\to\infty} \sum_{k=1}^{n} V_k(x) \\ &= \lim_{n\to\infty} v_n(x), \\ w(x) &:= \sum_{k=1}^{\infty} W_k(x) \\ &= \lim_{n\to\infty} \sum_{k=1}^{n} W_k(x) \\ &= \lim_{n\to\infty} w_n(x). \end{aligned}$$

The functions $v(x)$ and $w(x)$ are holomorphic in $\mathcal{D}(r)$ and satisfy

$$xv'(x) = w(x),$$

because we have

$$\begin{aligned} xv'(x) &= x\sum_{k=1}^{\infty} V_k'(x) \\ &= \sum_{k=1}^{\infty} xV_k'(x) \\ &= \sum_{k=1}^{\infty} W_n(x) = w(x). \end{aligned}$$

By (2.3.11) and (2.3.12), there is a constant $M > 0$ independent of

n such that

$$|v(x) - v_n(x)| \le \sum_{k=n+1}^{\infty} |V_k(x)| \le M(|x| + |e^{-\kappa} x^{\omega}| + |e^{\kappa} x^{1-\omega}|)^n,$$

$$|w(x) - w_n(x)| \le \sum_{k=n+1}^{\infty} |W_k(x)| \le M(|x| + |e^{-\kappa} x^{\omega}| + |e^{\kappa} x^{1-\omega}|)^n$$

in $\mathcal{D}(r)$, and hence using Lemmas 2.3.3 and 2.3.4, we have

$$\begin{aligned}
&|I_1(v_n, w_n; x) - I_1(v, w; x)| \\
&\le 2cM \int_{\Gamma(x)} |t|^{-1}(|t| + |e^{-\kappa} t^{\omega}| + |e^{\kappa} t^{1-\omega}|)^{n+1} |dt| \\
&\le \frac{2cMK(\omega)}{n+1} (|x| + |e^{-\kappa} x^{\omega}| + |e^{\kappa} x^{1-\omega}|)^{n+1}
\end{aligned}$$

and

$$\begin{aligned}
&|I_0(v_n, w_n; x) - I_0(v, w; x)| \\
&\le \frac{2cMK(\omega)^2}{(n+1)^2} (|x| + |e^{-\kappa} x^{\omega}| + |e^{\kappa} x^{1-\omega}|)^{n+1}
\end{aligned}$$

in $\mathcal{D}(r)$. Let $n \to \infty$ in both sides of recursive formula (2.3.9), and use the above estimates and the inequality

$$|x| + |e^{-\kappa} x^{\omega}| + |e^{\kappa} x^{1-\omega}| < 3r < 1.$$

Then we conclude that $(v(x), w(x))$ is a holomorphic solution of the integral equation (2.3.5). Furthermore, by (2.3.11) and (2.3.12), we have

$$\begin{aligned}
|v(x)| &\le \sum_{n=1}^{\infty} |V_n(x)| \\
&\le \exp(2cK(\omega)^2(|x| + |e^{-\kappa} x^{\omega}| + |e^{\kappa} x^{1-\omega}|)) - 1 \\
&= \mathcal{O}(|x| + |e^{-\kappa} x^{\omega}| + |e^{\kappa} x^{1-\omega}|),
\end{aligned}$$

and

$$\begin{aligned}|xv'(x)| &= |w(x)| \\ &\le \sum_{n=1}^{\infty} |W_n(x)| \\ &\le \exp(2cK(\omega)^2(|x| + |e^{-\kappa}x^{\omega}| + |e^{\kappa}x^{1-\omega}|)) - 1 \\ &= \mathcal{O}(|x| + |e^{-\kappa}x^{\omega}| + |e^{\kappa}x^{1-\omega}|)\end{aligned}$$

in $\mathcal{D}(r)$. Thus we have obtained the desired estimates of $v(x)$ and $xv'(x) = w(x)$. (It is easy to see that the dominating constants corresponding to the symbol $\mathcal{O}$ depend only on ω. This proves Remark 2.2.3.)
In order to show that $v(x)$ satisfies equation (2.3.1), we need the following lemma which will be proved later.

LEMMA 2.3.6 *If $f(x)$ is a holomorphic function satisfying*

$$f(x) = \mathcal{O}(|x| + |e^{-\kappa}x^{\omega}| + |e^{\kappa}x^{1-\omega}|) \tag{2.3.13}$$

in $\mathcal{D}(r)$, then

$$F(x) = \int_{\Gamma(x)} t^{-1} f(t)dt$$

is holomorphic in $\mathcal{D}(r)$ and satisfies $F'(x) = x^{-1}f(x)$.

Substituting $v(x)$ into (2.3.5) and using $xv'(x) = w(x)$, we have

$$\begin{aligned}xv'(x) = \int_{\Gamma(x)} t^{-1}\{&\Phi(t, e^{-\kappa}t^{\omega}, e^{\kappa}t^{1-\omega}) \\ &+ \Psi(t, e^{-\kappa}t^{\omega}, e^{\kappa}t^{1-\omega}, v(t), tv'(t))\}dt\end{aligned}$$

in $\mathcal{D}(r)$. Notice that, by (2.3.10),

$$|v(x)| < \frac{r'}{2}, \quad |xv'(x)| < \frac{r'}{2}.$$

Hence, by these inequalities and Lemma 2.3.3, the integrand of the above equality is of order $\mathcal{O}(|t|^{-1}(|t| + |e^{-\kappa}t^{\omega}| + |e^{\kappa}t^{1-\omega}|))$ in $\mathcal{D}(r)$. By virtue of Lemma 2.3.6, differentiating the both sides, we conclude that $v(x)$ satisfies equation (2.3.1), which proves Theorem 2.2.1.

Assuming Lemma 2.3.6, we prove Lemma 2.3.5.

Proof of Lemma 2.3.5. By Lemmas 2.3.3 and 2.3.6, $W_1(x) = w_1(x) = \varphi_1(x)$ and $V_1(x) = v_1(x) = \varphi(x)$ are holomorphic in $\mathcal{D}(r'/2)$ and satisfy $xv_1'(x) = w_1(x)$. Thus, by Lemma 2.3.4, we have

$$\begin{aligned}|W_1(x)| = |w_1(x)| &= |\varphi_1(x)| \\ &\leq \int_{\Gamma(x)} |t|^{-1}|\Phi(t, e^{-\kappa}t^{\omega}, e^{\kappa}t^{1-\omega})||dt| \\ &\leq K(\omega)c(|x| + |e^{-\kappa}x^{\omega}| + |e^{\kappa}x^{1-\omega}|)\end{aligned}$$

and

$$|V_1(x)| = |v_1(x)| = |\varphi(x)| \leq K(\omega)^2 c(|x| + |e^{-\kappa}x^{\omega}| + |e^{\kappa}x^{1-\omega}|)$$

in $\mathcal{D}(r'/2)$. Take the constant r so small that

$$r < \min\{r'/2, 1/3\},$$

$$\begin{aligned}&\sum_{n=1}^{\infty}(2c)^n K(\omega)^{2n}(3r)^n(n!)^{-1} \\ &= \exp(2cK(\omega)^2 \cdot 3r) - 1 < r'/3.\end{aligned} \tag{2.3.14}$$

Then inequalities (2.3.11) and (2.3.12) hold for $n = 1$ in the domain $\mathcal{D}(r)$ ($\subset \mathcal{D}(r'/2)$). (Note that $K(\omega) \geq 1$.) Moreover, since

$$\begin{aligned}|v_1(x)| &\leq cK(\omega)^2 \cdot 3r < r'/3, \\ |w_1(x)| &\leq cK(\omega) \cdot 3r \leq cK(\omega)^2 \cdot 3r < r'/3\end{aligned}$$

in $\mathcal{D}(r)$, inequalities (2.3.10) hold for $n = 1$.

Suppose that (2.3.10), (2.3.11) and (2.3.12) hold for $n \leq N - 1$. By Lemmas 2.3.3, 2.3.4 and inequalities (2.3.10) with $n = N - 1$, the integrands of the recursive formula (2.3.9) with $n = N$ is of order $\mathcal{O}(|t|^{-1}(|t| + |e^{-\kappa}t^{\omega}| + |e^{\kappa}t^{1-\omega}|))$. Hence, by Lemma 2.3.6, $v_N(x)$ and $w_N(x)$ are holomorphic in $\mathcal{D}(r)$. Putting $n = N$ and $n = N - 1$ in the first equation in (2.3.9), and subtracting the latter from the former, we have

$$\begin{aligned}|W_N(x)| &= |I_1(v_{N-1}, w_{N-1}; x) - I_1(v_{N-2}, w_{N-2}; x)| \\ &\leq \int_{\Gamma(x)} |t|^{-1}|\Psi(t, e^{-\kappa}t^{\omega}, e^{\kappa}t^{1-\omega}, v_{N-1}(t), w_{N-1}(t)) \\ &\quad - \Psi(t, e^{-\kappa}t^{\omega}, e^{\kappa}t^{1-\omega}, v_{N-2}(t), w_{N-2}(t))||dt|\end{aligned}$$

in $\mathcal{D}(r)$. This inequality together with Lemma 2.3.3 leads to

$$|W_N(x)| \le \int_{\Gamma(x)} c|t|^{-1}(|t| + |e^{-\kappa}t^{\omega}| + |e^{\kappa}t^{1-\omega}|) \times (|V_{N-1}(t)| + |W_{N-1}(t)|)|dt|. \tag{2.3.15}$$

In a similar way we obtain

$$|V_N(x)| \le \int_{\Gamma(x)} |s|^{-1} \int_{\Gamma(s)} c|t|^{-1}(|t| + |e^{-\kappa}t^{\omega}| + |e^{\kappa}t^{1-\omega}|) \times (|V_{N-1}(t)| + |W_{N-1}(t)|)|dt||ds|. \tag{2.3.16}$$

By virtue of inequalities (2.3.11) and (2.3.12) with $n = N - 1$ and Lemma 2.3.4, the above estimates (2.3.15) and (2.3.16) imply

$$\begin{aligned} &|W_N(x)| \\ &\le (2c)^N K(\omega)^{2(N-1)}((N-1)!)^{-1} \\ &\qquad \times \int_{\Gamma(x)} |t|^{-1}(|t| + |e^{-\kappa}t^{\omega}| + |e^{\kappa}t^{1-\omega}|)^N |dt| \\ &\le (2c)^N K(\omega)^{2N-1}(N!)^{-1}(|x| + |e^{-\kappa}x^{\omega}| + |e^{\kappa}x^{1-\omega}|)^N \end{aligned}$$

and

$$\begin{aligned} &|V_N(x)| \\ &\le \int_{\Gamma(x)} |s|^{-1}|W_N(s)||ds| \\ &\le (2c)^N K(\omega)^{2N}(N!)^{-1}(|x| + |e^{-\kappa}x^{\omega}| + |e^{\kappa}x^{1-\omega}|)^N \end{aligned}$$

for $x \in \mathcal{D}(r)$, which implies that (2.3.11) and (2.3.12) hold for $n = N$. Moreover, by (2.3.14) and (2.3.11), (2.3.12) for $n \le N$, we have

$$\begin{aligned} |v_n(x)| &\le \sum_{n=1}^{N} |V_n(x)| \\ &< \exp(6cK(\omega)^2 r) - 1 \\ &< r'/3, \end{aligned}$$

and

$$|w_n(x)| \le \sum_{n=1}^{N} |W_n(x)| < r'/3,$$

which implies that (2.3.10) holds for $n = N$. Notice that the holomorphy of $v_n(x)$ and $w_n(x)$ has been already proved. Thus, by Lemma 2.3.6 and (2.3.9), we see readily that $xv_n'(x) = w_n(x)$. Once the statement of the lemma for v_n and w_n is proved, the holomorphy of V_n and W_n and the equality $xV_n'(x) = W_n(x)$ follow from their definition. ∎

Finally we prove Lemma 2.3.6.

Proof of Lemma 2.3.6. It is sufficient to show that

$$F(x + \Delta x) - F(x) = \int_x^{x+\Delta x} t^{-1} f(t) dt,$$

namely

$$\left(\int_{\Gamma(x+\Delta x)} - \int_{\Gamma(x)} - \int_x^{x+\Delta x} \right) t^{-1} f(t) dt = 0, \tag{2.3.17}$$

where x is an arbitrarily fixed point in $\mathcal{D}(r)$, $x + \Delta x$ is a point in the neighbourhood of x and the path of the third integral is taken to be the line-segment joining x to $x + \Delta x$.

Assume that $\Im\omega \neq 0$. Let ε be an arbitrary small positive constant. Let $t = c(\varepsilon)$ and $t = c'(\varepsilon)$ be points on the paths $\Gamma(x)$ and $\Gamma(x + \Delta x)$, respectively, such that

$$|c(\varepsilon)| = |c'(\varepsilon)| = \varepsilon.$$

By the condition $c(\varepsilon) \in \Gamma(x)$ and (2.3.7), we have

$$\arg c(\varepsilon) = \arg x + \frac{\Re\omega - 1/2}{\Im\omega} \log \frac{\varepsilon}{|x|}.$$

If $t \in \Gamma(c(\varepsilon))$, then

$$\begin{aligned} t &= \tau \exp\left((\arg c(\varepsilon) + \frac{\Re\omega - 1/2}{\Im\omega} \log \frac{\tau}{\varepsilon}) \sqrt{-1} \right) \\ &= \tau \exp\left((\arg x + \frac{\Re\omega - 1/2}{\Im\omega} \log \frac{\tau}{|x|}) \sqrt{-1} \right) \quad (\tau = |t| < \varepsilon < |x|), \end{aligned}$$

which implies $t \in \Gamma(x)$.

Thus we have $\Gamma(c(\varepsilon)) \subset \Gamma(x)$ and similarly $\Gamma(c'(\varepsilon)) \subset \Gamma(x+\Delta x)$. Hence, for any $\varepsilon > 0$, we have

$$\int_{\Gamma(x+\Delta x)} - \int_{\Gamma(x)} - \int_{x}^{x+\Delta x} = \int_{\Gamma(c'(\varepsilon))} - \int_{\Gamma(c(\varepsilon))} - \int_{c(\varepsilon)}^{c'(\varepsilon)}.$$

In order to show (2.3.17), it is sufficient to prove

$$\lim_{\varepsilon \to 0} \left(\int_{\Gamma(c'(\varepsilon))} - \int_{\Gamma(c(\varepsilon))} - \int_{c(\varepsilon)}^{c'(\varepsilon)} \right) t^{-1} f(t)dt = 0. \tag{2.3.18}$$

Since $c(\varepsilon) \in \Gamma(x)$, $c'(\varepsilon) \in \Gamma(x+\Delta x)$, we have

$$|c(\varepsilon)^{\omega}| = |x^{\omega}| \left| \frac{\varepsilon}{x} \right|^{1/2}$$

and

$$|c'(\varepsilon)^{\omega}| = |(x+\Delta x)^{\omega}| \left| \frac{\varepsilon}{x+\Delta x} \right|^{1/2}.$$

Then, by Lemma 2.3.4 and (2.3.13), we have estimates for the first and the second integrals in (2.3.18) as follows:

$$\begin{aligned} \left| \int_{\Gamma(c(\varepsilon))} t^{-1} f(t)dt \right| &= \mathcal{O}(|c(\varepsilon)| + |e^{-\kappa} c(\varepsilon)^{\omega}| + |e^{\kappa} c(\varepsilon)^{1-\omega}|) \\ &= \mathcal{O}(\varepsilon^{1/2}) \end{aligned} \tag{2.3.19}$$

and

$$\left| \int_{\Gamma(c'(\varepsilon))} t^{-1} f(t)dt \right| = \mathcal{O}(\varepsilon^{1/2}). \tag{2.3.20}$$

We estimate the third integral. By (2.3.7), if $\Gamma(x)$ and $\Gamma(x+\Delta x)$ meet at some point, then

$$\arg x + \frac{\Re\omega - 1/2}{\Im\omega} \log \frac{1}{|x|} = \arg(x+\Delta x) + \frac{\Re\omega - 1/2}{\Im\omega} \log \frac{1}{|x+\Delta x|}$$

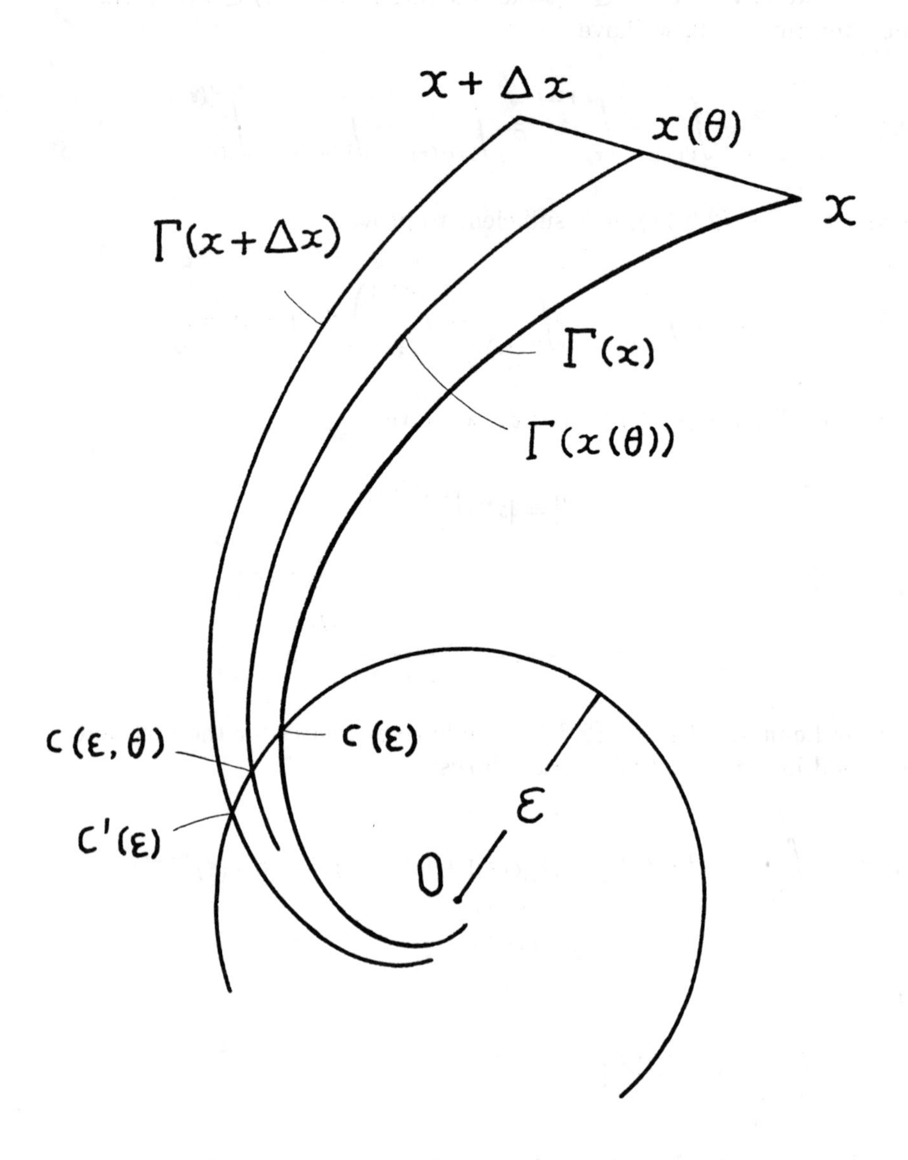

Figure 2.3.2

which implies that one of the two paths includes the other. In such a case the third integral vanishes and the lemma is proved. We consider the case where $\Gamma(x)$ and $\Gamma(x+\Delta x)$ do not meet in $\mathcal{D}(r)$. The segment $[x, x+\Delta x]$ is parameterized as follows

$$t = x(\theta) = x + \theta\Delta x \qquad (0 \le \theta \le 1).$$

Let $t = c(\varepsilon, \theta)$ be the point on the path $\Gamma(x(\theta))$ such that

$$|c(\varepsilon, \theta)| = \varepsilon.$$

When $x(\theta)$ moves on the segment $[x, x+\Delta x]$ from x to $x+\Delta x$, the point $c(\varepsilon, \theta)$ moves continuously on the arc $A(c(\varepsilon), c'(\varepsilon))$ (included in the circle $|t| = \varepsilon$) from $c(\varepsilon)$ to $c'(\varepsilon)$. Since $c(\varepsilon, \theta) \in \Gamma(x(\theta))$, we have

$$|c(\varepsilon, \theta)^\omega| = |x(\theta)^\omega| \left|\frac{\varepsilon}{x(\theta)^\omega}\right|^{1/2} = \mathcal{O}(\varepsilon^{1/2}) \qquad (0 \le \theta \le 1).$$

Hence, if $t \in A(c(\varepsilon), c'(\varepsilon))$ then

$$f(t) = \mathcal{O}(\varepsilon^{1/2}). \tag{2.3.21}$$

Moreover, by (2.3.7), we have

$$\begin{aligned} &|\arg c(\varepsilon) - \arg c'(\varepsilon)| \\ &\quad = \left|\arg x - \arg(x+\Delta x) + \frac{\Re\omega - 1/2}{\Im\omega}\log\left|1 + \frac{\Delta x}{x}\right|\right| \\ &\quad = \mathcal{O}(\Delta x) = \mathcal{O}(1) \end{aligned}$$

and hence

$$\text{the length of } A(c(\varepsilon), c'(\varepsilon)) = \mathcal{O}(\varepsilon). \tag{2.3.22}$$

Therefore, by (2.3.21) and (2.3.22), we have

$$\left|\int_{c(\varepsilon)}^{c'(\varepsilon)} t^{-1} f(t)dt\right| = \mathcal{O}(\varepsilon^{1/2}). \tag{2.3.23}$$

From (2.3.19), (2.3.20) and this estimate, (2.8.18) follows immediately. In case $0 < \omega < 1$, the path $\Gamma(x)$ is the line-segment joining 0 to x. The estimates corresponding to (2.3.19), (2.3.20) and (2.3.23) are obtained in a similar way and are of order $\mathcal{O}(\varepsilon^{\min\{\omega, 1-\omega\}})$. Thus the lemma is proved. ∎

2.4. Solutions of Painlevé equations

By Proposition 2.1.1 and Theorem 2.2.1, we obtain solutions of Painlevé equations near the fixed singular points of regular type.

THEOREM 2.4.1. *Let*

$$\omega \in \mathbb{C} - (-\infty, 0] - [1, +\infty)$$

be a constant and let κ *be an arbitrary complex constant. Then*

(1) *the third Painlevé equation* ($\mathrm{P_{III}}$) *admits a holomorphic solution with the properties*

$$\lambda_{\mathrm{III}}(\omega, \kappa; t) = e^{-\kappa} t^{2\omega-1}(1 + \mathcal{O}(|t|^2 + |e^{-\kappa} t^{2\omega}| + |e^{\kappa} t^{2(1-\omega)}|))$$

and

$$\lambda'_{\mathrm{III}}(\omega, \kappa; t) = (2\omega - 1)e^{-\kappa} t^{2\omega-2}(1 + \mathcal{O}(|t|^2 + |e^{-\kappa} t^{2\omega}| + |e^{\kappa} t^{2(1-\omega)}|))$$

in the domain $\{t \in \mathcal{R}_0; t^2 \in \mathcal{D}(r)\}$,

(2) *the fifth Painlevé equation* ($\mathrm{P_V}$) *admits a holomorphic solution with the properties*

$$\lambda_{\mathrm{V}}(\omega, \kappa; t) = 1 - e^{-\kappa} t^{\omega}(1 + \mathcal{O}(|t| + |e^{-\kappa} t^{\omega}| + |e^{\kappa} t^{1-\omega}|))$$

and

$$\lambda'_{\mathrm{V}}(\omega, \kappa; t) = -\omega e^{-\kappa} t^{\omega-1}(1 + \mathcal{O}(|t| + |e^{-\kappa} t^{\omega}| + |e^{\kappa} t^{1-\omega}|))$$

in the domain $\mathcal{D}(r)$,

(3) *the sixth Painlevé equation* ($\mathrm{P_{VI}}$) *admits a holomorphic solution with the properties*

$$\lambda_{\mathrm{VI}}(\omega, \kappa; t) = e^{-\kappa} t^{\omega}(1 + \mathcal{O}(|t| + |e^{-\kappa} t^{\omega}| + |e^{\kappa} t^{1-\omega}|))$$

and

$$\lambda'_{\mathrm{VI}}(\omega, \kappa; t) = \omega e^{-\kappa} t^{\omega-1}(1 + \mathcal{O}(|t| + |e^{-\kappa} t^{\omega}| + |e^{\kappa} t^{1-\omega}|))$$

in the domain $\mathcal{D}(r)$.

Here r is a sufficiently small positive constant depending on ω, $\mathcal{R}_0$ is the universal covering of $\mathbb{C} - \{0\}$, and

$$\mathcal{D}(r) = \{t \in \mathcal{R}_0; |t| < r,\ |e^{-\kappa} t^{\omega}| < r,\ |e^{\kappa} t^{1-\omega}| < r\}.$$

Proof. (1) Recall that by the change of variables

$$\lambda = t^{-1} e^{-u}, \quad t^2 = x,$$

equation $(\mathrm{P_{III}})$ is transformed into the normal form obtained in Section 2.1. Substituting a solution of the normal form expressed as

$$u(\omega, \kappa; x) = -\omega \log x + \kappa + \mathcal{O}(|x| + |e^{-\kappa} x^{\omega}| + |e^{\kappa} x^{1-\omega}|)$$

$(x \in \mathcal{D}(r))$ into $\lambda = t^{-1} e^{-u}$ and going back to the original variable $x (= t^2)$, we obtain the solution of equation $(\mathrm{P_{III}})$ expressed as

$$\begin{aligned}\lambda_{\mathrm{III}}(\omega, \kappa; t) &= t^{-1} \exp(-u(\omega, \kappa; t^2)) \\ &= t^{-1} \exp(\omega \log t^2 - \kappa + \mathcal{O}(|t^2| + |e^{-\kappa} t^{2\omega}| + |e^{\kappa} t^{2(1-\omega)}|)) \\ &= e^{-\kappa} t^{2\omega - 1}(1 + \mathcal{O}(|t^2| + |e^{-\kappa} t^{2\omega}| + |e^{\kappa} t^{2(1-\omega)}|))\end{aligned}$$

which is holomorphic for $t \in \mathcal{R}_0$ satisfying $t^2 \in \mathcal{D}(r)$.
The derivative $\lambda'_{\mathrm{III}}(\omega, \kappa; t)$ is computed in the following way. Note that

$$\begin{aligned}&\lambda'_{\mathrm{III}}(\omega, \kappa; t) \\ &= -\, t^{-2} \exp(-u(\omega, \kappa; t^2)) - 2u_x(\omega, \kappa; t^2) \exp(-u(\omega, \kappa; t^2)) \\ &= -\, t^{-1}(1 + 2t^2 u_x(\omega, \kappa; t^2)) \lambda_{\mathrm{III}}(\omega, \kappa; t^2).\end{aligned} \tag{2.4.1}$$

Substituting

$$t^2 u_x(\omega, \kappa; t^2) = -\omega + \mathcal{O}(|t^2| + |e^{-\kappa} t^{2\omega}| + |e^{\kappa} t^{2(1-\omega)}|),$$

which is obtained from (2.2.2), into (2.4.1), we obtain

$$\begin{aligned}&\lambda'_{\mathrm{III}}(\omega, \kappa; t) \\ &= -\, t^{-1}(1 - 2\omega + \mathcal{O}(|t^2| + |e^{-\kappa} t^{2\omega}| + |e^{\kappa} t^{2(1-\omega)}|)) \lambda_{III}(\omega, \kappa; t^2) \\ &= (2\omega - 1) e^{-\kappa} t^{2\omega - 2}(1 + \mathcal{O}(|t^2| + |e^{-\kappa} t^{2\omega}| + |e^{\kappa} t^{2(1-\omega)}|)).\end{aligned}$$

(2) Equation $(\mathrm{P_V})$ is transformed into normal form by the change of variables

$$\lambda = \tanh^2 \frac{u}{2} = \left(\frac{1-e^{-u}}{1+e^{-u}}\right)^2, \qquad t = x. \tag{2.4.2}$$

By Theorem 2.2.1, the normal form admits a solution expressed as

$$u(\omega,\kappa;x) = -\omega \log x + \kappa + \mathcal{O}(|x| + |e^{-\kappa}x^{\omega}| + |e^{\kappa}x^{1-\omega}|)$$

in the domain $\mathcal{D}(r)$. Hence equation $(\mathrm{P_V})$ admits a solution

$$\lambda_{\mathrm{V}}(\omega,\kappa;t) = \tanh^2(u(\omega,\kappa;t)/2)$$

in $\mathcal{D}(r)$. If $t \in \mathcal{D}(r)$, the absolute value of

$$\exp(-u(\omega,\kappa;t)) = e^{-\kappa}t^{\omega}(1 + \mathcal{O}(|t| + |e^{-\kappa}t^{\omega}| + |e^{\kappa}t^{1-\omega}|))$$

is sufficiently small. Substituting $u(\omega,\kappa;t)$ into (2.4.2), we obtain

$$\begin{aligned}\lambda_{\mathrm{V}}(\omega,\kappa;t) &= \left(\frac{1-\exp(-u(\omega,\kappa;t))}{1+\exp(-u(\omega,\kappa;t))}\right)^2 \\ &= 1 - 4\exp(-u(\omega,\kappa;t)) \cdot [1 + \mathcal{O}(\exp(-u(\omega,\kappa;t)))] \\ &= 1 - 4e^{-\kappa}t^{\omega}(1 + \mathcal{O}(|t| + |e^{-\kappa}t^{\omega}| + |e^{\kappa}t^{1-\omega}|))\end{aligned}$$

in $\mathcal{D}(r)$. Since κ is arbitrary, replacing $4e^{-\kappa}$ by $e^{-\kappa}$, we obtain a solution in the theorem. The derivative $\lambda'_{\mathrm{V}}(\omega,\kappa;t)$ is obtained as follows:

$$\begin{aligned}&\lambda'_{\mathrm{V}}(\omega,\kappa;t) \\ &= u'(\omega,\kappa;t)\tanh(u(\omega,\kappa;t)/2)\cosh^{-2}(u(\omega,\kappa;t)/2) \\ &= 4u'(\omega,\kappa;t)\exp(-u(\omega,\kappa;t)) \cdot [1 + \mathcal{O}(\exp(-u(\omega,\kappa;t)))]\end{aligned}$$

in $\mathcal{D}(r)$.

(3) The solution $\lambda_{\mathrm{VI}}(\omega,\kappa;t)$ of equation $(\mathrm{P_{VI}})$ is obtained in a similar way. ∎

By the asymptotic expressions of solutions of Painlevé equations and of their normal forms, we can see the behavior of solutions. For example, consider a solution of equation $(\mathrm{P_V})$ expressed as

$$\lambda_{\mathrm{V}}(\omega,\kappa;t) = 1 - e^{-\kappa}t^{\omega}(1+\mathcal{O}(|t|+|e^{-\kappa}t^{\omega}|+|e^{\kappa}t^{1-\omega}|))$$

in $\mathcal{D}(r)$.

1. Let κ be an arbitrary complex constant and ω be a constant satisfying $0<\omega<1$. Then the solution satisfies

$$\lim_{\substack{t\to 0\\ t\in\mathcal{D}(r)}} \lambda_{\mathrm{V}}(\omega,\kappa;t) = 1.$$

2. Let $\omega = \sigma\sqrt{-1}$ $(\sigma \neq 0)$ is a pure imaginary constant and κ be a complex constant such that $|e^{-\kappa}|$ is sufficiently small. Then the real interval $(0,r)$ is contained in $\mathcal{D}(r)$ and the solution admits the expression

$$\begin{aligned}&\lambda_{\mathrm{V}}(\sigma\sqrt{-1},\kappa;t)\\ &\quad = 1 - e^{-\kappa}(\cos(\sigma\log t)+\sqrt{-1}\sin(\sigma\log t))(1+\mathcal{O}(|e^{-\kappa}|+|e^{\kappa}t|))\end{aligned}$$

on the real interval $(0,r)$. By Remark 2.2.3, if $0<t<r\le|e^{-2\kappa}|$, we can replace the remainder term by $\mathcal{O}(|e^{-\kappa}|)$. This expression shows that $\lambda_{\mathrm{V}}(\sigma\sqrt{-1},\kappa;t)$ oscillates on the real interval $(0,r)$, if r is sufficiently small.

3. When $\alpha=\beta=0$, the normal form of equation $(\mathrm{P_V})$ can be written as

$$t(tu')' = \frac{\gamma}{4}(te^{u}-te^{-u}) - \frac{\delta}{8}(t^2e^{2u}-t^2e^{-2u}). \tag{2.4.3}$$

By the change of variables $u = v+\log t$, $t^2 = x$, this equation is transformed into

$$x(xv_x)_x = \frac{\gamma}{16}(xe^{v}-e^{-v}) - \frac{\delta}{32}((xe^{v})^2-(e^{-v})^2) \tag{2.4.4}$$

which is also in normal form. Applying Theorem 2.2.1 to equation (2.4.4), we obtain a solution of equation (2.4.3) expressed as

$$\begin{aligned}u = U(\omega,\kappa;t) &= (1-2\omega)\log t+\kappa\\ &+\mathcal{O}(|t|^2+|e^{-\kappa}t^{2\omega}|+|e^{\kappa}t^{2(1-\omega)}|)\end{aligned}$$

if $t^2 \in \mathcal{D}(r)$, where κ is an arbitrary complex constant and ω is a constant satisfying (2.2.1). Hence we obtain a solution of equation $(\mathrm{P_V})$ expressed as

$$\begin{aligned}\lambda &= \Phi(\omega, \kappa; t) \\ &= \tanh^2\left(\left(\frac{1}{2} - \omega\right)\log t + \frac{\kappa}{2} + \mathcal{O}(|t|^2 + |e^{-\kappa}t^{2\omega}| + |e^{\kappa}t^{2(1-\omega)}|)\right)\end{aligned}$$

if $t^2 \in \mathcal{D}(r)$. If we put $\omega = (1/2) - c\sqrt{-1}$ and $\kappa = 2c'\sqrt{-1}$ $(c, c' \in \mathbb{R})$, then, for any $A > 0$, the domain $\mathcal{D}(r)$ includes the sector

$$|\arg t| < A, \quad |t| < r_0,$$

where r_0 is a sufficiently small positive constant depending on c, c' and A. Thus we have

THEOREM 2.4.2 *Assume that $\alpha = \beta = 0$. Then, for any real constants c and c', equation* $(\mathrm{P_V})$ *admits a solution expressed as*

$$\begin{aligned}\lambda &= \phi(c, c'; t) \\ &= -\tan^2(c\log t + c' + \mathcal{O}(t))\end{aligned}$$

in the sector $|\arg t| < A$, $|t| < r_0$. Here A is an arbitrary positive constant and r_0 is a sufficiently small positive constant.

REMARK The solution $\phi(c, c'; t)$ has infinitely many poles and zeros near the positive real axis.

Bibliography

[Aom.1] Aomoto, K., Une remarque sur la solution des équations de L. Schlesinger et Lappo-Danilevski, J. Fac. Sci. Univ. Tokyo, **17**(1970) 341-354.

[Aom.2] Aomoto, K., Un théorème du type Matsushima-Murakami concernant l'intégrale des fonctions multiformes, J. Math. Pures Appl., **52**(1973), 1-11.

[Aom.3] Aomoto, K., Les équations aux différences linéaires et les intégrales des fonctions multiformes, J. Fac. Sci. Univ. Tokyo, **22**(1975), 271-297; Une correction et complément à l'article "Les équations aux différences linéaires et les intégrales des fonctions multiformes", ibid., **26**(1979),519-523.

[Aom.4] Aomoto, K., On vanishing of cohomology attached to certain many valued meromorphic functions, J. Math. Soc. Japan, **27**(1975), 248-255.

[Aom.5] Aomoto, K., On the structure of integrals of power product of linear functions, Sci. Papers College Gen. Ed. Univ. Tokyo, **27**(1977), 49-61.

[AKF] Appell, P et Kampé de Fériet, J. Fonctions hypergéométriques et hypersphériques, Gauthier Villars, Paris, 1926.

[AW] Askey, R.A. and Wilson, J.A. Some basic hypergeometric orthogonal polynomials that generalize Jacobi polynomials, Memoirs of the AMS, AMS, 1985.

[BV] Babbitt, D.G. and Varadrajan, V.S. Local moduli for meromorphic differential equations, Astérisque, 169-170, 1989.

[Bail] Bailey, W.N. Generalized hypergeometric series. Cambridge Mathematical Tract No.32. Cambridge Univ. Press, 1935.

[BJL.1] Balser, W., Jurkat, B. and Lutz, D.A., Birkhoff invariants and Stokes multipliers for meromorphic linear differential equations, J. Math. Anal. Appl., **71**(1979) 48-94.

[BJL.2] Balser, W., Jurkat, B. and Lutz, D.A., A general theory of invariants for meromorphic differential equations; Part I, formal invariants; Part II, proper invariants, Funkcial. Ekvac., **22**, 197-221, 257-283 (1979); Part III, Houston J. Math., **6**(1980) 149-189.

[BJL.3] Balser, W., Jurkat, B. and Lutz, D.A., On the reduction of connection problems for differential equations with an irregular singular point to ones with only regular singularities, I., SIAM J. Math. Anal., **12**(1981) 691-721.

[Bern] Bernstein, I.N., The analytic continuation of generalised functions with respect to a parameter, Func. Anal. Akademia Nauk CCCR **6** (4)(1972), 26-40.

[Bieb] Bieberbach, L., Theorie der gewöhnlichen Differentialgleichungen, Springer-Verlag, 1953.

[Birk] Birkhoff, G. (Editor), A source book in classical analysis, Harvard Univ. Press, Cambridge, Massachusetts, 1973.

[Bor] Borel, É., Leçon sur les séries divergentes, Gauthier Villars, Paris, 1928.

[Bout] Boutroux, P., Recherches sur les transcendantes de M.Painlevé et l'étude asymptotique des équations différentielles du second ordre, Ann Sci. Ecole Norm. Sup., (3) **30**(1913), 255-375; (3) **31**(1914), 99-159.

[Brj] Brjuno, A.D., Analytic form of differential equations, Trans. Moscow Math. Soc., **25**(1971), 131-288.

[Bur.1] Bureau, F.J., Differential equations with fixed critical points, I. Annali di Matematica, **64**(1964), 229-364; II ibid., **66**(1964), 1-116.

[Bur.2] Bureau, F.J., Équations différentielles du second ordre en Y et du second degré en $\ddot{Y}$ dont l'integrale générale est à points critiques fixes, ibid., **91**(1972), 163-281.

[Chaz] Chazy, J., Sur les équations différentielles du troisième ordre et d'ordre supérieur dont l'intégrale a ses points critiques fixes, Acta. Math., **34**(1911), 317-385.

[CL] Coddington, E.A. and Levinson, N., Theory of Ordinary Differential Equations, McGraw-Hill, New York, 1955.

[Del] Deligne, P., Équations différentielles à points singuliers réguliers, Lec. Notes in Math., **163**, Springer-Verlag, 1970.

[Eca] Ecalle, J., Les fonctions resurgentes, Tome I,II,III, Publications Mathématiques D'Orsay, 85-05, Université de Paris-Sud.

[Erd] Erdélyi, A. (Editor), Higher transcendental functions, I, II and III, MacGraw Hill, New York, 1953.

[Ext.1] Exton, H. Multiple HGF's and applications, John Wiley, 1976.

[Ext.2] Exton, H. Handbook of HG integrals, John Wiley, 1978.

[FucR] Fuchs, R., Über lineare homogene Differentialgleichungen zweiter Ordnung mit drei im Endlichen gelegenen wesentlich singulären Stellen, Math., **63**(1907), 301-321.

[Gamb] Gambier, B., Sur les équations différentielles du second ordre et du premier degré dont l'intégrale générale est à points critiques fixes, Acta. Math. Ann., **33**(1910), 1-55.

[Gar.1] Garnier, R., Sur les équations différentielles du troisième ordre dont l'intégrale générale est uniforme et sur une classe d'équations nouvelles d'ordre supérieur dont l'intégrale générale a ses points critiques fixes, Ann. Ecole Norm. Sup., **29** (1912), 1-126.

[Gar.2] Garnier, R., Etude de l'intégrale générale de l'équation VI de M. Painlevé, Ann. Sci. Ecole Norm. Sup. 3e série, **34**(1917), 239-353.

[Gar.3] Garnier, R., Contribution à l'étude des solutions de l'équation (V) de Painlevé , J. Math. Pures Appl., **46**(1968), 353-413.

[Gel] Gelfand, I.M. General theory of hypergeometric functions, Soviet Math. Dokl. **33**(1986), 573-577.

[GG] Gelfand, I.M. and Gelfand, S.I., Generalized hypergeometric equations, Soviet Math. Dokl. **33**(1986), 643-646.

[GGR.1] Gelfand, I.M. and Graev, M.I., Hypergeometric functions associated with the Grassmannian $G_{3,6}$, Soviet Math. Dokl. **35**(1987), 298-303.

[GGR.2] Gelfand, I.M. and Graev, M.I., Generalized hypergeometric functions on the Grassmannian $G_{3,6}$ (in Russian), Preprint **123** Keldysh Inst. of Appl. Math. (1987).

[GGR.3] Gelfand, I.M. and Graev, M.I., Strata in $G_{3,6}$ and the associated hypergeometric functions (in Russian), Preprint **127** Keldysh Inst. of Appl. Math. (1987).

[Gér.1] Gérard, R., Une classe d'équations différentielles non linéaires à singularité régulière, Funkcial. Ekvac., **29**(1986), 55-76.

[Gér.2] Gérard, R., Etude locale des équations différentielles de la forme $xy' = f(x,y)$ au voisinage de $x = 0$, J. Fac. Sci. Univ. Tokyo **36**(1989), 729-752

[GL.1] Gérard, R. and Levelt, A.H.M., Etude d'une classe particulière de systèmes de Pfaff du type de Fuchs sur l'espace projectif complexe, J. Math. pures et appl., **51**(1972), 189-217.

[GL.2] Gérard, R. and Levelt, A.H.M., Sur les connexions à singularités régulières dans le cas de plusieurs variables, Funkcial. Ekvac., **19**(1976), 149-173.

[GR] Gérard, R. and Ramis, J.-P. (editors), Equations Différentielles et Systèmes de Pfaff dans le Champ Compexe, I, II, Lec. Note in Math., Springer, **712**, 1979; ibid **1015**, 1983.

[GO] Gérard, R. and Okamoto, K. (editors), Équations différentielles dans le champ complexe (colloque franco-japonais 1985), I, II, III, Publications IRMA,(1989).

[GS] Gérard, R. and Sibuya, Y., Étude de certains systèmes de Pfaff avec singularités, Lec. Note in Math., Springer, **712**, 1979.

[Gol] Golubew, W.W., Vorlesungen über Differentialgleichungen im Komplexen, Deutscher Verlag der Wiss. Berlin, 1958.

[Gol] Golubeva, V.A., Some problems in the analytic theory of Feynman integrals, Russian Math. Surveys **31** (1976), 139-207.

[Gray] Gray, J., Linear differential equations and group theory from Riemann to Poincaré, Birkhäuser, 1986.

[Gun] Gunning, R.C. Lectures on Vector Bundles over Riemann Surfaces, Mathematical Notes vol. 6 Princeton, Princeton University Press, 1967.

[HSY] Hara, M., Sasaki, T. and Yoshida, M., Tensor products of linear differential equations, Funkcial. Ekvac., **32**(1989), 455-477.

[Har.1] Haraoka, Y., Theorems of Sibuya-Malgrange type for Gevrey functions of several variables, Funkcial. Ekvac., **32**(1989), 365-388.

[Har.2] Haraoka, Y., G-primitive extensions for linear ordinary differential equations, Kumamoto J. Math., **3**(1990).

[Har.3] Haraoka, Y., The Galois theory for linear homogeneous partial differential equations of the first order, Funkcial. Ekvac., **33**(1990).

[HK] Hattori, A. and Kimura, T., On the Euler integral representations of hypergeometric functions in several variables, J. of Math. Soc. of Japan, **26**(1974), 1-16.

[Hil] Hille, E., Ordinary differential equations in the complex domain, Wiley-Interscience Publication, New York-London-Sydney-Tronto, 1978.

[Huk.1] Hukuhara, M., Sur les points singuliers des équations différentielles linéaires: domain réel, J. Fac. Sci. Hokkaido Imp. Univ., **2**(1934), 13-88.

[Huk.2] Hukuhara, M., Sur les points singuliers des équations différentielles linéaires II, J.Fac.Sci. Hokkaido Univ., **5** (1937), 157-166.

[Huk.3] Hukuhara, M., Sur les points singuliers d'une équation différentielle ordinaire du premier ordre, I, II, III, IV, Mem. Fac. Eng. Kyushu Imp. Univ., **8**(1937), 203-247; Proc. Phys.-Math. Soc. Japan, **20** (1938), 157-189; ibid **20**(1938), 409-441; ibid **20**(1938), 865-907.

[Huk.4] Hukuhara, M., Intégration formelle d'un système d'equations différentielles non linéaires dans le voisinage d'un point singulier, Ann. Mat. Pura Appl., **19**(1940), 35-44.

[Huk.5] Hukuhara, M., Sur les points singuliers des équations différentielles linéaires III, Mem. Fac. Sci. Kyushu Univ., **2** (1941), 125-137.

[Huk.6] Hukuhara, M., Quelques remarques sur le mémoire de P. Painlevé sur les équations différentielles dont l'intégrale générale est uniforme, Pub. Res. Inst. Math. Sci., A**3**(1967), 139-159.

[Huk.7] Hukuhara, M., Theory of ordinary differential equations, (in Japanese), Iwanami-Zensho, 1950, 2-nd edition, 1980.

[HKM] Hukuhara, M., Kimura, T., and Matuda, T., Équations différentielles ordinaires du premier ordre dans le champ complexe, Math. Soc. Jap., Tokyo, 1961.

[Inc] Ince, E.L., Ordidnary Differential Equations, Dover, 1956.

[Inui] Inui, T., Special functions (in Japanese), Iwanami, 1962.

[IN] Its, A.R. and Novokshenov, V,Yu., The isomonodromic deformation method in the theory of Painlevé equations, Lec. Notes in Math. **1191**, Springer, 1986.

[Iwn.1] Iwano, M., Intégration analytique d'un système d'équations différentielles non linéaires dans le voisinage d'un point singulier, I, II, Ann. Mat. Pura Appl., **44**(1957), 261-292, **47**(1959), 91-150.

[Iwn.2] Iwano, M., On a singular point of Briot-Bouquet type of a system of ordinary nonlinear differential equations, Comment. Math. Univ. St. Paul., **11**(1963), 37-78.

[Iwn.3] Iwano, M., On general solution of a first-order non-linear differential equation of the form $x(dy/dx) = y(\lambda + f(x,y))$ with negative rational λ, Ann. Mat. Pura Appl., **126**(1980), 19-80.

[Iwn.4] Iwano, M., On a general solution of a non-linear 2-system of the form $x^2 dw/dx = \Lambda w + xh(x,w)$ with a constant matrix Λ of signature (1.1), Tôhoku Math. J., **32**(1980), 453-486.

[Iwn.5] Iwano, M., On an n-parameter family of solutions of a nonlinear n-systems with an irregular type singularity, Ann. Mat. Pura Appl., **140**(1985), 57-132.

[Iwn.6] Iwano, M., Schwartz Theory , Math. Seminar Notes, Tokyo Metropolitan Univ., 1989.

[Iwn.7] Iwano, M., A general solution of a system of nonlinear ordinary differential equations $xy' = f(x,y)$ in the case when $f_y(0,0)$ is the zero matrix, Ann. Mat. Pura. Appl., **83**(1969), 1-42.

[Iwn.8] Iwano, M., Applications of Nagumo-Hukuhara theory on the boundary value problems for nonlinear ordinary differential equations, Ann. Mat. Pura. Appl., **113**(1977), 303-392.

[Iwn.9] Iwano, M., Analytic expressions for bounded solutions of non-linear ordinary differential equations with an irregular type singular point, Ann. Mat. Pura. Appl., **LXXXII**(1969), 189-256.

[Iwn.10] Iwano, M., Analytic integration of a system of nonlinear ordinary differential equations with an irregular type singularity I,II, Ann. Mat. Pura. Appl., **94**(1972) 109-160, **99** (1974) 221-246.

[Iws.1] Iwasaki, K., On the Riemann-Hilbert-Birkhoff problem for ordinary differential equations containing a parameter, J. Fac. Sci. Univ. Tokyo **35**(1988), 251-312.

[Iws.2] Iwasaki, K., Riemann equation of harmonic equation and Appell's F_4, SIAM J. Math. Anal., **19**(1988), 902-917.

[Iws.3] Iwasaki, K., Moduli and deformation for Fuchsian projective connections on a Riemann surface, Univ. Tokyo preprint 89-16.

[Jimb] Jimbo, M., Monodromy problem and the boundary condition for some Painlevé equations, Publ. Res. Inst. Math. Sci., **18**(1982), 1137-1161.

[JMU] Jimbo, M., Miwa, T. and Ueno, K., Monodromy preserving deformation of linear ordinary differential equations with rational coefficients, I, Physica D, **2**(1981), 306-352.

[JM] Jimbo, M. and Miwa, T., Monodromy preserving deformation of linear ordinary differential equations with rational coefficients, II, Physica D, **2**(1981), 407-448.

[Jou] Jouanolou, J.P., Equations de Pfaff Algébriques, Lec. Note in Math., Springer-Verlag, **708**(1979).

[Kam.1] Kametaka, Y., On the Euler-Poisson-Darboux equation and the Toda equation I, Proc. Japan. Acad. **60**A(1984), 145-148, II, ibid 181-184.

[Kam.2] Kametaka, Y., Hypergeometric solutions of Toda equation, RIMS Kokyuroku **554**(1985), 26-46.

[KNFH] Kametaka, Y., Noda, M., Fukui, Y. and Hirano, S, A numerical approach to Toda equation and Painlevé II equations, Mem. Fac. Eng. Ehime Univ. **9**(1986), 1-26.

[Kas] Kashiwara, M., b-functions and holonomic systems, rationality of roots of b-functions, Inv. Math. **38**(1976), 33-53.

[KimH.1] Kimura, H., On the isomonodromic deformation of linear ordinary differential equations of the third order, Proc. Jap. Acad., **57**(1981), 446-449.

[KimH.2] Kimura, H., The construction of a general solution of a Hamiltonian system with regular type singularity and its application to Painlevé equations, Ann. Mat. Pura Appl., **134**(1983), 363-392.

[KimH.3] Kimura, H., The degeneration of the two dimensional Garnier system and the polynomial Hamiltonian structure, Ann. Mat. Pura. Appl., (to appear in 1989).

[KimH.4] Kimura, H., Feuilletage uniforme associé au systéme de Garnier, Tokyo Univ., preprint (1988).

[KO.1] Kimura, H. and Okamoto, K., On the isomonodromic deformation of linear ordidnary differential equations of higher order, Funkcial. Ekvac., **26**(1983), 37-50.

[KO.2] Kimura, H. and Okamoto, K., On the polynomial Hamiltonian structure of the Garnier systems, J. Math. Pures et Appl., **63**(1984), 129-146.

[KO.3] Kimura, H. and Okamoto, K., On particular solutions of Garnier systems and the hypergeometric functions of several variables, Quarterly J. Math., **37**(1986), 61-80.

[KimT.0] Kimura, T., Ordinary differential equations (in Japanese), Kyoritu Publ., 1978.

[KimT.1] Kimura, T., Sur une généralisation d'un théorème de Malmquist, I, II, III, Commentarii Math. Univ. Sancti Pauli, **2**(1953), 23-28; ibid **3**(1955), 97-107; ibid **4**(1955), 25-41.

[KimT.2] Kimura, T., Sur les points singuliers des équations différentielles ordinaires, I, II, Commentarii Math. Univ. Sancti Pauli, **2**(1954), 47-53; ibid **3**(1954), 43-49.

[KimT.3] Kimura, T., Sur la propriété d'Iversen et l'équation différentielle ordinaire du second ordre, Commentarii Math. Univ. Sancti Pauli, **7**(1960-1), 63-70.

[KimT.4] Kimura, T., Sur la propriété d'Iversen et l'équation différentielle ordinaire du second ordre, II, Commentarii Math. Univ. Sancti Pauli, **8**(1961-2), 87-90.

[KimT.5] Kimura, T., Sur la direction de Julia au point singulier fixe d'une équation différentielle du premier ordre, Funkcial. Ekvac., **4**(1962), 1-27.

[KimT.6] Kimura, T., On Riemann's equations which are solvable by quadratures, Funkcial. Ekvac., **12**(1969-70), 269-281.

[KimT.7] Kimura, T., On Fuchsian differential equations reducible to hypergeometric equations by linear transformations, Funkcial. Ekvac., **13**(1970), 213-232.

[KimT.8] Kimura, T, Hypergeometric functions of two variables, Seminar Note in Math. Univ. of Tokyo, 1973.

[KimT.9] Kimura, T., Ordinary differential equations II, Kisosugaku-koza, Iwanami, (in japanese), 1976.

[KimT.10] Kimura, T., Analytic theory of ordinary differential equations, IV. Global theory of nonlinear differential equations, Recent Progress of Natural Sciences in Japan, Science Council of Japan, **1**(1976), 47-55.

[KimT.11] Kimura, T., On the isomonodromic deformation for linear ordinary differential equations of the second order, I, II, Proc. Jap. Acad.,

57(1981), 285-290; ibid **58**(1982), 294-297.

[KS] Kimura, T. and Shima, K., On the monodromy of the hypergeometric differential equation (in preparation)

[Kin] Kinosita, K., On the system of Pfaffian equations of Briot- Bouquet type I, J. Fac. Sci. Univ. Tokyo, 24(1977), 341-356.

[Kit] Kita, M., The Riemann-Hilbert problem and its application to analytic functions of several complex variables, Tokyo J. of Math., **2** (1979), 1-27.

[KN] Kita, M. and Noumi, M., On the structure of cohomology groups attached to the integral of certain many-valued anaytic functions, Jap. J. Math. **9**(1983), 113-157.

[Kle] Klein, F., Vorlesungen über die hypergeometrische Funktion, Springer-Verlag, 1933.

[Koh.1] Kohno, M., An extended Airy function of the first kind, Hiroshima Math. J. **9**(1979), 473-489.

[Koh.2] Kohno, M., A two point connection problem, Hiroshima Math. J. **9**(1979), 61-135.

[Koh.3] Kohno, M., A simple reduction of single linear differential equations to Birkhoff and Schlesinger's canonical systems, Kumamoto J. of Math. **2**(1989), 1-18.

[KY] Kohno, M. and Yokoyama, T., A central connection problem for a normal system of linear differential equations, Hiroshima Math. J. **14** (1984), 257-263.

[LM] Léauté, B. and Marcilhacy, G., Sur certaines solutions particuliéres transcendentes des équations d'Einstein, Ann. Inst. H.Poincaré Sect. A, **31**(1979), 363-375.

[MTW] MacCoy, B.M., Tracy, C.A. and Wu, T.T., Painlevé functions of the third kind, J. Math. Phys., **18**(1977), 1058-1092.

[Majm] Majima, H., Asymptotic Analysis for Integrable Connections with Irregular Singular Points, Lec. Note in Math., Springer-Verlag, **1075** (1984).

[Malg.1] Malgrange, B., Sur les polynômes de I.N.Bernstein, Uspekhi Math. Nauk 24-9 (1974), 81-88.

[Malg.2] Malgrange, B., Sur les déformation isomonodromiques, I. Singularités régulierès, séminaire E.N.S. Birkhäuser, 1982.

[Mal.1] Malmquist, J., Sur les fonctions d'un nombre fini de branches définies par les équations différentielles du premier ordre, Acta Math., **36**(1913), 297-343.

[Mal.2] Malmquist, J., Sur les équations différentielles du second ordre dont l'intégrale générale a ses points critiques fixes, Arkiv. Math. Astr.

Fys., **17**(1922-23), 1-89.

[Mal.3] Malmquist, J., Sur l'étude analytique des solutions d'un système d'équations différentielles dans le voisinage d'un point singulier d'indétermination, I, II, III, Acta Math., **73** (1940), 87-129, **74**-(1941), 1-64, **74**(1941), 109-128.

[MR] Martinet, J. and Ramis, J.P., Problèm de modules pour des équations différentielles non-linéaires du premier ordre, Publ. I.H.E.S. **55** (1982) 63-164.

[MatM] Matsuda, M. First order algebraic differential equations, Lec. Note in Math., Springer, **804**(1980).

[MatT] Matuda, T., Étude de l'équation différentielle ordinaire sur une surface de Riemann, Funkcial. Ekvac.,**3**(1961), 75-103.

[MSY] Matsumoto, K., Sasaki, T. and Yoshida, M., The period map of a 4-parameter family of $K3$ surfaces and the Aomoto-Gelfand hypergeometric function of type $(3,6)$, Proc. J. Acad. **64**(1988), 307-310.

[Mill.1] Miller, W.Jr., Lie theory and the Lauricella functions F_D, J. Math. Phys. **13**(1972), 1393-1399.

[Mill.2] Miller, W.Jr., Lie theory and generalized hypergeometric functions SIAM J. Math. Anal. **3**(1972), 31-44.

[Mill.3] Miller, W.Jr., Symmetry and separation of variables, Encyclopedia of Mathematics and its application, Vol.4, Addison-Wesley, MA, 1977.

[Mill.4] Miller, W.Jr., Symmetries of differential equations, the hypergeometric and Euler-Darboux equation, SIAM. J. Math. Annl. **4** (1973), 314-328.

[Mis] Misaki, N., Reducibility of the Pochhammer equation of order three, Master thesis, Tokyo Univ., (1973).

[Miw] Miwa, T., Painlevé property of monodromy preserving deformation equations and the analyticity of τ functions, Publ. RIMS, Kyoto Univ., **17**(1981), 703-721.

[Mur.1] Murata, Y., Rational solutions of the second and the fourth Painlevé equations, Funkcial. Ekvac., **28**(1985), 1-32.

[Mur.2] Murata, Y., On fixed and movable singularities of systems of rational differential equations of order n, J. Fac. Sci. Univ. Tokyo, **35**(1988), 436-506.

[Mur.3] Murata, Y., The Picard type theorem for essential singularities of solutions of systems of n rational differential equations., J. Diff. Eq., **82**(1989), 174-190.

[Mur.4] Murata, Y., Classical solutions of the third Painlevé equation, Nagasaki Univ., preprint (1989).

[Nag] Nagumo, M., Degree of mapping in convex linear topological spaces, Amer. J. Math. **73**(1951), 497-511.

[Nism] Nishimoto, T., Global disperation relation for density waves in a certain simplified model of disk shaped galaxy, Studies in Appl. Math., **60**(1979), 11-26.

[Nis.1] Nishioka, K., General solutions of algebraic differential equations (in Japanese), Seminar on Math. Sci. Keio Univ. No. **11**, 1987.

[Nis.2] Nishioka, K., A note on the transcendency of Painlevé first transcendent, Nagoya Math. J., **109**(1988), 63-67.

[Nis.3] Nishioka, K., Differential algebraic function fields depending rationally on arbitrary constants, Nagoya Math. J., **113**(1989), 173-179.

[Nis.4] Nishioka, K., General solutions depending rationally on arbitrary constants, Nagoya Math. J., **113**(1989), 1-6.

[Noum] Noumi, M., Wronskian determinants and the Gröbner representation of a linear differential equation, in Algebraic Analysis. Papers Dedicated to Prof. Mikio Sato on the Occasion of His 60th Birthday. Ed. by M. Kashiwara and T. Kawai. Academic Press, 1989.

[Oht.1] Ohtsuki, M., A residue formula for Chern classes associated with logarithmic connections, Tokyo J. of Math., **5**(1982), 13-21.

[Oht.2] Ohtsuki, M., On the number of apparent singularities of a linear differential equations, Tokyo J. of Math. **5**(1982), 23-29.

[Okm.1] Okamoto, K., Sur les feuilletages associés aux équations du second ordre à points critiques fixes de P. Painlevé, Jap. J. Math., **5**(1979), 1-79.

[Okm.2] Okamoto, K., Déformation d'une équation différentielle linéaire avec une singularité irréguliére sur un tore, J. Fac. Sci. Univ. Tokyo Sec. IA, **26**(1979), 501-518.

[Okm.3] Okamoto, K., Polynomial Hamiltonians associated with Painlevé equations, I, II, Proc. Japan Acad., 56, Ser. A (1980), 264-268; ibid, 367-371.

[Okm.4] Okamoto, K., On the τ-function of the Painlevé equations, Physica D, **2**(1981), 525-535.

[Okm.5] Okamoto, K., Introduction to the Painlevé equations (in Japanese), Sophia Kokyuroku in Math. 19, Sophia Univ.,1985.

[Okm.6] Okamoto, K., Isomonodromic deformation and Painlevé equations and the Garnier system, J. Fac. Sci. Univ. Tokyo Sec. IA, Math. **33** (1986), 575-618.

[Okm.7] Okamoto, K., Studies on the Painlevé equations, I, Ann. Mat. Pura Appl., **146**(1987), 337-381; II, Jap.J.Math., **13**(1987), 47-76; III, Math. Ann., **275**(1986), 221-256;IV, Funkcial. Ekvac., **30**(1987),

305-332.

[Okm.8] Okamoto, K., Sur les échelles associées aux fonctions spéciales et l'équation de Toda, J. Fac. Sci. Univ. Tokyo, Sect. IA **34**(1987), 709-740.

[Okm.9] Okamoto, K., The Hamiltonian structure derived from the Holonomic Deformation of the linear ordinary differential equations on an elliptic curve, Sci. Papers College Arts Sci. Univ. Tokyo, **37**(1987), 1-11.

[Okm.10] Okamoto, K., Echelles et l'équation de Toda, Univ. Tokyo preprint 87-04.

[Okm.11] Okamoto, K., Elliptic Garnier systems, Univ. Tokyo preprint 87-01.

[Okm.12] Okamoto, K., Bäcklund transformations of classical orthogonal polynomials, (Colloque franco-japonais 1985), Publ. IRMA, Univ. de Strasbourg, 1989.

[Okm.13] Okamoto, K., The Painlevé equations and the Dynkin diagrams, Proc. of the NATO workshop, (to appear in 1991).

[Okub.1] Okubo, K., A global representation of a fundamental set of solutions and a Stokes phenomenon for a system of linear ordinary differential equations, J. Math. Soc. Japan **15**(1963), 268-288.

[Okub.2] Okubo, K., Connection problems for systems of linear differential equations, Proc. of Jap.-U.S. seminar on ordinary differential and functional equations, Lec. Notes in Math., **243**, Springer, 1971.

[Okub.3] Okubo, K., Group of Fuchsian equations, Seminar report of Tokyo Metropolitan Univ., 1988.

[OTY] Okubo, K., Takano, K. and Yoshida, S., A connection problem for the generalized hypergeometric equation, Funkcial. Ekvac., **31** (1988), 483-495.

[Pain] Painlevé, P., Œuvres t. I, II, III, SNRS, Paris, 1976.

[Pic] Picard, E., Remarques sur les équations différentielles, Acta. Math., **17**(1893), 297-300.

[Plem] Plemelj, J., Problems in the sense of Riemann and Klein, Interscience, 1964.

[Ram.1] Ramis, J.P., Dévissage Gevrey, Astérisque **59-60**(1978), 173-204.

[Ram.2] Ramis, J.P., Les séries k-sommables et leurs applications, Lec. Note in Phys., Springer **126**(1980), 178-199.

[Ram.3] Ramis, J.P., Phénomène de Stokes et filtration Gevrey sur le groupe de Picard-Vessiot, C.R. Acad. Sc. Paris **301**(1985), 165-167.

[Ram.4] Ramis, J.P., Phénomène de Stokes et resommation, C.R. Acad. Sc. Patis **301**(1985), 99-102.

[Ram.5] Ramis, J.P., Filtration Gevrey sur le groupe de Picard-Vessiot d'une équation différentielle irrégulière, preprint, Instituto de Matematica Pura a Aplicada, Rio de Janeiro, 45(1985), 38 pages.

[Röh] Röhrl, H., Das Riemann-Hilbertsche Problem der Theorie der linearen Differntialgleichungen, Math. Ann., **133**(1957), 1-25.

[Sait.1] Saito, T., A note on the linear differential equation of Fuchsian type with algebraic coefficients, Kodai Math. Seminar Reports, **10**(1958), 58-63.

[Sait.2] Saito, T., On Fuchs' relation for the linear differential equation with algebraic coefficients, ibid., **10**(1958), 101-104.

[Sasai] Sasai, T., On a monodromy group and irreducibility conditions of a fourth order Fuchsian differential system of Okubo type, J. Reine Angew. Math. **299/300**(1978), 38-50.

[ST] Sasai, T. and Tsuchiya, S., On a fourth order Fuchsian differential equation of Okubo type, Funkcial. Ekvac., **33**(1990).

[Sski] Sasaki, T., Contiguity relations of Aomoto-Gelfand hypergeometric functions and their application to Appell's system F_3 and Goursat's system ${}_3F_2$, to appear in SIAM Math. Analysis (1991).

[SY.1] Sasaki, T. and Yoshida, M., Linear differential equations in two variables of rank 4, Math. Ann. **282**(1988), 69-111.

[SY.2] Sasaki, T. and Yoshida, M., Linear differential equations modeled after hyperquadrics, Tohoku Math. J. **41**(1989), 321-348.

[SY.3] Sasaki, T. and Yoshida, M., Tensor products of linear differential equations II, Funkcial. Ekvac.**33**(1990), 527-549.

[Schl] Schlesinger, L., Über eine Klasse von Differentialsystemen beliebliger Ordnung mit festen kritischen Punkten, J. für Math., **141** (1912), 96-145.

[Shim.1] Shimomura, S., Painlevé transcendents in the neighbourhood of fixed singular points, Funkcial. Ekvac., **25**(1982), 163-184.

[Shim.2] Shimomura, S., Series expansions of Painlevé transcendents in the neighbourhood of a fixed singular point, Funkcial. Ekvac., **25** (1982), 185-197.

[Shim.3] Shimomura, S., Supplement to "Series expansions of Painlevé transcendents in the neighbourhood of a fixed singular point", Funkcial. Ekvac., **25**(1982), 363-371.

[Shim.4] Shimomura, S. Analytic integration of some nonlinear ordinary differential equations and the fifth Painlevé equation in the neighbourhood of an irregular singular point, Funkcial. Ekvac., **26** (1983), 301-338.

[Shim.5] Shimomura, S., On solutions of the fifth Painlevé equation on the positive real axis I, II, Funkcial. Ekvac., **28**(1985), 341-370, **30** (1987), 203-224.

[Shim.6] Shimomura, S., A family of solutions of a nonlinear ordinary differential equation and its application to Painlevé equations (III),(V) and (VI), J. Math Soc. Japan, **39**(1987), 649-662.

[Sib.1] Sibuya, Y., Sur un système des équations différentielles ordinaires linéaires à coefficients périodiques et contenant des paramètres, J. Fac. Sci. Univ. Tokyo, **7**(1954), 229-241.

[Sib.2] Sibuya, Y., Sur réduction analytique d'un système d'équations différentielles ordinaires linéaires contenant un paramètre, J. Fac. Sci. Univ. Tokyo, **7**(1958), 527-540.

[Sib.3] Sibuya, Y., Simplification of a system of linear ordinary differential equaitons about a singular point, Funkcial. Ekvac., **4**(1962), 29-56.

[Sib.4] Sibuya, Y., Asymptotic solutions of a system of linear ordinary differential equaitons containing a parameter, Funkcial. Ekvac., **4** (1962), 83-113.

[Sib.5] Sibuya, Y., Formal solutions of a linear ordinary differential equation of the n th order at a turning point, Funkcial. Ekvac., **4**(1962), 115-139.

[Sib.6] Sibuya, Y., Global theory of second order linear ordinary differential equations with a polynomial coefficient, Math. Studies **18**, North-Holland, 1975.

[Sib.7] Sibuya, Y., Linear ordinary differential equations in the complex domain, Kinokuniya-shoten (1976), (in Japanese).

[Sie.1] Siegel, C.L., Topics in complex function theory I, Wiley-Interscience, 1969.

[Sie.2] Siegel, C.L., Über die Normalform analytischer Differentialgleichungen in der Nähe einer Gleichgewichtslösung, Nachr, Akad. Wiss. Göttingen, Math.-Phys. Kl. (1952), 21-30.

[Sie.3] Siegel, C.L., Über die Existenz einer Normalform analytischer Hamiltonischer Differentialgleichungen in der Nähe einer Gleichgewichtslösung, Math. Ann., **128**(1954), 144-170.

[Slat] Slater, L.J., Generalized hypergeometric functions, Cambridge University Press, 1966.

[Tkn.1] Takano, K., A 2-parameter family of solutions of Painlevé equation(V) near the point at infinity, Funkcial. Ekvac., **26** (1983),79-113.

[Tkn.2] Takano, K., Reduction for Painlevé equations at the fixed singular points of the first kind, Funkcial. Ekvac., **29**(1986), 99-119.

[Tkn.3] Takano, K., Reduction for Painlevé equations at the fixed singular points of the second kind, J. Math. Soc. of Japan, (to appear in 1990).

[TSY] Takano, K., Shimomura, S. and Yoshida, S., On the fixed singular points of Painlevé equations, Sugaku, **39**(1987), 289-304, Jap. Math. Soc. (in Japanese).

[TB] Takano, K. and Bannai, E, A global study of Jordan-Pochhammer differential equations, Funkcial. Ekvac., **19**(1976), 85-99.

[Tky.1] Takayama, N., Gröbner basis and the problem of contiguous relations, Japan J. Appl. Math., **6**(1989), 147-160.

[Tky.2] Takayama, N., Holonomic solutions of Weisner's operator, Funkcial. Ekvac., **32**(1989), 323-341.

[Tky.3] Takayama, N., An approach to the zero recognition problem by Buchberger algorithm, J. Symbolic Computation (to appear 1991).

[Tky.4] Takayama, N., Monodromy transformation formula of Euler-Darboux equation, Kobe Univ. preprint (1989).

[Tky.5] Takayama, N., An algorithm of constructing the integral of a module — an infinite dimensional analog of Gröbner basis, Proc. of ISSAC'90, ACM Press (to appear 1990).

[Trji] Trjitzinsky, W.J., Analytic theory of nonlinear singular differential equations, Mém. Sci. Math., Gauthier Villars, 1938.

[Tsu.1] Tsutsui, T., Linear partial differential equations with regular singularities in all variables, Japan J. Math., **11**(1985), 131-143.

[Tsu.2] Tsutsui, T., On the Cauchy problem with ramified initial data whose singular loci are intersecting hypersurfaces, Chiba Uni., preprint (1987).

[Tsu.3] Tsutsui, T., Fuchsian initial value problem on $\mathbb{P}^2(\mathbb{C})$ with hypergeometric functions as data along $\mathbb{P}^1(\mathbb{C})$, preprint (1988).

[Tur] Turrittin, H.L., Convergent solutions of ordinary linear homogeneous differential equaitons in the neighborhood of an irregular singular point, Acta Math., **93**(1955), 27-66.

[Uen.1] Ueno, K., Hypergeometric series formulas through operator calculus, Funkcial. Ekvac.(to appear in 1990).

[Uen.2] Ueno, K., Hypergeometric series formulas generated by the Chu-Vandermonde convolution, Mem. Fac. Sci. Kyushu Univ. (to appear in 1990).

[Uen.3] Ueno, K., Umbral calculus and special functions, Adv. in Math. **67**(1988), 174-229.

[Ume.1] Umemura, H., Algebro-geometric problems arising from Painlevé's work, Algebraic and Topological Theories, Kinokuniya Tokyo (1985), 467-495.

[Ume.2] Umemura, H., Birational automorphism groups and differntial equations, in [GO].

[Ume.3] Umemura, H., On the irreducibility of the first differential equation of Painlevé, Algebraic geometry and Commutative algebra in honor of Masayoshi Nagata, (1987), 101-119.

[Ume.4] Umemura, H., On the second proof of the irreducibility of the first differential equation of Painlevé, Kumamoto Univ. preprint (1988).

[Vil] Vilenkin, N.J., Special functions and the theory of group representations, Transl. of Math. monographs, vol. **22**, AMS Providence, Rhode Island, 1968.

[Waso] Wasow, W., Asymptotic expansions for ordinary differential equations, Interscience, 1965.

[WW] Whittaker, E.T. and Watson, A., A course of modern analysis, Cambridge Univ. Press, 1972.

[Witt] Wittich, H., Neuere Untersuchungen über Eindeutige Analytische Functionen, Springer, Berlin-Göttingen- Heidelberg, 1955.

[Yag] Yagami, T., On Gevrey asymptotic solutions of linear Pfaffian equations, Master's Thesis, Univ.Tokyo (1983).

[Yok.1] Yokoyama, T., On connection formulas for a fourth order hypergeometric system, Hiroshima Math. J., **15**(1985), 297-320.

[Yok.2] Yokoyama, T., Characterization of connection coefficients for hypergeometric system, Hiroshima Math. J., **17**(1987), 225-239.

[Yok.3] Yokoyama, T., On the structure of connection coefficients for hypergeometric systems, Hiroshima Math. J., **18**(1988), 309-239.

[Yok.4] Yokoyama, T., A system of total differential equations of two variables and its monodromy group, submitted to Funkcial. Ekvac.(1989).

[YosM] Yoshida, M., Fuchsian differential equations, Vieweg Verlag, Wiesbaden, 1987.

[YosS.1] Yoshida, S., A 2-parameter family of solutions of Painlevé equations (I)-(V) at an irregular singular point, Funkcial. Ekvac., **28**(1985), 233-248.

[YosS.2] Yoshida, S., A general solution of a nonlinear 2-system without Poincaré's condition at an irregular singular point, Funkcial. Ekvac., **27** (1984), 367-391.

[Zei] Zeilberger, D., A holonomic systems approach to special functions identities, J. of Computational and Applied Math. (to appear in 1991).

[ZT.1] Zograf, P.G. and Takhtadzhyan, L.A., On Liouville's equation, accessary parameters and the geometry of Teichmüller space for Riemann surfaces of genus 0, Mat. USSR Sbornik (AMS), **60-1**(1988), 143-161.

[ZT.2] Zograf, P.G. and Takhtadzhyan, L.A., On uniformization of Riemann surfaces and the Weil-Petersson metric on Teichmüller and Schottky spaces, Mat. USSR Sbornik (AMS), **60-2**(1988), 297-313.

Notes on the chapter titlepage illustrations

The titlepage illustration of Chapter 1 shows a solution of the Toda equation expressed by rational solutions of the second Painlevé equation [KNFH].

The titlepage illustration of Chapter 2 shows a solution of the Toda equation expressed by the Bessel function, as drawn by N. Takayama.

The titlepage illustration of Chapter 3 shows a group of symmetries of the fourth Painlevé equation [Mur.1].

The titlepage illustration of Chapter 4 shows the distribution of poles of a rational solution of the second Painlevé equation [KNFH].

Index of symbols

General

Chapter 2

$F(\alpha,\beta,\gamma;x)$: hypergeometric series . . . 31
$E(\alpha,\beta,\gamma)$: hypergeometric differential equation . . . 30
$R(\alpha,\beta,\gamma)$: Riemann scheme of $E(\alpha,\beta,\gamma)$ . . . 36
$L(\alpha,\beta,\gamma)$: hypergeometric differential operator . . . 44
$RE(\rho,\sigma,\tau)$: Riemann equation . . . 55
D_a^α: Euler transform . . . 55
$\mathcal{D}^\lambda$: hypergeometric Euler transform . . . 62
F_{pq}: solution of $E(\alpha,\beta,\gamma)$ given by an integral over the arc $\overline{pq}$ 65, 96
$\langle\xi,\eta\rangle$: solution of $E(\alpha,\beta,\gamma)$ given by an integral
over the double loop $[\xi,\eta]$ . . . 101
$f_0(x;0), f_0(x;1-\gamma), f_1(x;0), f_1(x;\gamma-\alpha-\beta), f_\infty(x;\alpha), f_\infty(x;\beta)$:
Kummer's solutions . . . 38, 39
$H_j(\alpha,\beta,\gamma)$: step-up operators . . . 42, 43
$B_j(\alpha,\beta,\gamma)$: step-down operators . . . 42, 43
$c_1(\alpha,\beta,\gamma) := -\alpha(\alpha-\gamma+1)$
$c_2(\alpha,\beta,\gamma) := -\beta(\beta-\gamma+1)$
$c_3(\alpha,\beta,\gamma) := (\gamma-\alpha)(\gamma-\beta)$
$\delta_x := xd/dx$
$\mathcal{R}$: ring of linear differential operators . . . 41
$\mathcal{S}(\alpha,\beta,\gamma)$: linear space of solutions at a point $(\neq 0,1)$
of $E(\alpha,\beta,\gamma)$ . . . 42
$\mathcal{H}_n$: set of n-th order differential operator
of hypergeometric type . . . 62
$\varepsilon(\cdot) := \exp(2\pi i\cdot)$
$\langle m\rangle := \{1,\dots,m\}$
$\gamma_* f$: analytic continuation of f along a curve γ

Chapter 3

$\sum_i$: summation over $i=1,\dots,n$
$\sum_i^j$: summation over $i=1,\dots,j-1,j+1,\dots,n$
$\sum_{(i)}$: summation over $i=1,\dots,n+2$
$\sum_{(i)}^j$: summation over $i=1,\dots,j-1,j+1,\dots,n+2$
$\prod_i$: product over $i=1,\dots,n$

Chapter 4

Index

INDEX

Aspects of Mathematics

* Eine Veröffentlichung des Max-Planck-Instituts für Mathematik, Bonn